Iceland Geodynamics

Crustal Deformation and Divergent Plate Tectonics

Freysteinn Sigmundsson

Iceland Geodynamics

Crustal Deformation and Divergent Plate Tectonics

Springer

Published in association with
Praxis Publishing
Chichester, UK

PRAXIS

Dr. Freysteinn Sigmundsson
Geophysicist at the Nordic Volcanological Centre
Institute of Earth Sciences
University of Iceland
Reykjavik
Iceland

SPRINGER–PRAXIS BOOKS IN GEOPHYSICAL SCIENCES
SUBJECT *ADVISORY EDITOR*: Dr. Philippe Blondel, C.Geol., F.G.S., Ph.D., M.Sc., Senior Scientist, Department of Physics, University of Bath, Bath, UK

ISBN 10: 3-540-24165-5 Springer-Verlag Berlin Heidelberg New York

Springer is part of Springer-Science + Business Media (springeronline.com)

Bibliographic information published by Die Deutsche Bibliothek

Die Deutsche Bibliothek lists this publication in the Deutsche Nationalbibliografie; detailed bibliographic data are available from the Internet at http://dnb.ddb.de

Library of Congress Control Number: 2005930645

Printed in Germany

Cover design: Jim Wilkie
Project management: Originator Publishing Services, Gt Yarmouth, Norfolk, UK

Printed on acid-free paper

Contents

Preface

Iceland, a land in continuous motion and deformation, has inspired my study and work on geodynamics for the last 20 years. Consequences of crustal deformation are not only evident in the zones of active volcanoes and earthquake fractures in Iceland; outside of these zones the interiors of dyke swarms and ancient volcanoes are revealed by erosion, the lava pile is regionally tilted because of loading from lava flows on top, and glacial rebound has left raised beaches high above the current relative sea level. Precise geodetic measurements, including the use of space-geodetic techniques, of current crustal movements have been used to provide constraints on active deformation processes. Recent and extensive results from Global Positioning System (GPS) geodesy and satellite radar interferometry (InSAR) complement earlier results from levelling and electronic distance measurements in Iceland, providing a long time series of deformation. Research on crustal deformation is carried out by a number of scientists in Iceland at different institutes, in extensive collaboration with scientists from other countries. Cooperation and collaboration between the many persons involved has been key to gathering extensive new knowledge on deformation fields in Iceland and their interpretation. This book is intended as an overview of some of the recent work and would have been impossible to write without extensive help from many individuals actively involved in the study of Iceland geodynamics.

This book rests on personal experience gained during the last 20 years and interaction with a large number of scientists during this time. My initial mentor in Iceland and now a long term collaborator, Páll Einarsson, was the one who got me started working on crustal movements. Sveinbjörn Björnsson who co-supervised my M.Sc. study together with Páll Einarsson on viscosity under Iceland was also instrumental in raising my interest in geophysics. My Ph.D. study was then conducted at the University of Colorado in Boulder, U.S.A., under the supervision of Roger Bilham. Thank you Roger for your guidance, extraordinary enthusiasm and scientific motivation that still guides me through most days.

Interaction with other members of the geophysics group in Boulder 1988–1992 was important, as well as extensive support from UNAVCO[1] to GPS projects in Iceland.

I was at the Nordic Volcanological Institute 1992–2004 and I am now at the Nordic Volcanological Centre, Institute of Earth Sciences, University of Iceland after the merging of academic geoscience groups in Iceland. Everyone at these institutes is acknowledged for stimulating discussions and taking part in cooperative projects. My time as the director of the Nordic Volcanological Institute 1999–2004 expanded my view of volcanology and geoscience, helping me to put crustal deformation results into a broader context. Interaction with other geoscience groups in Iceland has been extensive and important, including the Icelandic Meteorological Office, Iceland Geosurvey (previously the National Energy Authority), the Icelandic Institute of Natural History, and the National Land Survey of Iceland. In particular I want to thank the late Guðmundur Sigvaldason, the director of the Nordic Volcanological Institute until 1999. He provided unique inspiration and continuous support to deformation studies. Special thanks also to Eysteinn Tryggvason, the father of crustal deformation studies in Iceland, who introduced me to the techniques of optical levelling and tilt measurements, as well as electronic distance measurements and emphasized the dedication needed for those measuring crustal movements. Halldór Ólafsson has been instrumental in carrying out fieldwork for crustal deformation projects, and Anna Eiríksdóttir and Rósa Ólafsdóttir have provided various support in the office. Through the years, discussions and interaction with numerous scientists in Iceland has been enlightening and important for my understanding of Iceland geodynamics. Some of these are Níels Óskarsson, Karl Grönvold, Þóra Árnadóttir, Erik Sturkell, Amy Clifton, and Reidar Trönnes at Nordvulk, Páll Einarsson, Bryndís Brandsdóttir, and Magnús Tumi Guðmundsson at the Science Institute, Ragnar Stefánsson, Sigurður Rögnvaldsson, Steinunn Jakobsdóttir, and Kristín Vogfjörð at the Icelandic Meteorological Office, and Ólafur Flóvenz and Kristján Sæmundsson at Iceland Geosurvey. Most important of all have been research fellows at Nordvulk and various graduate students I have had the fortune to work with, some of them now being long-term collaborators. They have done much of the work behind results presented here. These include Erik Sturkell, Rikke Pedersen, Halldór Geirsson, Carolina Pagli, Elske de Zeeuw-van Dalfsen, Pete La Femina, Sverrir Guðmundsson, Sigurjón Jónsson, Sigrún Hreinsdóttir, Ingrid Anell, Dominique Richard, Johan Camitz, and Malou Blomstrand Stinessen.

Interaction with scientists outside of Iceland has also been extensive. In particular I want to mention Alan Linde, Christof Völksen, Thierry Villemin, Virginie Pinel, Claude Jaupart, Gillian Foulger, Wolfgang Niemeier, Wolfgang Jacobi, Hazel Rymer, Tim Dixon, John Sinton and Bob Detrick, and the Nordic board of directors for the Nordic Volcanological Institute. In Toulouse, France, I was taught the InSAR technique by Hélène Vadon and Didier Massonnet at the French Space Agency and by Kurt Feigl at the CNRS. They have all been

[1] UNAVCO stands for University Navstar Consortium.

instrumental for InSAR studies in Icleand, and Kurt Feigl is a long-term collaborator.

Funding for research I have been involved in has come from various sources. The Nordic Council of Ministers has been the main sponsor of the Nordic Volcanological Institute and Centre, with contributions as well from Icelandic authorities. Project funding in Iceland has also come from the Icelandic Centre for Research (*Rannís*), the National Power Company of Iceland (*Landsvirkjun*) and the Icelandic Road Authority. International cooperative projects have been many, and I acknowledge support from the European Union through participation in numerous projects, including the projects on European Laboratory Volcanoes, Prenlab-1, Prenlab-2, Retina, and Prepared. The National Science Foundation, U.S.A., has also provided support to enable work on geodynamical projects in Iceland.

The writing of this book would have been impossible without tremendous help from a large number of individuals. Many have provided artwork as detailed in the acknowledgements. Earlier versions of parts of this book have been read by the following individuals: Þóra Árnadóttir, Amy Clifton, Páll Einarsson, Sigmundur Freysteinsson, Áslaug Geirsdóttir, Magnús Tumi Guðmundsson, Bill Menke and Kristján Sæmundsson. Extensive advice from these and others during writing is acknowledged. Thanks also to Oddur Sigurðsson and Ágúst Guðmundsson for providing photographs. Excuses to those I have forgotten to mention but have contributed, they are acknowledged as well. I also want to acknowledge the publishers. Support from Clive Horwood at Praxis has been unfailing, and the flexibility offered has allowed the completion of this book in harmony with other undertakings. Philippe Blondel (University of Bath, UK) read the manuscript and was instrumental in shaping it into final form, as well as providing encouragement throughout all the writing. The team at Originator did the copy-editing in an excellent manner.

My hope is that this book will provide a useful overview of selected aspects of Iceland geodynamics and crustal deformation, provide insights into the physical processes of plate spreading and related processes in general, and stimulate further research on how the Earth deforms.

Reykjavík, October 2005
Freysteinn Sigmundsson

To my family

Acknowledgements

The following publishers and authors are acknowledged for giving permission to use original or redrawn figures based on illustrations in journals and books for which they hold copyright. Figure captions include citations to original authors. Attempts have been made to secure permission for use of all copyrighted materials. Apologies if there are any errors or omissions.

Publishers	**Figure No.**
American Association for the Advancement of Science (AAAS)	
Science	2.9
American Geophysical Union	
EOS, Transactions American Geophysical Union	3.21
Geophysical Research Letters	4.12, 4.13, 5.14, 5.15, 5.19, 5.20, 6.12, 7.14, 7.15, 8.5, 8.10, 8.14, 8.15, 8.16
Geochemistry, Geophysics, Geosystems	8.11, 8.13
Journal of Geophysical Research	2.4, 2.5, 2.13, 4.3a, 4.9, 4.10, 5.5a, 5.5b, 5.12, 5.16, 6.2, 6.3, 6.4, 6.5, 6.11, 6.16, 6.17, 6.19, 7.7, 8.12
Reviews of Geophysics	2.6, 2.15
Arnold Publishers Ltd.	
The Holocene	3.14
Blackwell Publishing Ltd	
Geophysical Journal International	4.2, 4.3b, 4.5, 5.6, 8.18

Publishers	**Figure No.**
Elsevier	
Earth and Planetary Science Letters	2.16, 2.17, 4.6, 7.11, 7.12
Tectonophysics	4.8, 6.20, 7.4a, 7.10
Marine Geology	8.1
J. Volcanology and Geotherm. Res.	5.17
Geological Society of America	
The Geology of North America	2.2, 2.3, 2.8, 3.20, 4.15
Iceland Glaciological Soc. and Geoscience Soc. of Iceland	
Jökull	3.1, 3.3, 3.16, 3.17, 3.18, 3.19, 8.4
ÍSOR, Iceland Geosurvey	2.10, 4.4
Icelandic Institute of Natural History	3.2, 3.8
Macmillan Magazines Ltd	
Nature	2.7, 2.12, 5.18, 7.16, 8.4
Springer	
Geologische Rundschau	3.4, 3.7b
Taylor and Francis Ltd.	
Boreas (*www.tandf.no/boreas*)	8.2, 8.6

Authors

Þóra Árnadóttir	7.5
Bryndís Brandsdóttir	5.5
Hjálmar Eysteinsson	2.1, 2.10, 4.4
Halldór Geirsson	2.11, 6.1, 6.18, 7.2, 8.17
Ágúst Guðmundsson, JTS Geotechnical Services	5.3, main cover figure
Gunnar Guðmundsson	3.19, 7.1, 7.4b, 7.8
Ólafur Guðmundsson and Bryndís Brandsdóttir	5.6
Ólafur Flóvenz and Kristján Sæmundsson	4.7
Sigrún Hreinsdóttir	7.6
Peter La Femina	6.3, 6.4, 6.5
Rafaella Montelli	2.14
Halldór Ólafsson	5.2c
Carolina Pagli	6.7
Guðmundur Sigvaldason	5.2a
Oddur Sigurðsson	3.5, 3.7a, 5.1a, 5.1b, 5.1c, 5.1d, 8.7
Sigurjón Sindrason	5.2d
Erik Sturkell	5.17
Eysteinn Tryggvason	6.9

I also acknowledge those individuals who have contributed figures to this book (held copyright by publishers) to facilitate figure reproduction. These include: Kristján Ágústsson, Richard Allen, Hjálmar Eysteinsson, Ólafur Flóvenz, Gunnar Guðmundsson, Sigrún Hreinsdóttir, Garrett Ito, Rosa Ólafsdóttir, Rikke Pedersen, Erik Sturkell, and Cecily Wolfe.

Figures

Tables

Abbreviations

AAAS	American Association for the Advancement of Science
EDM	Electronic Distance Measurement
EVFZ	Eastern Volcanic Flank Zone
EVRZ	Eastern Volcanic Rift Zone
EVZ	Eastern Volcanic Zone
FIRE	Faroes–Iceland Ridge Experiment
GPS	Global Positioning System
InSAR	SAR interferometry
LILE	Large Ion Lithophile Element
MAR	Mid-Atlantic Ridge
Moho	Mohorovicic discontinuity
MORB	Mid-Ocean Ridge Basalt
NUVEL	Global plate motion model by DeMets et al. (1990, 1994)
NVZ	Northern Volcanic Zone
OIB	Ocean Island Basalt
REVEL	Global plate motion model by Sella et al. (2002)
RISE	Reykjanes–Iceland Seismic Experiment
RP	Reykjanes Peninsula
RSL	Relative Sea Level
SIL	South Iceland Lowland
SIST	South Iceland Seismic Tomography
SISZ	South Iceland Seismic Zone
TFZ	Tjörnes Fracture Zone
UNAVCO	University NAVstar COnsortium
WVZ	Western Volcanic Zone

1

Introduction

Iceland is the largest portion of the mid-ocean ridge system emerged above sea level, a consequence of excessive volcanism caused by interaction of a mantle plume and a mid-oceanic ridge. As it is above sea level, it is a unique site to study the physical processes of divergent plate tectonics and the consequences of plume–ridge interaction. This book summarizes extensive new knowledge on geodynamics in this natural laboratory that has been collected in the last decades. The emphasis is on geophysical results—in particular, crustal deformation studies. The aim is to put geodynamical results in a broad context and discuss in general the physical processes of divergent plate tectonics.

The book consists of nine chapters, the first being this introduction and the last a summary and discussion of some future research topics. The other seven chapters each cover a special topic and are relatively independent from each other. Readers having some familiarity with Icelandic geology can read the chapters of their interest and skip others, without suffering from lack of continuity. Geodynamics is the focus of the book, and frequent reference is given to crustal deformation results that have provided a variety of constraints on geodynamic processes.

Most available geodetic techniques have been used in Iceland, including triangulation, trilateration, and electronic distance measurements, precise optical levelling, Global Positioning System (GPS) geodesy, and interferometric analysis of synthetic aperture radar images acquired by satellites (InSAR), as well as continuous observations of strain, tilt, and displacements. Crustal deformation studies in Iceland were initiated by German geodesists, inspired by Wegener's ideas of continental drift. They installed a geodetic network in 1938 to detect widening across the rift zone in the northern part of the country. With the acceptance of plate tectonics in the 1960s, programs of electronic distance measurements (led by Bob Decker) and precise optical levelling measurements (led by Eysteinn Tryggvason) were initated. Initial GPS measurements were conducted in 1986 shortly after the introduction of that technique; initial SAR interferometry (InSAR) studies were conducted in the middle of the 1990s.

Reference to locations in Iceland is frequent; most of these are shown on maps in Chapter 3. There is some confusion regarding spelling of Icelandic words, as the Icelandic language has ten special letters different from Latin letters (á, ð, é, í, ó, ú, ý, þ, æ, and ö). These letters are often transliterated into equivalent Latin characters as outlined in Appendix 1 that also has for reference a list of Icelandic words in this book written with and without usage of the special characters. In general, Icelandic spelling is used in this book, except names in references are always spelled out as in the original publications (Icelandic authors with names including special Icelandic characters often modify their names, sometimes differently, when authoring articles in the international literature).

Chapter 2 aims at placing Iceland in context with other parts of the North Atlantic, describe the underlying mantle plume and how it interacts with the Mid-Atlantic Ridge, and how it influences a large part of the North Atlantic region. An alternative to the plume theory is also discussed.

Chapter 3 discusses the surface geology and describes how it is governed by the physical processes of divergent plate tectonics and plume–ridge interaction. It also gives an overview of geologic activity in historic times going back to 874 AD. A broad overview of seismic activity is also included.

The fourth chapter takes the reader from the surface to deeper levels and describes the layered crust/mantle structure under Iceland as identified by seismology, gravity, earthquake distribution, temperature conditions, and rheology.

Volcanology is the focus of Chapter 5, and in particular how geodetic measurements have helped to understand plumbing systems of volcanoes, magma migration, and eruption dynamics. Some theoretical background to interpretation of volcano deformation data is given.

Details of the plate-spreading process are the topic of Chapter 6. It attempts to provide answers to such questions as: What governs the width of the plate boundary deformation zone, and why is there time variability in spreading rates as measured at short distances across the plate boundary? The post-rifting style of deformation observed in North Iceland after a rifting episode in the Krafla volcanic system 1975–1984 is discussed, as well as inter-rifting and co-rifting deformation.

Seismology and recent earthquake activity is the focus of Chapter 7. Seismic activity is focused in two transform zones in Iceland, and in one of them, the South Iceland Seismic Zone, two M_S6.6 earthquakes happened in the year 2000. These events triggered widespread seismic activity along a large part of the plate boundary in South Iceland. Seismic and geodetic observations have provided important knowledge about these events.

Chapter 8 is on glacio-isostatic adjustments and vertical movements. During the last glaciation Iceland was fully covered by ice, and post-glacial rebound occurred when it melted. The rebound was much faster than in most parts of the world as it was completed in about 1,000 years, because of low viscosity. One of the exceptional features of Iceland geology is that in the time period of the few thousand years during and following the deglaciation, the volcanic production rates were extremely high. The deglaciation affected mantle melting and volcanic systems as described in this chapter. Furthermore, historical ice volume changes have caused land elevation changes, currently at a rate of up to 1–2 cm/year, and these can be used to study rheology. Global warming in the future and associated ice melting may provide a still new experiment in rheology.

The final chapter of the book then provides a summary. and discusses the role of Iceland as the geo-laboratory of the future.

2

Mantle plume–mid-ocean ridge interaction in the North Atlantic

Opening of the North Atlantic began about 60 million years ago, with massive basaltic volcanism from that time now found on both sides of the Atlantic. Divergence of the North American plate and the Eurasian plate since that time has formed the ocean floor in the North Atlantic, with the Mid-Atlantic Ridge (MAR) marking the present day plate boundary. The history of spreading is well documented by regular magnetic lineaments, with magnetic observations from the ocean floor south of Iceland being used in the early development of the ideas of plate tectonics (e.g., Vine and Matthews, 1963; Vine, 1966). The North Alantic area is also dominated by the Iceland Hotspot and excessive magmatic activity that has built up Iceland. A mantle plume under Iceland was suggested by Jason Morgan (1971), but geophysical models for the region still differ widely. Many of the characteristics of the North Atlantic can be attributed to the interaction of a mantle plume under Iceland and the MAR, as reviewed, for example, by Ito et al. (2003).

2.1 GEOLOGY OF THE NORTH ATLANTIC: THE ICELAND HOTSPOT

Bathymetric maps of the North Atlantic reveal a huge topographic anomaly centred on Iceland, with decreasing ocean depths towards Iceland (Figure 2.1, see colour plates). This anomalous topography is the swell associated with the Iceland Hotspot, about 1,000 km in radius. Superimposed on this radial anomaly is the broad MAR in the middle of the Atlantic Ocean. It is an integral part of the submarine system of mid-ocean ridges, the longest mountain chain on Earth. Perpendicular to it lies the Greenland–Scotland Ridge, a topographic high across the Atlantic. Other more subtle but significant topographic features in the North Atlantic include so-called V-shaped ridges found on each side of the MAR, particularly well expressed in the area south of Iceland. Each of these ridges has one limb west of the MAR with a strike a few degrees less than the MAR, and another limb east of the MAR, with a strike a few degrees larger than the MAR. The two limbs of each V-shaped ridge meet at the ridge crest.

2.1.1 The Mid-Atlantic Ridge

The MAR is composed of a series of spreading centres marking the ridge crest, offset in a number of places by transform faults that mark the seismicially active parts of fracture zones (Figure 2.2). South of Iceland, the largest offset is the Charlie–Gibbs Fracture Zone at 53°N where the ridge is offset about 350 km. To the north of the Charlie–Gibbs Fracture Zone the ridge is relatively straight and water depths decrease steadily towards Iceland. Here the MAR is called the Reykjanes Ridge. It extends all the way to Iceland where it comes onshore at the southwestern tip of the Reykjanes Peninsula. Immediately north of Iceland, the ridge is offset about 150 km to the west at the Tjörnes Fracture Zone. The ridge segment just north of Iceland is called the Kolbeinsey Ridge, which owes its name to the small Kolbeinsey Island 100 km off the north coast of Iceland. The Kolbeinsey Ridge extends towards Jan Mayen, where the MAR is offset again by the Jan Mayen Fracture Zone.

Earthquakes occur in a narrow zone along the entire length of the northern MAR (Figure 2.3), outlining well the central axis of the plate boundary between the North American and Eurasian plates. Seismicity along the spreading portions of the ridge (ridge crests) and along the transform faults is fundamentally different. Earthquakes on the transform faults are mostly strike–slip events occurring in mainshock–aftershock sequences, whereas a large majority of earthquakes on the ridge crests are normal faulting earthquakes occurring in swarms (Einarsson, 1986, 1987). This seismic behaviour shows that extensional tectonics dominate along the ridge crests, whereas the transforms are zones of horizontal shearing caused by lateral offsets in spreading.

2.1.2 The North Atlantic Large Igneous Province

One of the largest volcanic events on Earth in the last 200 Myr was the eruption of huge volumes of flood basalts during the opening of the North Atlantic 55–60 million years ago. Up to 10 million km^3 of igneous rocks were produced on the associated rifted margins during as little as 2–3 million years (e.g., White et al., 1987; White and McKenzie, 1989). Volcanic formations from this period are today found on both sides of the North Atlantic and include extensive, submarine, volcanic, rifted margins as well as the onshore Tertiary igneous provinces of Britain, Northern Ireland, the Faeroes, Greenland, and Baffin Island (Figure 2.4). White and McKenzie (1989) conclude that these volcanic provinces form a well-documented example of the influence of a mantle plume on igneous activity when the overlying lithosphere is stretched and rifted. A thermal anomaly in the mantle underlying stretched and rifted lithosphere is the cause of excessive volcanism.

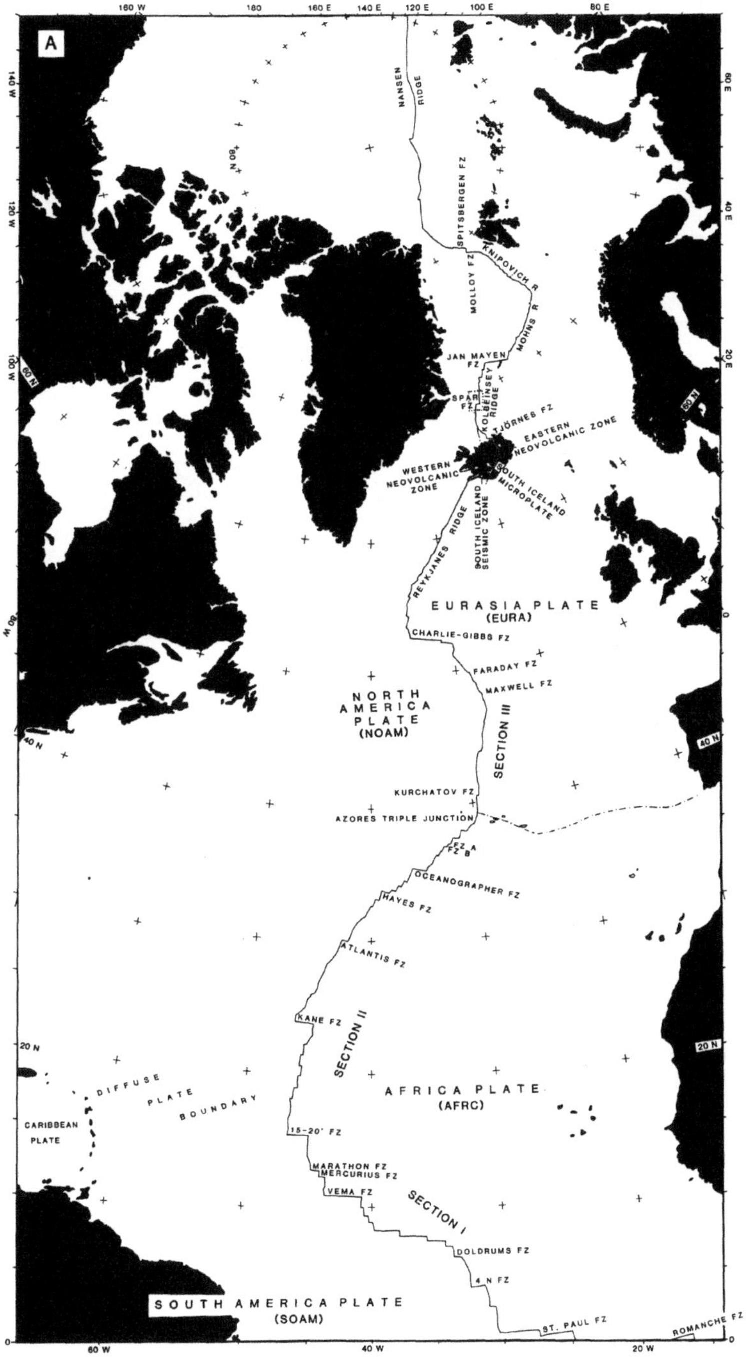

Figure 2.2. The Mid-Atlantic Ridge from bathymetric and magnetic data.

Reproduced from Vogt (1986a) with permission of the Geological Society of America.

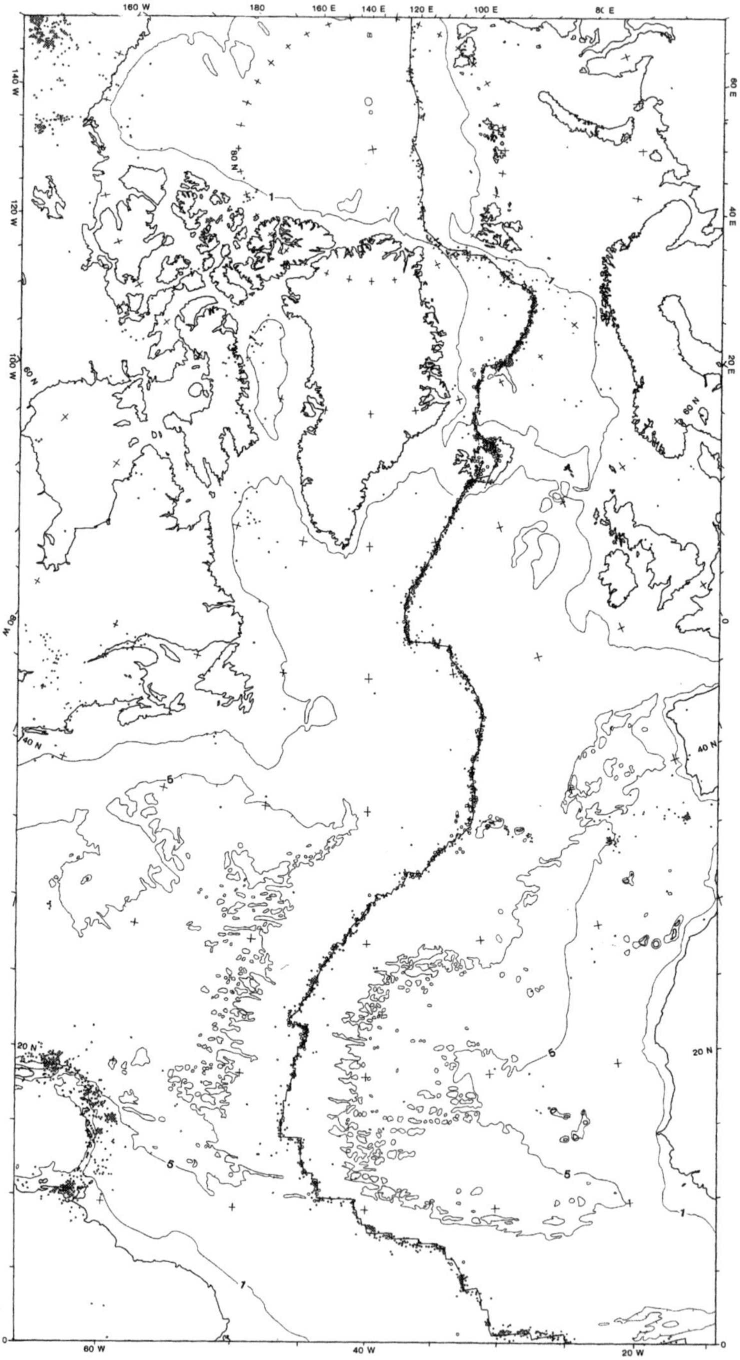

Figure 2.3. Earthquakes in the Atlantic Ocean 1961–1982 (dots).
Reproduced from Vogt (1986a) with permission of the Geological Society of America.

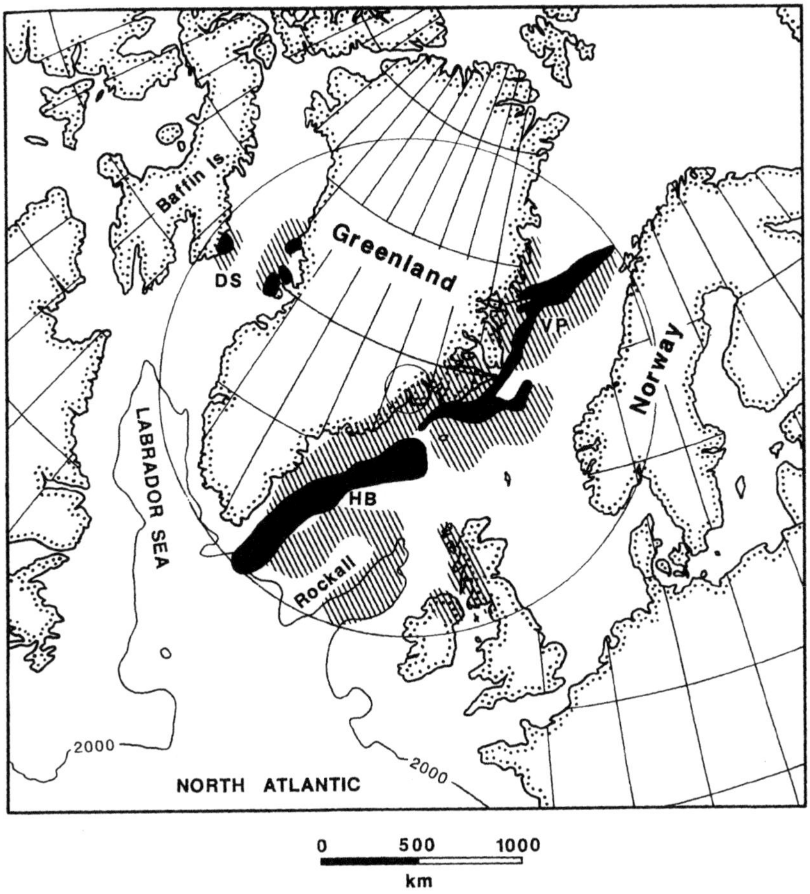

Figure 2.4. Reconstruction of the northern North Atlantic region at magnetic-anomaly-23 time (about 52 Myr ago), just after the onset of ocean spreading. Black shading shows position of extrusive rocks, with hatching showing the extent of early Tertiary igneous activity in the region. Inferred position of a mantle plume under eastern Greenland (small circle) and the extent of the plume head (larger circle). Also shown are the Vøring Plateau (VP), Hatton Bank (HB), and the Davies Strait (DS).

2.1.3 Geochemical variations

Geochemistry reveals the process of melt generation in the mantle. The amounts of rare earth and trace elements are dependent on the degree of melting and its depth extent, as well as on mantle sources. The concentration of rare earth elements can be used to derive partial melt distributions in the mantle (e.g., White and McKenzie, 1995). Such studies suggest that melting under Iceland begins at over the 100-km depth and extends upward towards the lithosphere, with the maximum percentage of melting being around 20% (Figure 2.5).

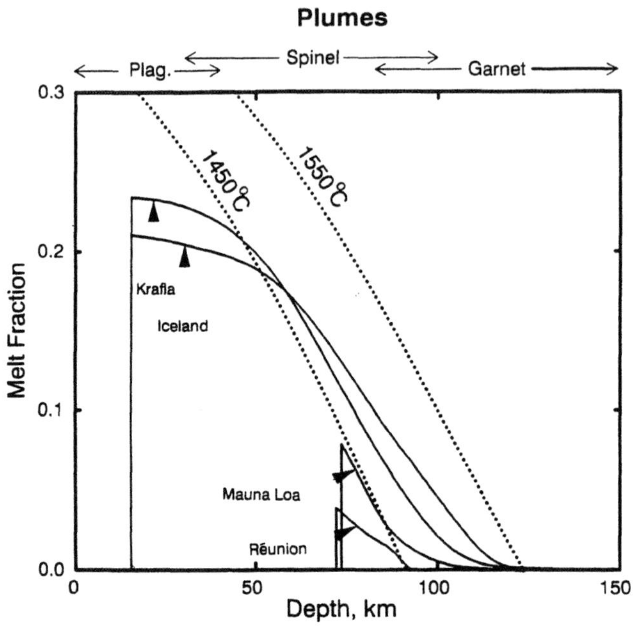

Figure 2.5. Curves of inferred melt distribution in mantle plumes under oceanic plates. Decompression melting beneath Iceland reaches the base of the crust as Iceland lies above a spreading centre; mantle melting is stopped at greater depth by thick old plates that overlie the Hawaiian and Réunion Plumes. Dotted curves show mantle potential temperature.

Isotopic ratios show significant changes along the MAR, correlating with the locations of hotspots as well as fracture zones. In particular, there are clear gradients in some of the ratios along the MAR axis for several hundred kilometres away from the centre of the Icelandic hotspot (e.g., Schilling, 1973a, b, 1986; Ito et al., 2003). Geochemical anomalies centred over Iceland include elevated $^{87}Sr/^{86}Sr$ and $^{3}He/^{4}He$ isotopic ratios, as well as an excessive La/Sm ratio (Figure 2.6). The common explanation for such systematic variation along ridges is the mixing of distinct sources for Mid-Ocean Ridge Basalts (MORBs) and Ocean Island Basalts (OIBs). A binary mixing model calling for mixing of melts from a mantle plume source and an upper asthenospheric source depleted in Large Ion Lithophile Elements (LILEs, e.g., light rare earth elements) was proposed by Schilling (1973a, b) to explain the observations and was further supported by lead isotope studies (Sun et al., 1975). Various types of mantle topology may conform to the observed isotope gradients, including a plume with central upwelling under Iceland spreading laterally along the MAR (Schilling, 1986). Geochemical discontinuities occur across the fracture zones in the North Atlantic and may be explained by damming of flow along the ridge because of older and colder lithosphere opposite the fracture zones.

Within Iceland, the geochemical signatures are further complicated by reworking of old crust, caused by eastward rift jumps in response to westward

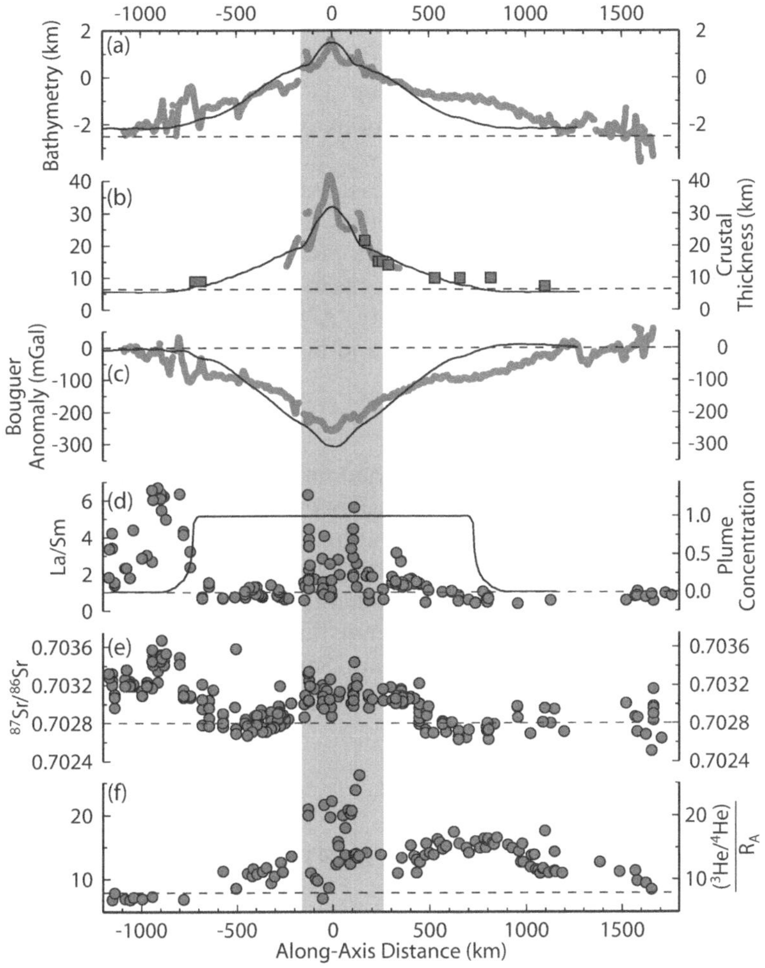

Figure 2.6. Profiles along the Mid-Atlantic Ridge centred on Iceland of (a) bathymetry, (b) crustal thickness, (c) Bouguer gravity, (d) La/Sm ratio, (e) $^{87}Sr/^{86}Sr$, and (f) $^{3}He/^{4}He$ normalized by atmospheric ratio. Compilation by Ito et al. (2003). Shading marks the extent of Iceland. Thin curves in a–c are model predictions from a three-dimensional geodynamic model of Ito et al. (1999).

migration of the plate boundary relative to the mantle plume (see Sections 2.6 and 3). Compositional and isotopic variations along the rift zones in Iceland are influenced by remelting of crust and extinct volcanic centres. Assimilation of partial melts from the crust into ascending mantle-derived melts has been suggested to be a feedback process contributing to isotopic ratios and resulting in accumulation of LILEs in the crust, most extensively in silicic volcanic centres (Óskarsson et al., 1982, 1985; Sigvaldason et al., 1974).

2.1.4 Gravity and geoid anomalies

Important information on crust and mantle can be derived from the gravity field of the Earth. The free air gravity field over the oceans is particularly sensitive to topography and crustal thickness variations at the ocean floor (Figure 2.7, see colour plates), and the shape of the geoid provides information as well. Geoid anomalies, the difference between the measured geoid and a reference geoid taken as an ellipsoid of revolution, provide clues to the compensation of the Earth's topography. In particular, gradients in the ratio of geoid and topography heights have been used to infer the depth of compensation of hotspot swells. With an ocean basin as a reference, a geoid anomaly ΔN associated with an isostatically compensated topography, h, is (Turcotte and Schubert, 1982; Schubert et al., 2001):

$$\Delta N = \frac{\pi G(\rho_0 - \rho_w)}{g} hW \tag{2.1}$$

where ρ_0 is the reference density corresponding to zero elevation, ρ_w is seawater density, g is the acceleration of gravity, G is the gravitational constant, and W is the depth of compensation of the topography. One of the largest long-wavelength, positive geoid anomalies on Earth has a centre in the North Atlantic close to Iceland (Figure 2.8). Despite this, the $\Delta N/h$ gradients around Iceland are relatively low because the dimensions of the geoid anomaly are much larger than the dimensions of the Iceland Hotspot swell. For Iceland, Sandwell and MacKenzie (1989) find $\Delta N/h \approx 1.5\,\text{m/km}$. They derive shallow compensation depths for most hotspot swells in the range of 75–125 km.

2.2 OPENING OF THE NORTH ATLANTIC

2.2.1 Magnetic recording of sea floor spreading

The Atlantic was a key site in early studies of magnetic anomalies at ocean ridges. A magnetic profile at the MAR was one of the examples used by Vine and Matthews (1963) when explaining magnetic lineaments on the sea floor in terms of normally and reversely magnetized crust. They realized that if the oceanic crust was formed over a "convective upcurrent in the mantle at the centre of an oceanic ridge" and crustal spreading of the ocean floor would take place, the crust would have alternating normally and reversely magnetized material blocks parallel to the ridge, dependent on the magnetic field polarity at the time of formation. The Vine–Matthews hypothesis and its application to the North Atlantic was further considered by Vine (1966), using a detailed aeromagnetic survey of the Reykjanes Ridge south of Iceland made by the U.S. Naval Oceanographic Office. The survey revealed linear magnetic anomalies parallel or subparallel to the ridge (Figure 2.9) that could indeed be explained by a model utilizing a reversal timescale for the magnetic field, and invoking a full spreading rate across the Reykjanes Ridge of close to 2 cm/year.

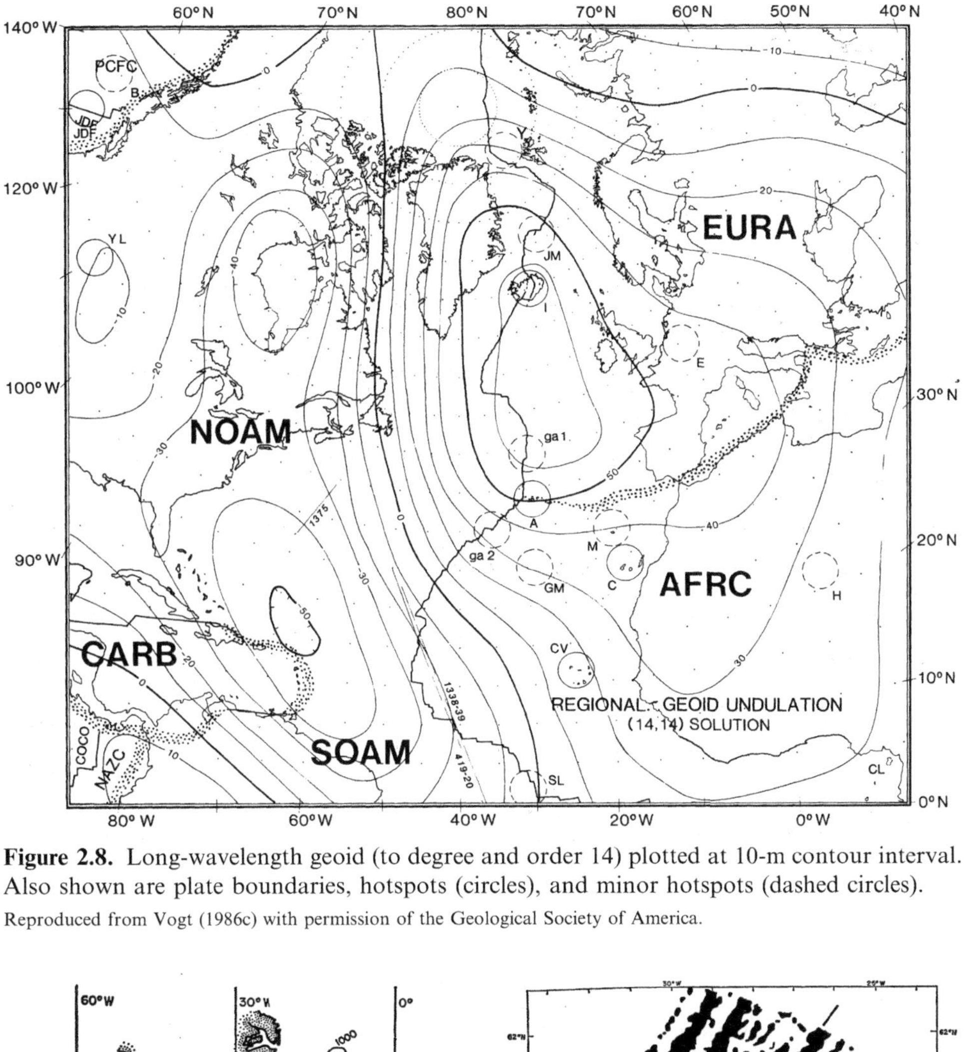

Figure 2.8. Long-wavelength geoid (to degree and order 14) plotted at 10-m contour interval. Also shown are plate boundaries, hotspots (circles), and minor hotspots (dashed circles).
Reproduced from Vogt (1986c) with permission of the Geological Society of America.

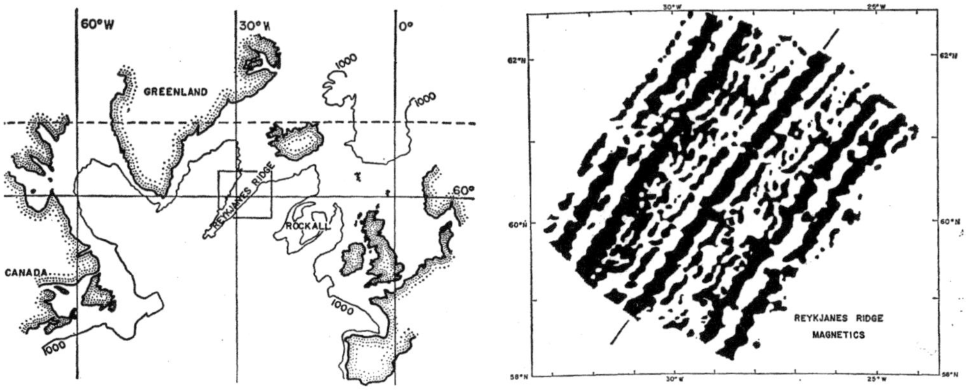

Figure 2.9. Magnetic anomalies along the Reykjanes Ridge south of Iceland.
Reproduced from Vine (1966) with permission of *Science*. Copyright AAAS.

The magnetic field in the Iceland region has been revealed in detail by extensive marine magnetic surveys, as well as aeromagnetic surveys conducted over Iceland. Efforts in Iceland include the extensive work of Þorbjörn Sigurgeirsson (1970–1985) and Leó Kristjánsson and coworkers (e.g., Kristjánsson et al., 1989; Jonsson et al., 1991). The geomagnetic field anomalies over Iceland are irregular, whereas clear lineaments parallel to the MAR are observed both south and north of Iceland. Spreading has been restricted to a single axis south of Iceland, but rift relocations have occurred north of Iceland. Spreading across the currently active Kolbeinsey Ridge began about 24 Myr ago, but a prominent extinct ridge, the Aegir Ridge in the Norway Basin, was active before that. Magnetic anomalies are clear on each side of the Kolbeinsey Ridge (Figure 2.10, see colour section), revealing a spreading rate of about 2 cm/yr for the last 12 Myr (Vogt et al., 1980). Spreading rates inferred from marine magnetic anomalies form one set of observations used as constraints on global plate motion models. For example, the spreading rate north of Iceland inferred by Vogt et al. (1980) is one of the 277 globally distributed spreading rates used by DeMets et al. (1990, 1994) to constrain the NUVEL-1 and NUVEL-1A models described below.

2.2.2 Geologic and geodetic plate motion models

Plate tectonics describes plate motion on the surface of the Earth in mathematical terms with the help of Euler's theorem, giving the motion of two rigid plates on a spherical surface by their pole of rotation and an associated angular velocity (e.g., Fowler, 2005). The relative velocity, u, between plates at a plate boundary is given by:

$$u = \omega \cdot a \sin \Delta \tag{2.2}$$

where ω is the angular velocity of rotation, a is the radius of the Earth, and Δ is the angle subtended at the centre of the Earth by the pole of rotation and a particular location on a plate boundary. The relative velocity along a plate boundary thus increases as the surface distance from a pole of rotation, $a \sin \Delta$, increases.

Various observations can be used to constrain relative plate motion and construct plate motion models. The best constrained global plate motion model based on geologic evidence is the NUVEL-1A model (DeMetz et al., 1994). It is based on the earlier NUVEL-1 model (DeMetz et al., 1990) with velocities scaled to accommodate revisions in the geomagnetic timescale. The NUVEL-1A model is based on spreading rates from marine magnetic anomalies, spreading directions from the azimuth of transform faults, and earthquake slip data. An extensive dataset is inverted to define the Euler pole of rotation for Earth's lithospheric plates. The location of the pole of rotation and angular velocity describing the relative motion of the Eurasian and the North American Plates is given in Table 2.1. According to the NUVEL-1A model, the full spreading velocity in central Iceland (64.5°N, 18°W) is 18.3 mm/yr in direction N105°E. The variation in spreading rate across Iceland due to different distance from pole of rotation is less

Table 2.1. Euler poles and relative angular velocities for the Eurasian and North American plates.

Model	Latitude	Longitude	ω	Error ellipse		ξ	σ_ω
	(°N)	(°E)	(°/Myr)	$\sigma_{\max}$	$\sigma_{\min}$		
NUVEL-1A	62.4	135.8	0.21	4.1	1.3	−11	0.01
REVEL	68.05	136.42	0.245	1.5	0.8	−38	0.004

Error ellipses are one-sigma angular lengths in degrees of the semimajor and semiminor axes of the pole of rotation, and ξ is the azimuth of the semimajor ellipse axis in degrees clockwise from north.

than 2 mm/yr. A compilation of earlier plate motion models describing the relative motion between the North American and Eurasian Plates is given by Vogt (1986b).

Plate motion models can also be derived from geodetic data. A model based on space-geodetic data from 1993 to 2000, primarily observations from continuous Global Positioning System (GPS) stations distributed around the globe, was inferred by Sella et al. (2002). Their REVEL model gives plate motion as Euler poles of rotation and angular velocities, in the same manner as plate motion models based on geologic evidence. The model incorporates only GPS data from stable plate interiors when determining angular velocities. GPS stations at or close to plate boundaries, like in Iceland, are excluded. This model gives a full spreading rate in central Iceland as 19.7 mm/yr in a direction N103°E. The agreement with the NUVEL-1A model is good despite the fact that the NUVEL-1A model corresponds to average motion in the last 3 Myr, whereas the REVEL model describes plate motion in the 1993–2000 period.

2.2.3 Geodetic measurements in Iceland

The current divergence rate across the MAR in Iceland can also be inferred from geodetic measurements. The data have to be interpreted with care, as geodetic stations within plate boundary deformation zones show spatial and temporal variation relating to various processes. Only stations outside the main plate boundary deformation zones directly give the divergence rate. A network of continuous GPS stations in Iceland (Geirsson et al., submitted) contains some stations outside these zones. Stations with the longest observation span are the REYK station in Reykjavík, on the North American Plate, and the HOFN station, located in Höfn, on the Eurasian Plate (Figure 2.11). The relative velocity between these two stations in 1999–2004 inferred by Geirsson et al. (submitted) is 21.9 mm/yr in direction N102°E, slightly larger than the NUVEL-1A and REVEL velocities. The rate may reflect minor contributions from local processes, such as ongoing glacio-isostatic movements around the Vatnajökull Icecap (see Section 8.4). The observed relative velocity between the REYK and HOFN stations allows, however, the conclusion that essentially all of the spreading across the MAR is accommodated within the width of Iceland. Extensive network GPS measurements

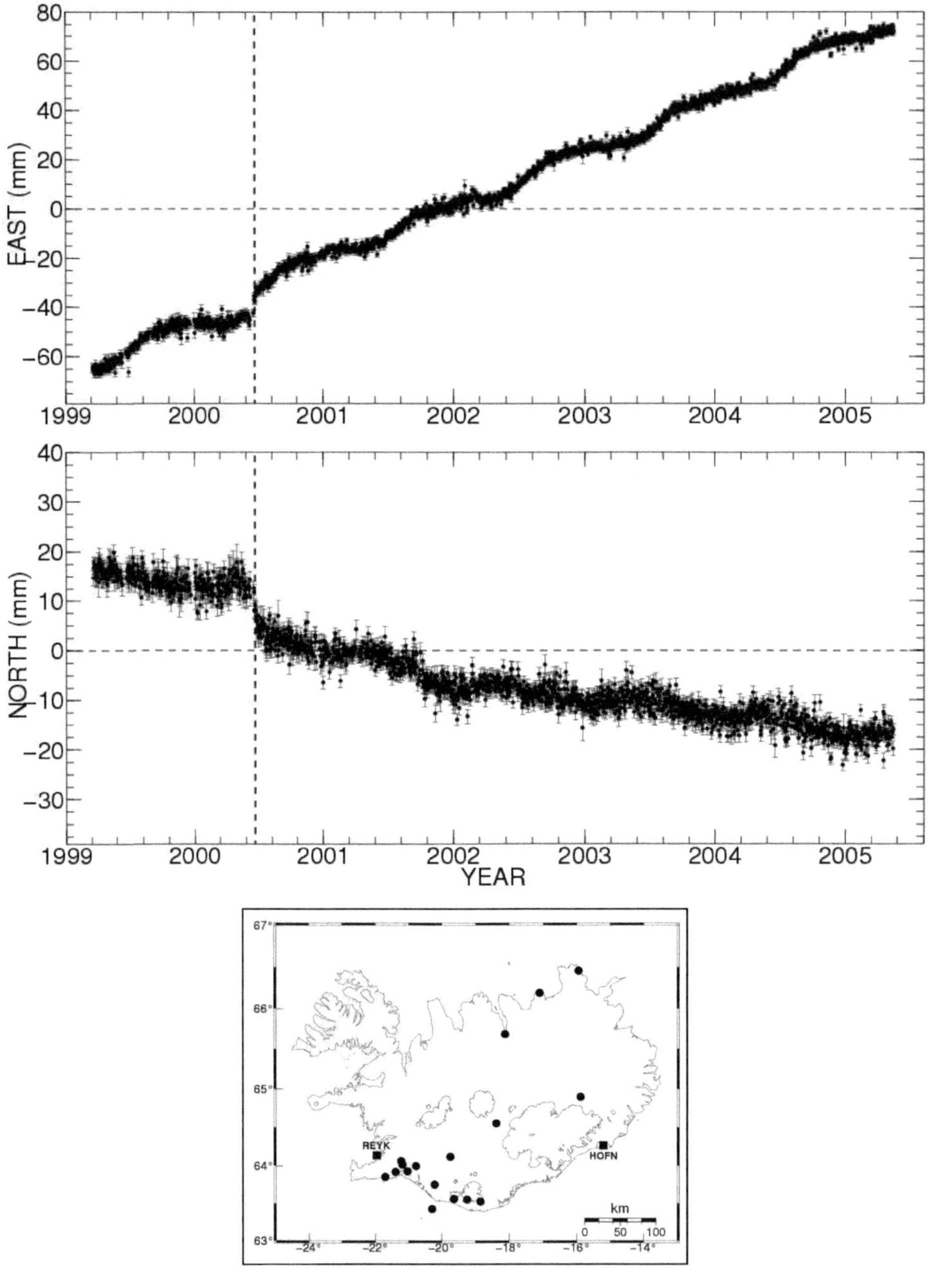

Figure 2.11. Spreading across Iceland inferred from continuous GPS measurements at the HOFN station in SE Iceland and the REYK station in SW Iceland. East (upper panel) and north (lower panel) relative displacements. Map shows location of the REYK and HOFN stations (squares), and other continuous GPS sites (circles).

Courtesy of Halldór Geirsson, Icelandic Metorological Office.

in Iceland conducted since 1986 constrain further the style of spreading and are the topic of Chapter 6.

2.3 SEISMIC STRUCTURE OF THE ICELAND MANTLE PLUME

The Iceland Hotspot is commonly thought to be the surface expression of a mantle plume—a buoyant convection plume of anomalously hot material rising from deeper levels in the mantle. This idea extends back to the original suggestion of mantle plumes by Jason Morgan (1971), who argued that Iceland was formed due to a ridge-centred mantle plume under the island. Numerous seismic experiments have aimed at detecting the mantle plume under Iceland. They provide relatively similar conclusions regarding the structure of the uppermost few hundred kilometres, but their resolution is, in general, poor below a depth of about 400 km.

2.3.1 Plume structure in the upper mantle

The upper mantle under Iceland is characterized by anomalously low seismic velocities, as initially pointed out by Eysteinn Tryggvason (1964), and first mapped out in a pioneering seismic tomography study by Kristján Tryggvason et al. (1983). Data from the ICEMELT network of broadband seismic stations operated in Iceland (1993–1996) then allowed Wolfe et al. (1997) to resolve these low-velocity seismic anomalies much further. They found low P- and S-wave velocities extending from a 100-km to at least a 400-km depth beneath central Iceland, and concluded that Iceland is underlain by a hot, narrow plume of upwelling mantle with a radius of ~150 km (Figure 2.12, see colour plates).

A study by Allen et al. (2002b) used a combination of body wave and surface wave data, primarily from deployment of 30 broadband seismometers in 1996–1998 (the HOTSPOT experiment), supplemented with other datasets. Prior to inversion, the crustal portion of the travel time anomalies were removed using a crustal model (see Section 4.1). Three datasets were used to calculate three independent velocity models for Iceland. These were S-velocity structure as sampled at 0.03–0.1 Hz, and P-velocity structure as sampled at 0.03–0.1 Hz and 0.8–2.0-Hz, yielding three independent but similar velocity models for Iceland. The favoured model is the S-velocity model, ICEMAN-S (Figure 2.13, see colour plates). This shows a cylindrical low-velocity anomaly extending from the maximum depth of resolution at 400 km up toward the surface, where it spreads out beneath the lithosphere. The results are interpreted as a vertical plume conduit at a 400- to ~200-km depth, and a horizontal plume head above 200 km. In the plume conduit, the cylindrical anomaly has a radius of ~100 km and peak v_p and v_s anomalies of −2% and −4%, respectively. Recent work suggests these velocity anomalies may be even more pronounced (Hung et al., 2004).

In the top 250 km under Iceland, Foulger et al. (2000, 2001) also infer a cylindrical seismic anomaly. However, at greater depths it becomes a tabular anomaly oriented N–S along the plate boundary. Another study of S-wave velocity heterogeneity and anisotropy beneath the North Atlantic from regional surface wave tomography (Pilidou et al., 2004) reveals a 5–7% negative anomaly in the mantle above the 200-km depth under Iceland. Low velocities in this model extend along the ridges adjacent to Iceland, being more pronounced beneath the Reykjanes Ridge. The model only resolves structures in the uppermost mantle and has a horizontal resolution of a few hundred kilometres, extending to about a 400-km depth.

At deeper levels in the mantle, studies of the conversion of P to S waves from primary discontinuities at 410- and 660-km depths argue for the presence of upwelling mantle at a 400–700-km depth beneath Iceland (Shen et al., 1996, 1998). The mantle transition zone between 410 and 660 km under Iceland has been inferred to be anomalously thin, and this observation is taken as an indication of mantle upwelling. Excessive temperature within a mantle plume influences the 410- and 660-km phase boundaries in a different manner, causing upward shift of the 660-km discontinuity and downward shift in the 410-km discontinuity. The mantle transition zone beneath Iceland has been inferred to be ~19 km thinner than beneath surrounding areas, with the centre of the zone lying at least 100 km south of the upper mantle low-velocity anomaly (Shen et al., 2002). This lateral shift has been interpreted as evidence for a tilted mantle plume under Iceland, with an inferred tilt angle of 9° from vertical.

The upper mantle structure under Iceland has further been addressed in a series of papers aimed at resolving the whole mantle structure under Iceland (see below).

2.3.2 Plume structure in the lower mantle: a resolution problem

The size of Iceland limits onland seismic station distribution and causes a narrow aperture of seismic networks relative to the mantle plume under the island. Good seismic resolution is achieved only in the uppermost few hundred kilometres under Iceland, and seismic tomography studies generally have poor resolution at greater depths. This is one of the reasons for widely different tomographic results regarding the lower mantle structure under the Iceland region. Seismic anomalies under the Iceland region in different models range from having a depth extent all the way to the core–mantle boundary to being entirely focused in the uppermost few hundred kilometres. Bijwaard and Spakman (1999) argue that tomography provides evidence for a narrow whole mantle plume under Iceland extending all the way to the core–mantle boundary. On the other hand, Foulger et al. (2000, 2001) argue that the seismic anomaly and mantle upwelling under Iceland is confined to the upper mantle. Their main argument is based on inferred morphological change in seismic anomalies with depth. Their inferred change from a cylindrical seismic anomaly to a tabular anomaly oriented N–S along the spreading plate boundary occurs at about a 250-km depth. It is taken as evidence for mantle upwelling under Iceland extending no deeper than the mantle transition zone. Numerical models of convection suggest such a transition in the shape of buoyant upwelling near their base (see discussion by Foulger et al., 2000).

The widely different tomographic results are not only caused by poor resolution of structures in the lower mantle, but are also a consequence of simplifying assumptions made in most seismic tomography analyses. Montelli et al. (2004a, b) present a new tomography technique based on evaluation of finite-frequency travel time tomography for seismic waves. Previous tomographic models were all based on ray theory for transmission of seismic waves through the Earth, valid only in the high-frequency limit of the elastodynamic equations of motion. The new technique is based on a wave approach, considering that travel time of a finite-frequency wave is sensitive to anomalies in a hollow, banana-shaped region surrounding the unperturbed ray path. Depending on the depth and size of anomalies, amplitudes of velocity perturbations in the finite-frequency tomographic images can be 30–50% larger than in other tomography analyses. The results of Montelli et al. (2004a, b) from P-wave studies demonstrate that only a limited number of hotspots are fed by plumes causing P-wave anomalies in the lower mantle. A number of major hotspots do not have a P-wave anomaly in the lower mantle, including Iceland. A study of S-wave anomalies with the same technique (Montelli et al., 2004c and in preparation) reopens, however, the question of the depth extent of the Iceland Plume. These studies confirm a weak plume structure in the mid-mantle around the 1,000-km depth, but a clear S-wave anomaly is observed beneath it, at greater depth in the lowermost mantle (Figure 2.14, see colour plates). Montelli et al. (in preparation) take these observations as an indication for a pulsating plume under Iceland, consistent with surface features in the North Atlantic (see Section 2.5.2), and conclude that if the plume under Iceland extends to the core–mantle boundary, it must be pulsating. This could explain the lack of clear seismic signatures relating to the plume in the mid-mantle at around 1,000 km deep.

2.3.3 An alternative to the plume model

An alternative model which does not invoke a mantle plume has been proposed to explain the existence of the Iceland Hotspot. The discussion, led by Gillian Foulger and coworkers, originates from their seismic tomography observations indicating that mantle upwelling beneath Iceland is confined to the upper mantle. This is taken as evidence for shallow upper-mantle origin for the processes responsible for the Iceland Hotspot. Fundamental differences between a mantle plume model and this alternative model are the depth extent of the anomalous mantle structure beneath Iceland, as well as mantle temperatures. Most plume models for Iceland require high mantle temperatures, whereas the new alternative model calls for only a modest increase in mantle temperatures. The seismic anomalies under Iceland are attributed to both elevated temperatures and the presence of partial melt. By taking into account the presence of partial melt, lower temperatures are needed to explain the anomalies than if only elevated temperatures are considered as the cause of seismic anomalies.

The alternative model attributes enhanced magmatism in the Iceland region to high local mantle fertility leading to anomalously large volumes of melt on this part of the ridge (e.g., Foulger and Anderson, 2005; Foulger et al., 2005). The source of the high local mantle fertility in this model is subducted ocean crust associated with the Caledonian collision around 440–400 Myr ago, when an earlier ocean in the North Atlantic region, the Iapetus Ocean, closed. In addition to fitting the upper mantle structure from seismic tomography, Foulger et al. (2005) argue that major-, trace-, and rare-earth-element compositions, as well as the isotopic characteristics of primitive Icelandic tholeiite, can all be explained by fractional remelting of abyssal gabbro. The diversity of Icelandic basalts may be caused by an enriched component already present in recycled crustal section. According to the model, compositions ranging from ferrobasalt to olivine basalt are produced by various degrees of partial melting of ecologite. Although Foulger et al. (2005) demonstrate that this alternative model can explain the petrology and geochemistry of Iceland, they do not argue that the geochemistry of Iceland is inconsistent with contemporary plume theory.

2.4 PLUME MODELS: EXCESS TEMPERATURES AND ENERGETICS

Mantle plumes are anomalously hot material rising from deeper levels in the mantle. Their excess temperature causes buoyant convection in the mantle that carries material towards the surface of the Earth (Figure 2.15, see colour plates). Various observations constrain the properties of mantle plumes, including geochemistry, seismic results, topography, gravity, geoid, and heat flow. Fluid-dynamical models have been applied to understand plume dynamics, as detailed in the comprehensive book by Schubert et al. (2001). Primary parameters of mantle plume models are the width of plumes and their temperature anomaly. Flux of material in mantle plumes is immense. Sleep (1990) gives an estimate of volume flux of the Iceland Mantle Plume based on the kinematics of spreading, by considering that the plume needs to supply the oceanic lithosphere at least down to the depth of extensive melting, about 80 km, and assuming that the plume flux balances the flow at great distance. The suggested volume flux in the Iceland Plume is 63 m^3/s or about 2 km^3/yr. Another parameter used to quantify mantle plumes is the so-called buoyancy flux. Denoting the plume volume flux by Q_v, then the buoyancy flux, $\dot{M}_p$, is defined as (Sleep, 1990; Schubert et al., 2001):

$$\dot{M}_p = Q_v \Delta\rho = (\rho_m \alpha \Delta T) Q_v \tag{2.3}$$

Here $\Delta\rho$ is the mean plume density deficit relative to the mantle, assumed to be due to thermal expansion of mantle material with density ρ_m, thermal expansion coefficient α, and an average excessive temperature of ΔT. For $\rho_m = 3{,}400$ kg/m^3, $\alpha = 3 \times 10^5\ ^\circ\mathrm{C}^{-1}$, and $\Delta T = 200^\circ\mathrm{C}$, the buoyancy flux for the Iceland Mantle Plume is estimated as 1,400 kg/s. This parameter can be related to the rate of hotspot swell formation, through the assumption that the excess mass of a swell is compensated by an equal mass deficit at depth.

The heat transported by a mantle plume, Q, is related to the volume flux by:

$$Q = (\rho_m c_p \Delta T) Q_v \tag{2.4}$$

where c_p is the specific heat at constant pressure. For $c_p = 1{,}250$ J/(kg °C) and the other parameters as above, the heat transported by the Icelandic mantle plume is estimated as 58 GW (Schubert et al., 2001).

2.5 PLUME–RIDGE INTERACTION AND THE ICELAND HOTSPOT SWELL

Fluid-dynamical models appropriate for the Iceland Mantle Plume are those considering a ridge-centred plume. The interaction of a mantle plume and a mid-ocean ridge has to be considered. A series of models has been calculated for Iceland, including those of Ito et al. (1996, 1999), Conrad et al. (2004), and Marquart and Schmeling (2004). An overview of observations and models of mantle-plume–MAR interaction is given by Ito et al. (2003).

2.5.1 Topography and gravity

A number of earlier models had problems fitting all observational constraints of plume–ridge interaction in the North Atlantic, including crustal structure, bathymetry, gravimetry, and width of geochemical anomalies. A range of models had been suggested, with one end-member consisting of a plume with relatively broad radius (~300 km) and a moderate temperature anomaly (ΔT ~75°C) and the other end-member consisting of a relatively narrow plume (radius less than 100 km) and a greater temperature anomaly (ΔT >150°C) (e.g., Ribe et al., 1995; Ito et al., 1996). These earlier models did not consider rheological changes associated with extraction of water from the mantle during partial melting, but Ito et al. (1999) have shown that consideration of this effect is of primary importance in explaining the spreading of plume heads.

Onset of mantle melting is associated with dehydration and a consequent increase in viscosity. The rheological effects of extracting water from the mantle during partial melting depend on the initial water content and temperature conditions. Based on the analysis of Hirth and Kohlstedt (1996), the viscosity increase associated with loss of water is likely to dominate over any viscosity reduction due to retention of melt. Ito et al. (1999) argue that above the dry solidus where most melting takes place, plume viscosity may be 50 times greater than below the dry solidus. The relative importance of various effects during melting is also estimated by Braun et al. (2000) who demonstrate that viscosity may increase up to two orders of magnitude from dehydration effects, with this effect dominating over other contributions to changes in viscosity (Figure 2.16). The dehydration effect fundamentally modifies upwelling rates above the dry solidus (Figure 2.17). Inclusion of the viscosity dehydration effect makes a model of a plume with relatively high excess temperature (180°C) and narrow radius (100 km) capable of reproducing the observed along-axis crustal thickness, bathymetry and gravity variations in the North Atlantic (Figure 2.6).

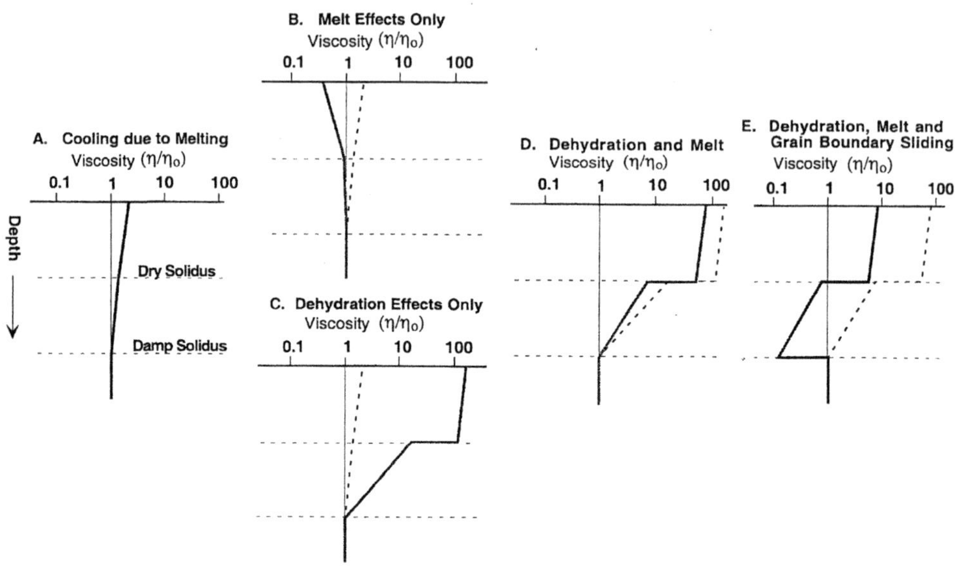

Figure 2.16. Normalized viscosity versus depth profiles showing the effects of melting on the viscosity of an initially damp mantle.

Reproduced from Braun et al. (2000) with permission of Elsevier.

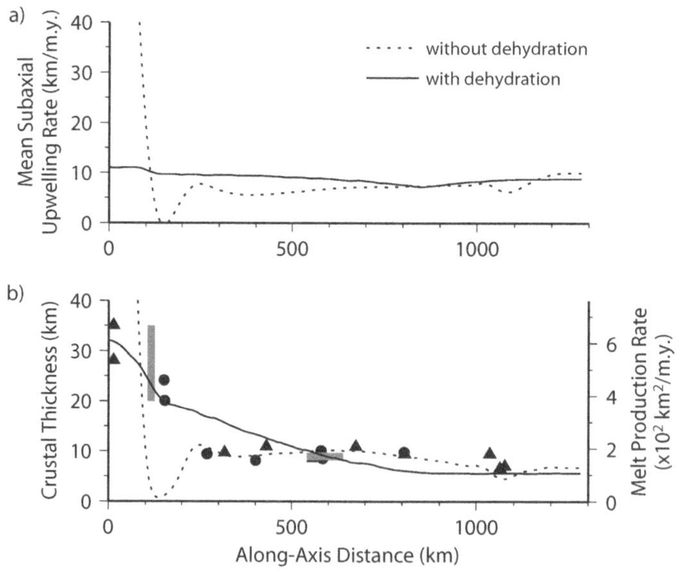

Figure 2.17. Mantle-upwelling rate above the dry solidus (a) and crustal thickness (b) in the North Atlantic versus distance from Iceland's centre, according to the geodynamic model of Ito et al. (1999), with and without viscosity dehydration effect.

Reproduced from Ito et al. (1999) with permission of Elsevier.

2.5.2 V-shaped ridges

The V-shaped ridges around Iceland (Figure 2.7, see colour plates) have stimulated various ideas about a pulsating mantle plume under Iceland since an initial suggestion by Vogt (1971). Pulses of activity travelling southward along the Reykjanes Ridge, when superimposed on plate spreading, can explain the formation of these V-shaped structures. The propagation velocity of these anomalies records the rate of lateral plume flow along the ridge and can be inferred from the shape of the V-shaped ridges. If a ridge axis is perpendicular to the spreading direction, the component of asthenospheric flow along the spreading axis, v_a, is given by:

$$v_a = S \cot \theta_R \tag{2.5}$$

where S is the spreading half rate and θ_R is the angle between either limb of the V-shaped ridge and the spreading axis. For the ocean floor south of Iceland S is about 1 cm/yr and θ_R is about 3–6°. Relation (2.5) is somewhat modified for oblique spreading, but in any case the inferred flow component along the ridge is 10–20 cm/yr, an order of magnitude faster than the plate-spreading rate (Vogt, 1971; Johansen et al., 1984). The V-shaped ridges around Iceland are further described and discussed, for example, by Jones et al. (2002)

The formation of V-shaped structures has been reproduced in a fully three-dimensional fluid-dynamical model of a pulsating and radially flowing mantle plume under a mid-ocean ridge. A model by Ito (2001) imposes variable flux in an upwelling plume by variation in the radius of the plume stem about a steady state as a periodic function of time. Large variations in the plume flux are needed to explain the observed structures around Iceland. Such pulses may influence topography and condition, including ocean circulation, over large parts of the North Atlantic (e.g, White and Lowell, 1997). Another view of the origin of V-shaped ridges is that they relate to relocation of the spreading axis in Iceland (rift jumps) as envisaged, for example, by Hardarson et al. (1997). Jones et al. (2002) conclude that the V-shaped ridges are probably generated by time-dependent flow in the Iceland Plume, with plume pulses eventually triggering the rift jumps in Iceland.

2.6 MOVEMENT OF THE MAR RELATIVE TO THE ICELAND MANTLE PLUME: THE HOTSPOT TRACK

Plate motion relative to Earth's hotspots can be estimated from studies of the ages and location of volcanoes at hotspot trails that define volcanic propagation rates and trends of hotspot paths. Ten hotspot datasets form the basis for the HS3–NUVEL1A model by Gripp and Gordon (2002), averaging plate motion over the last 5.8 Myr. The model gives angular velocities of plates relative to the hotspots, with the hotspots having insignificant relative motion (Table 2.2). If hotspots are the surface expression of plumes fixed in the mantle, this model gives the absolute plate motion. In Iceland, the inferred absolute plate motion is highly asymmetric.

Table 2.2. HS3–NUVEL1A Euler poles and angular velocities for the Eurasian and North American Plates.

Plate	Latitude	Longitude	ω	Error ellipse		ξ	σ_ω
	(°N)	(°E)	(°/Myr)	σ_{max}	σ_{min}		
Eurasia	−61.901	73.474	0.2047	27.38	17.52	3	0.0524
North America	−74.705	13.400	0.3835	15.59	8.7	56	0.0548

Error ellipses are one-sigma angular lengths in degrees of the semimajor and semiminor axes of the pole of rotation, and ξ is the azimuth of the semimajor ellipse axis in degrees clockwise from north.

According to the HS3–NUVEL1A model, the movement of the Eurasian Plate in central Iceland (64.5°N, 18°W) is 14 mm/yr in direction N218°E, and the North American Plate moves 27 mm/yr in direction N257°E. The Eurasian Plate has small movement relative to the plume, whereas the North American Plate is moving at more than the full plate-spreading velocity westward from the plume. As a consequence, the central axis of the plate boundary drifts also westward relative to the mantle plume underlying Iceland.

The HS3–NUVEL1A model averages plate motion for only the last few million years but models of hotspot tracks for the North Atlantic over longer time intervals show comparable trends, with the majority of the absolute motion (relative to the Iceland Plume) being taken up by the North American Plate. Jason Morgan initially studied Atlantic hotspot paths, and showed that during the early Tertiary the Iceland Hotspot was located under Greenland (e.g., Morgan, 1983). The Tertiary volcanism on both sides of the North Atlantic (Figure 2.4) is due to the Iceland Mantle Plume, and the Greenland–Scotland Ridge has been suggested to be the track of the Iceland Hotspot. This association is, however, not straightforward as the current hotspot location is about midway between their early Tertiary manifestations on both sides of the North Atlantic, but considering the asymmetric absolute plate motion the present Icelandic hotspot should be near the Eurasia continental margin (Vogt, 1983). In order to reconcile this apparent discrepancy, hotspot models for the North Atlantic presented by Morgan (1983) and Vink (1984) have the Greenland–Scotland Ridge formed at a point on the MAR accretion axis fed by mantle flow located some distance away from the axis. The Iceland Hotspot track has been further studied by Lawver and Müller (1994). Although some differences exist between models, they all have the Iceland Hotspot located under Greenland in the early Tertiary, being responsible for the formation of early Tertiary lavas in both eastern and western Greenland.

The different motion of the plates in Iceland relative to the underlying plume is reflected in a variety of structures both at the surface and at depths in the mantle. Seismic studies reveal differences in mantle structure east and west of the plate boundary, including observations of shear wave splitting. Bjarnason et al. (2002) study upper-mantle anisotropy from the splitting of teleseismic shear waves and relate the observations to a flow-induced, lattice-preferred orientation of olivine grains. A change in the fast polarization direction from eastern to western Iceland can be explained by the different absolute motion of the North American and Eurasian Plates, invoking also a background mantle flow of 3 cm/yr in a hotspot reference frame that governs the orientation of the anisotropy in the almost stationary Eurasian Plate.

On the surface, the area west of the current plate boundary in Iceland has several ancient rift zones. It is heavily fractured and characterized by extensive geothermal areas whereas the zone east of the boundary is much less fractured. The plate boundary drifts towards west from the plume centre and once it is sufficiently far away there is a tendency for the plume to break through the lithosphere again and form a new segment of the plate boundary. On the surface a rift jump towards east is observed, shifting the plate boundary again towards the mantle plume. These rift jumps and reorganisation of the plate boundary are the reasons for the complex tectonics of Iceland discussed in the following chapter.

3

Tectonic framework

The landscape of Iceland is shaped by volcanism and glaciations. Holocene and late glacial volcanic deposits are found in the neovolcanic zone that stretches across Iceland and are the onshore continuation of the Mid-Atlantic Ridge (MAR). The rocks exposed at the surface are up to 15 Myr old, with the oldest rocks occurring at the eastern and western extremities of Iceland. The onset of frequent glacial conditions in Iceland around 3.3 Myr ago marks a fundamental environmental change, with development of extensive subglacial volcanic products and more complex lithology than compared with the earlier Tertiary period. The currently active plate boundary consists of a series of volcanic and seismic zones that have developed and reorganized through time in a complex manner due to interaction of the MAR and the Iceland Mantle Plume. The building blocks of the volcanic zones are about 35 volcanic systems, typically consisting of a central volcano, often with a caldera and an associated fissure swarm.

3.1 GEOLOGY

Rocks in Iceland are divided into four stratigraphic groups based on climatic and paleomagnetic field conditions at the time of formation, and absolute age data (Figure 3.1). The stratigraphic geological division used in Iceland is somewhat modified from that used elsewhere, with primary epochs being the Tertiary, the Plio-Pleistocene (Upper Pliocene and Lower Pleistocene), the Upper Pleistocene and the Postglacial. This division of rock units is used conventionally for the geologic map of Iceland (Figure 3.2). An overview of the geology is given, for example, by Kristján Saemundsson (1978, 1979, 1986), Þorleifur Einarsson (1991), and Thordarson and Hoskuldsson (2002).

Iceland is mostly made up of basalts. They cover about 92% of the surface area of Postglacial volcanic zones, whereas 4% are basaltic andesites, 1% are andesites,

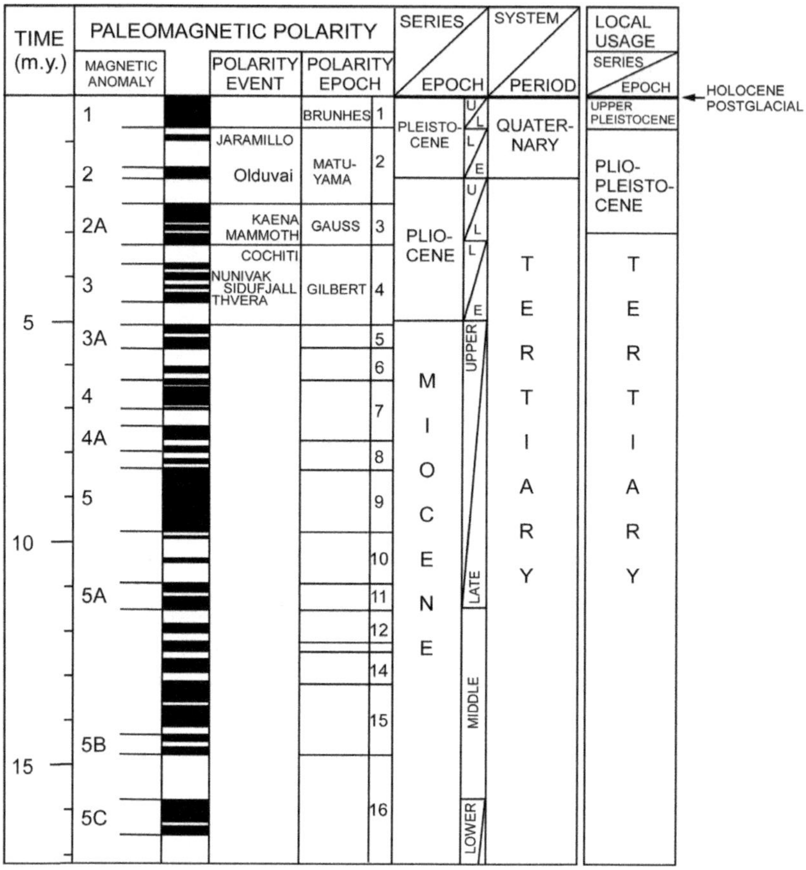

Figure 3.1. Stratigraphic timetable, with a modified version (far right) conventionally used in Iceland.

Modified from Saemundsson (1979) with permission of *Jökull*.

and 3% are dacite-rhyolites. An overview of the petrology of Iceland is, for example, given by Jakobsson (1979a, b). Geochemistry is discussed by Óskarsson et al. (1982, 1985) who argue that the geochemical characteristics of Icelandic volcanic rocks result from the interaction between the Iceland Hotspot and its trail. In response to eastward jumps of the plate boundary, new rift segments form in older crust.

3.1.1 The Tertiary

Rocks older than 3.3 Myr make up the Tertiary formation covering about half of Iceland. They occur in eastern, western, and northern Iceland, with ages increasing with distance from extinct and active spreading zones. The oldest rocks, 14–15 Myr old, are found in western and eastern Iceland (Figure 3.3), whereas rocks in north

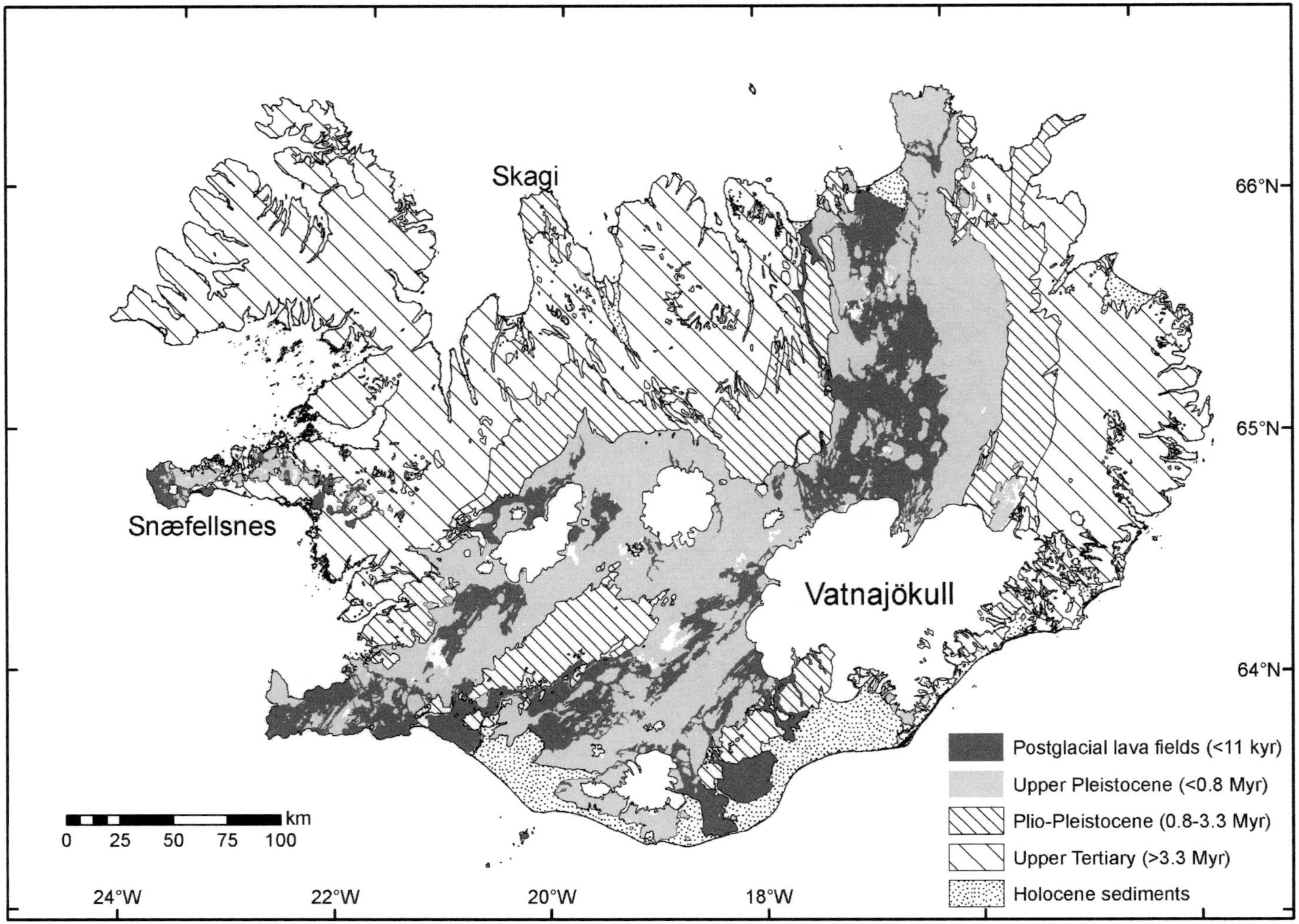

Figure 3.2. Geologic map of Iceland.

Modified from Jóhannesson and Sæmundsson (1998) with permission of Icelandic Institute of Natural History.

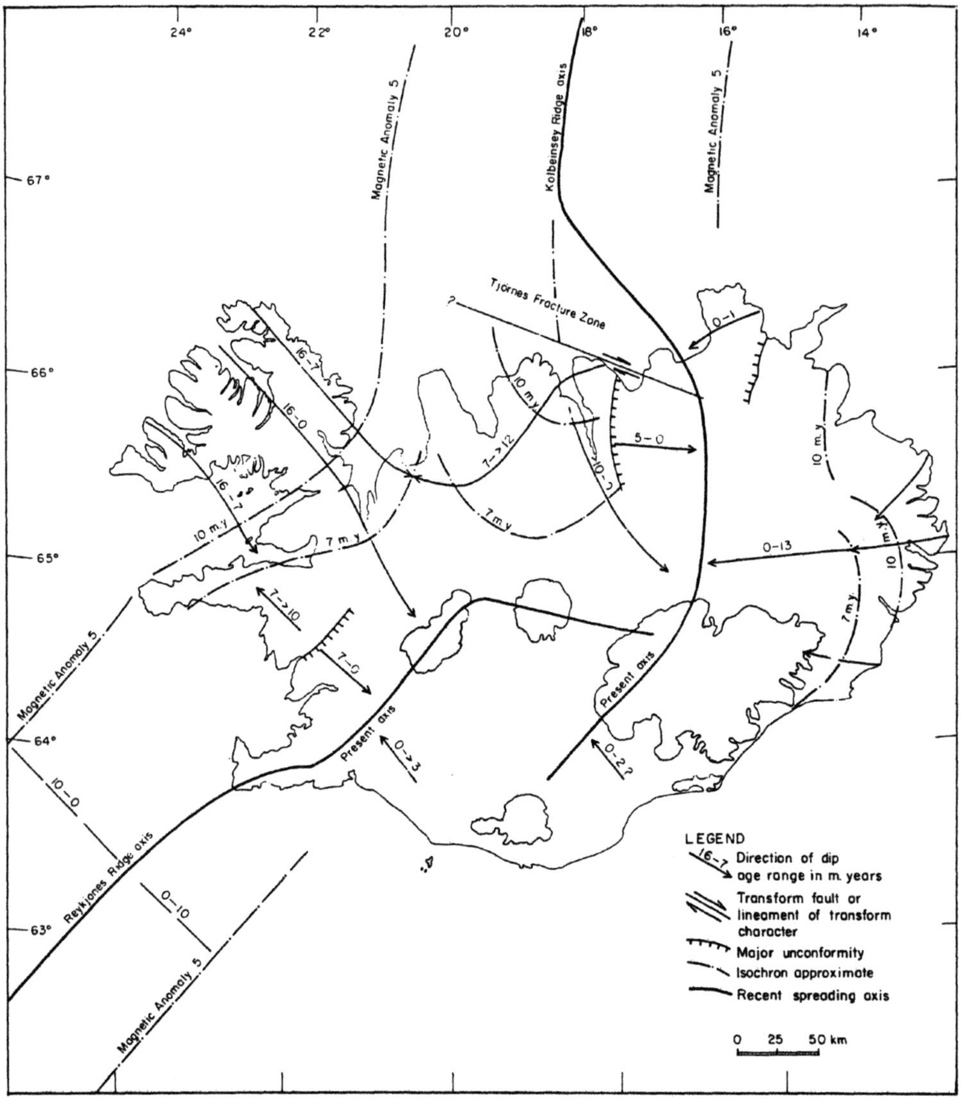

Figure 3.3. Development of the present geometry of the tectonically active zones in Iceland.
Reproduced from Saemundsson (1979) with permission of Jökull.

Iceland are up to 12 Myr old (Saemundsson, 1986). Tertiary rocks formed prior to extensive glaciations in Iceland, so glacial deposits and subglacial volcanic products are rare. Most of the Tertiary formation consists of a regular basaltic lava pile of uniform lithology. Subaerially erupted tholeiitic lavas about 5–15 m thick separated by minor clastic interbeds of volcanic origin form the bulk of the Tertiary lava pile (Saemundsson, 1979). Within this lava pile, the thickness of each flow is on the order

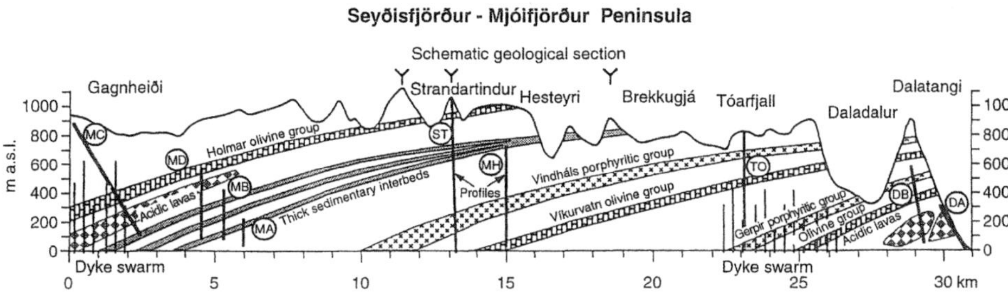

Figure 3.4. Schematic geological section in eastern Iceland showing regional tilting of the lava pile.

Reproduced from Kristjansson et al. (1995).

of 10 m, with each kilometre thickness of the lava pile spanning about 1 Myr. The average lava deposition rate was low, about one lava flow per 10,000 years. However, these rates vary over one order of magnitude, from about 360 to 4,000 metres per million of years (Saemundsson, 1986). The interbeds between lava layers in the lava pile are, most commonly, thin layers of red or red–brown clayey or tuffaceous material, believed to be mostly soil and windblown ash that has suffered chemical weathering (Saemundsson, 1979).

The regular structure of the Tertiary lava pile is interrupted by eroded volcanic centres associated with silicic rocks, extensive faulting, and diking. These ancient structures are analogues to the currently active volcanic centres found in the neovolcanic zone of Iceland. The rates of lava deposition in the Tertiary series referred to above apply to areas outside of these volcanic centres.

A characteristic feature of the Tertiary lava sequence outside the central volcanoes is a regional tilting of the lava pile (Figure 3.4). The lava layers have a distinct tilt towards the volcanic zones in which they originated, with dip varying from near zero at the highest exposed levels to about 5–10° at sea level (Figure 3.5). The lava layers were flat at the time of formation, with the regional tilt forming during growth of the lava pile (Saemundsson, 1979). Consequently, the load of the lava pile is responsible for flexure of the crust and development of associated syncline structures which are centred on rift areas. Anticline structures have formed in association with rift relocation, with lava loading occurring in two rift zones on each side of anticlines (Saemundsson, 1967).

The end of the Tertiary period and the beginning of the Quaternary period in Iceland is somewhat arbitrarily set at the end of the Mammoth paleomagnetic event at 3.3 million years, within the Gauss magnetic epoch. Around this time climate cooled with onset of frequent glaciations. The change was not abrupt as deposits of glacial origin are found in the Tertiary lava pile back to about 7 Myr, in southeastern and eastern Iceland (e.g., Friðleifsson, 1995; Hjartarson and Hafstað, 1997). However, the change around 3.3 Myr ago was drastic. Tuffs and volcanogenic sediments amount to only some 5% of the volume of the Tertiary series, whereas subglacial volcanics and glacially derived detrital beds gain in volume in later formations and may exceed 50% of its volume (Saemundsson, 1986). The onset of

Figure 3.5. View over Breiðdalsvík at the eastern coast of Iceland, showing well the regional tilt of lava layers.

Photo courtesy of Oddur Sigurðsson.

frequent glacial conditions marks the beginning of the Quaternary period which in Iceland is commonly divided up into three epochs described below.

3.1.2 The Plio-Pleistocene (Upper Pliocene and Lower Pleistocene)

The oldest part of the Quaternary series in Iceland formed during the Upper Pliocene and Lower Pleistocene epoch (termed Plio-Pleistocene in Iceland), beginning 3.3 million years ago and ending 0.8 Myr ago. The end of the epoch is marked by the transition from the Matuyama magnetic epoch of predominantly reversed magnetic polarity, to the Brunhes magnetic epoch of normal polarity. Rocks from the Plio-Pleistocene epoch bound the currently active rift zones and in most places lie conformably above the Tertiary sequence. Exceptions occur on the Skagi Peninsula in northern Iceland and on Snæfellsnes in western Iceland (Figure 3.2). In those locations an unconformity exists between the Plio-Pleistocene sequence and the Tertiary lava pile.

Rocks formed during the Plio-Pleistocene include extensive fluvioglacial and morainic deposits as well as hyaloclastites formed during subglacial eruptions. The structure of this rock series is therefore very different from the more uniform structure of the Tertiary lava pile, with many more irregularities. In subglacial volcanic eruptions, the erupted magma forms pillow lava, pillow breccia, or glassy tuffs at the eruptive site which are later transformed into hyaloclastites. Commonly the ice has confined the material, causing it to pile up at the eruptive site. Subaerially erupted lavas are found inbetween the subglacial formations, indicating that the Plio-Pleistocene was characterized by alternating warm and cold periods, with glaciation recurring every 100,000–120,000 years (e.g., Saemundsson and Noll, 1974). In southwestern Iceland, Kristjánsson et al. (1980) find evidence for 13 glaciations between 3.1 Myr and 1.8 Myr, in a 2.1-km-thick succession of lavas separated by glacial horizons.

3.1.3 Upper Pleistocene

The Upper Pleistocene rock series in Iceland consists of rocks formed during the Brunhes magnetic epoch which began 0.8 million years ago, excluding the Postglacial (Saemundsson, 1979). It is characterized by still more extensive hyaloclastite formations than the Plio-Pleistocene, as well as lavas erupted during interglacial times. The volcanic rocks of the Upper Pleistocene fall mostly into two primary types according to their mode of formation. One is subaerially erupted lava flows that have formed during interglacial times. The other type is the subglacial formations, subglacial pillow lavas, and hyaloclastite rocks referred to collectively as the "Palagonite formation" (Móberg in Icelandic). The increased proportion of subglacially formed rocks relative to subaerially erupted rocks in the Upper Pleistocene indicates more extensive glaciation occurred during that time. Furthermore, glacial erosional features are in general insignificant in the Upper Pleistocene sequence as volcanic accumulation appears to have dominated over glacial or fluvial erosion (Saemundsson, 1979). In addition to being little eroded, the formations of the Upper Pleistocene can frequently be related to currently active volcanic systems. The Upper Pleistocene rock series together with Postglacially erupted rocks are referred to as the neovolcanic zone of Iceland.

Rocks from the Upper Pleistocene reveal well the morphological relationship between subglacial volcanic landforms and the overlying ice sheets. Within the neovolcanic zones, the height of subglacially erupted landforms correlates directly with the thickness of the overlying ice sheet at the time of formation (Figure 3.6) (Walker, 1965). Recent studies (e.g., Werner et al., 1996) demonstrate that individual subglacial volcanic landforms can, however, be very complex formations, formed in a number of eruptions under different environmental conditions (Figure 3.7).

3.1.4 The Postglacial

Postglacial time in Iceland begins about 11,500 BP, corresponding to the time when outer parts of Iceland became deglaciated. Extensive fresh lava flows and pyroclastics as well as sediments and soil formed after deglaciation, and characterize this epoch. Lavas erupted during the Postglacial are glacially uneroded and cover the bulk of the neovolcancic zone of Iceland. The most

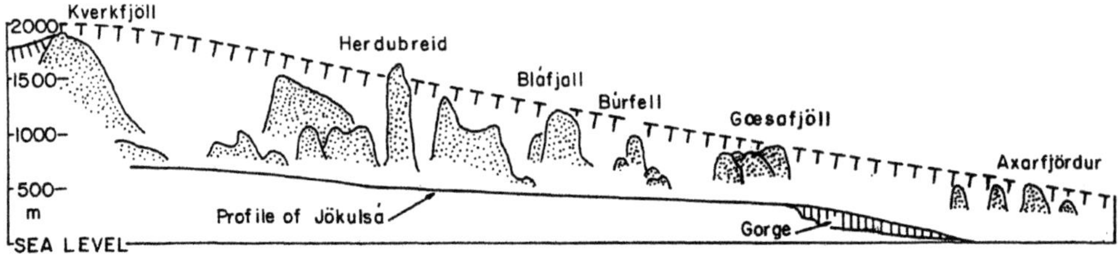

Figure 3.6. Elevation of subglacially erupted volcanoes in northern Iceland. The profile extends from the Kverkfjöll central volcano at the northern margin of Vatnajökull Icecap to the north coast of Iceland at Axarfjörður.
Reproduced from Walker (1965).

extensive postglacial sediments occur along the south coast of Iceland, formed in repeated glacial outburst floods ("jökulhlaup") associated with subglacial volcanic eruptions. Lavas erupted during the Holocene are divided into prehistorical lavas older than 1,100 years, and historical lavas younger than 1,100 years (Figure 3.8). This division is only for historic reasons; Iceland was settled around 900 AD and descriptions exist of some eruptions after that time. The intensity of volcanism during Postglacial time has varied, with much more extensive volcanic production in the initial millennia after deglaciation than is occurring currently (see Section 9.3).

3.2 THE PLATE BOUNDARY IN ICELAND

3.2.1 Volcanic zones

The neovolcanic zone in Iceland is divided into two types, depending on the amount of crustal spreading that has occurred within them. Volcanic flank zones are associated with little or no crustal spreading, whereas extensive crustal spreading characterizes the volcanic rift zones (Figure 3.9). There is also a fundamental difference in the composition of eruptive products within these zones. Tholeiites form in the volcanic rift zones, whereas alkali olivine basalts and transitional alkali basalts form in the flank zones (Jakobsson, 1972, 1979a, b). There are three volcanic flank zones, the Snæfellsnes Volcanic Zone, the South Iceland Volcanic Flank Zone (SIFZ), and the Öræfajökull–Snæfell Flank Zone.

The neovolcanic zone associated with extensive rifting forms the spreading plate boundary in Iceland. The neovolcanic zone along the Reykjanes Peninsula and in western Iceland from the Hengill area up through the Þingvellir area and to Langjökull has been termed the Reykjanes–Langjökull Volcanic Zone. Volcanic activity in this area forms a continuous zone that has been active during the same timespan, for the last 6–7 million years (Saemundsson, 1979, 1986). There is however considerable structural variability within this part of the neovolcanic zone, as its obliquity with respect to the general plate motion direction is very different north and south of the Hengill volcanic area. Furthermore, the Hengill area is a triple

(a)

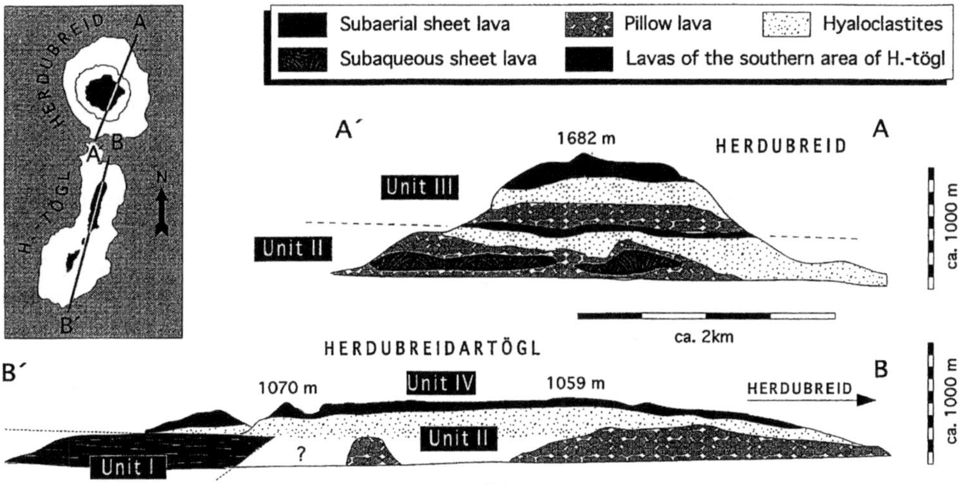

(b)

Figure 3.7. The Mt. Herðubreið table mountain: (a) photo; (b) stratigraphy according to Werner et al. (1996) indicating the mountain is a result of a series of eruptions under different environmental conditions. Dashed line marks a sharp mineralogical boundary.

(a) Courtesy of Oddur Sigurðsson. (b) Reproduced from Werner et al. (1996).

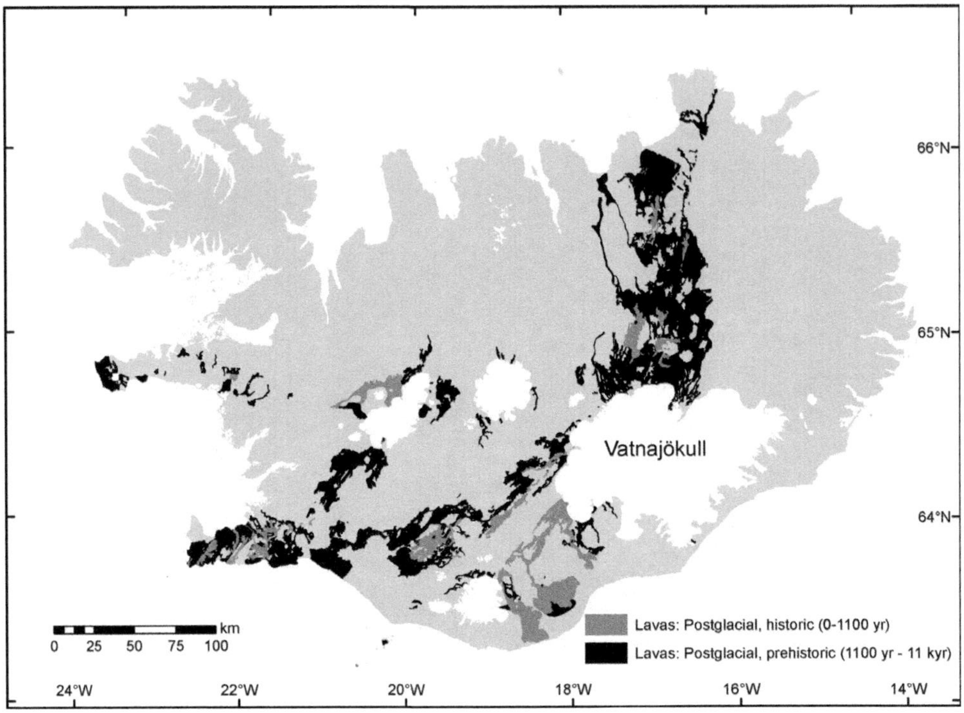

Figure 3.8. Postglacial lava fields, historic and prehistoric.

Modified from Jóhannesson and Sæmundsson (1998) with permission of the Icelandic Institute of Natural History.

point, with a third arm of a plate boundary being the transform zone in south Iceland, the South Iceland Seismic Zone (SISZ). Based on this difference, the Reykjanes–Langjökull Zone can be divided into two zones (Einarsson, 1991a), the Reykjanes Peninsula (RP) oblique rift, and the less oblique Western Volcanic Zone (WVZ).

Similarly, the eastern branch of the neovolcanic zone in Iceland extending from the north coast to the Vatnajökull Icecap and south towards the Westman Islands, can be classified as one or two volcanic zones (e.g., Oskarsson et al., 1985; Einarsson, 1991a). A division has been made at the Vatnajökull Icecap, as the volcanic zones north and south of it differ in age and are of somewhat different character. No lava shields are found in the zone south of Vatnajökull. North of Vatnajökull the zone has been active for 6–7 Myr, whereas south of Vatnajökull volcanism began 2–3 Myr ago (Saemundsson, pers. commun., 2005), and the rift appears to be propagating southwards. The volcanic rift zone north of Vatnajökull is termed the Northern Volcanic Zone (NVZ). South of Vatnajökull it is termed the Eastern Volcanic Zone (EVZ), to distinguish it from the overlapping WVZ which is also in south Iceland. A major structural change occurs in the EVZ in the Torfajökull area, at the junction of the EVZ with the SISZ (see below). Significant crustal spreading in

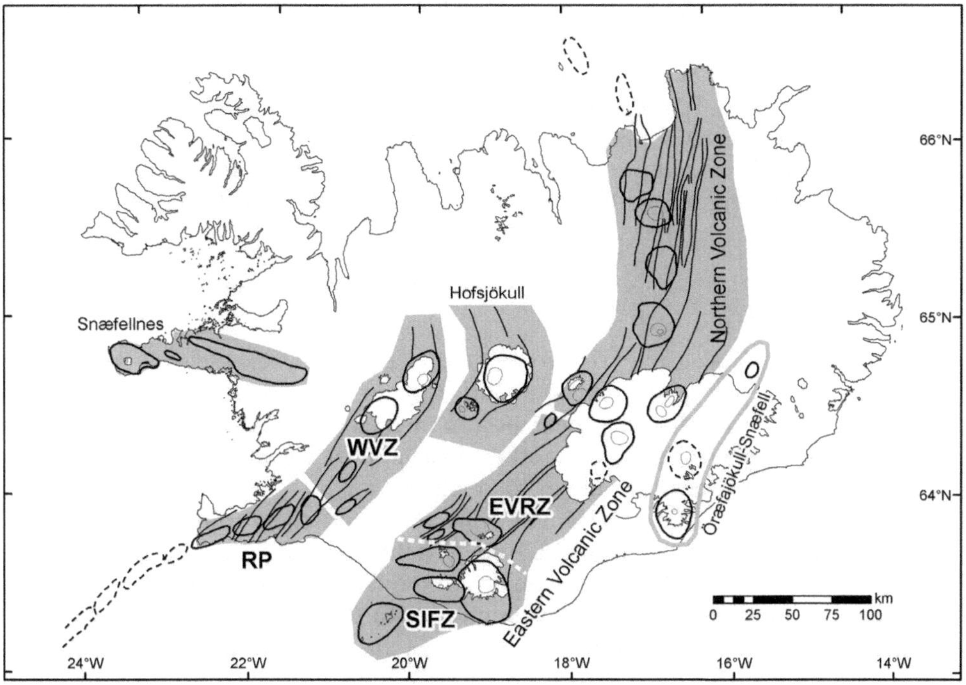

Figure 3.9. Volcanic zones of Iceland. The volcanic rift zones include the Northern Volcanic Zone (NVZ), the Western Volcanic Zone (WVZ), the Eastern Volcanic Rift Zone (EVRZ), and the Reykjanes Peninsula (RP) oblique rift. The volcanic flank zones (with little or no rifting) are the Snæfellsnes, Öræfajökull–Snæfell and the South Iceland Flank Zone (SIFZ). Together the SIFZ and the EVRZ are termed the Eastern Volcanic Zone (EVZ).

the EVZ has only developed north of the Torfajökull area. South of it, the EVZ may be termed the South Iceland Flank Zone (SIFZ), whereas north of it is termed the Eastern Volcanic Rift Zone (EVRZ). In addition to structural differences, the boundary is associated with a change in petrology from olivine tholeiites in the EVRZ to transitional and alkali olivine basalts in the SIFZ.

A majestic subglacial volcano in the middle of Iceland, Mt. Hofsjökull sits inbetween the northern ends of the main branches of the WVZ and the EVZ. This volcano has been classified as either a part of the WVZ or as a separate zone, the Middle Iceland Volcanic Zone or the Central Iceland Transform (e.g., Oskarsson et al., 1985). An active rhyolitic volcanic centre, Mt. Kerlingarfjöll, is located south of it. Because of its special tectonic setting, it is suggested to use the term Hofsjökull Volcanic Zone for this area. An extensive system of fractures extends from Hofsjökull north towards Skagafjörður, and forms a part of this zone. Currently the Hofsjökull Volcanic Zone accommodates little or no plate spreading. At this latitude, the plate spreading is currently focused on the EVZ (e.g., Geirsson et al., submitted).

The tectonic situation is more complex in south Iceland than in north Iceland, with two overlapping rifts, the WVZ and EVRZ. Activity within these overlapping rifts is very different. The WVZ has been the main locus of crustal spreading in south Iceland for the last 6–7 Myr, whereas activity began in the EVRZ 2–3 million years ago (Jóhannesson, 1980). Spreading activity seems, however, to be shifting from the WVZ to the EVRZ, and at present time the EVRZ is currently accommodating most of the plate spreading in South Iceland (e.g., Einarsson, 1991a; La Femina et al., in press).

3.2.2 Transforms

In addition to the volcanic rift zones, there are two main seismic zones that constitute the active plate boundary in Iceland, the SISZ, and the Tjörnes Fracture Zone (TFZ). Crustal shear occurs in these zones which transform plate movements from one rift zone to another. Instead of being simple transform faults, they are transform zones of complex character.

The SISZ extends from the Hengill triple junction to the Torfajökull area, where the EVZ changes character from rift zone to flank zone (Figure 3.9). It transforms plate motion from the RP Plate Boundary, to the EVRZ. These two zones currently accommodate most of the plate spreading in South Iceland, whereas little extension is currently occurring across either the WVZ or the South Iceland Volcanic Flank Zone. Rather than having a single east–west-trending left-lateral strike–slip fault along its length, the SISZ is characterized by an array of north–south-trending faults. This array accommodates the overall left-lateral shear across the zone by right-lateral slip on the north–south faults in a "bookshelf" style of faulting (see Section 8.2).

Near the north coast of Iceland the TFZ transforms plate motion from the NVZ to the submarine Kolbeinsey Ridge, about 150 km to the west-northwest. Because it is oriented obliquely to the spreading direction, the TFZ is a complex zone which accommodates both right-lateral shear as well as a component of extension (see Section 8.1).

3.3 SEGMENTATION OF THE VOLCANIC ZONES: VOLCANIC SYSTEMS

Volcanism within the active zones of Iceland differs widely in character. Intense volcanism has built up a number of volcanic edifices through repeated eruptions. These foci of volcanic production along the volcanic zones are termed central volcanoes. Many of them are associated with silicic rocks, high-temperature geothermal areas, and some have developed a caldera. Volcanism is frequent at the central volcanoes. Both inside and outside the central volcanoes, monogenetic crater rows, formed in fissure eruptions, often group together with an array of normal faults. Such zones of extensive fissuring and normal faulting have in Iceland been termed "fissure swarms". Central volcanoes within the volcanic rifts in Iceland are, as a rule, transected by fissure swarms. These features together, a central volcano and its associated fissure swarm, typically comprise a volcanic system (Saemundsson, 1979). In addition to their individual tectonic character, the volcanic systems also have their own petrographic and geochemical character (Jakobsson, 1979a, b).

The NVZ was initially divided into volcanic systems by Saemundsson (1974) who later divided as well all the volcanic zones into volcanic systems (Saemundsson, 1979). Focusing more on petrological and geochemical character, Jakobsson (1979a, b) outlined volcanic systems in the EVZ. Einarsson and Saemundsson (1987) defined over 30 volcanic systems in Iceland (Figure 3.10).

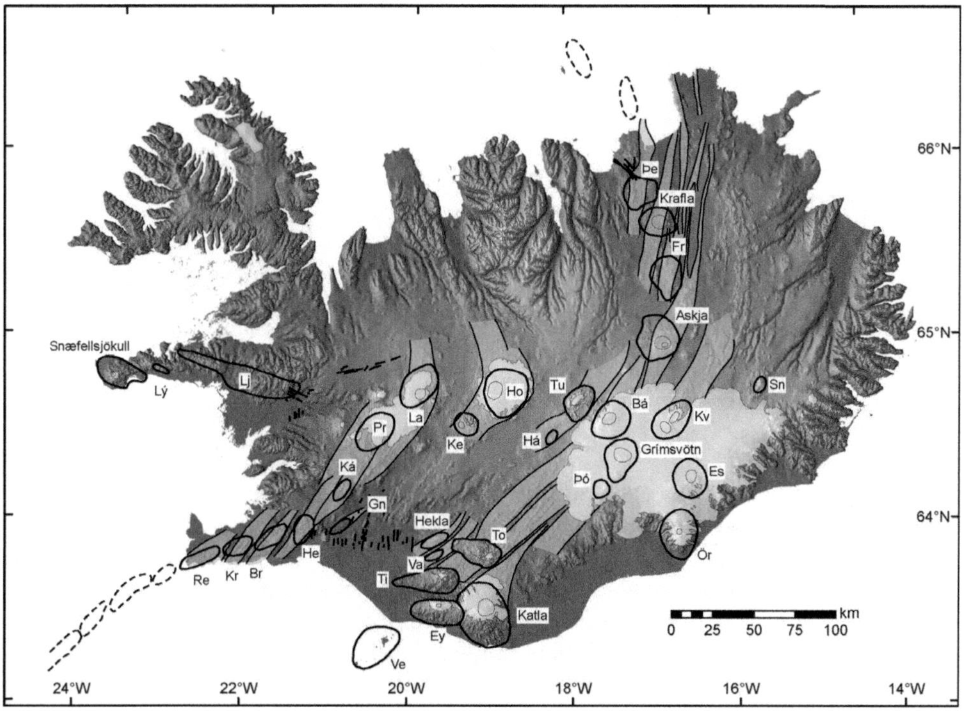

Figure 3.10. Volcanic systems in Iceland as mapped by Einarsson and Sæmundsson (1987). Background map shows shaded topography. The volcanic systems consist of fissure swarms (light shading with outlines), central volcanoes (thick oval outlines), and calderas at some of the central volcanoes (thin oval outlines). The volcanic systems are in alphabetical order: Askja, Bárðarbunga (Bá), Brennisteinsfjöll (Br), Esjufjöll (Es), Eyjafjallajökull (Ey), Fremri Námar (Fr), Grímsnes (Gn), Grímsvötn, Hágöngur (Há), Hekla, Hengill (He), Hofsjökull (Ho), Katla, Kálfstindar (Ká), Kerlingarfjöll (Ke), Krafla, Krísuvík (Kr), Kverkfjöll (Kv), Langjökull (La), Ljósufjöll (Lj), Lýsuskarð (Lý), Prestahnjúkur (Pr), Reykjanes (Re), Snæfellsjökull, Snæfell (Sn), Tindfjöll (Ti), Torfajökull (To), Tungnafellsjökull (Tu), Vatnafjöll (Va), Vestmannaeyjar-Westman Islands (Ve), Þeistareykir (Þe), Þórðarhyrna (Þó) and Öræfajökull (Ör).

Modified from Einarsson and Sæmundsson (1987).

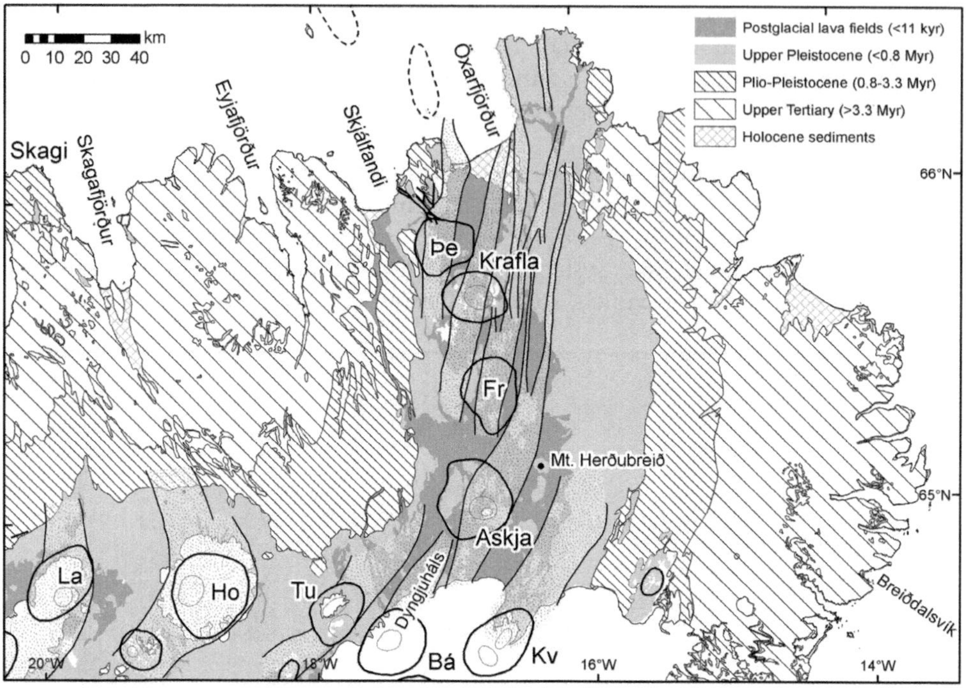

Figure 3.11. Tectonic map of northern Iceland. Volcanic systems as in Figure 3.10 and geologic boundaries as in Figure 3.2.

A number of large lava shields, inferred to be mostly monogenetic structures, are scattered throughout the volcanic zones, mainly in the NVZ and WVZ and some on the RP Plate Boundary. During the Postglacial, most lava shields formed in the initial millennia after deglaciation of Iceland, reflecting excessive magmatic production in that period compared with current conditions (see Section 9.3).

The NVZ north of Vatnajökull is simpler than other parts of the spreading plate boundary in Iceland as here the plate boundary has only one branch and it has been active for a long time. The zone is commonly divided into five volcanic systems: Þeistareykir, Krafla, Fremri-Námar, Askja, and Kverkfjöll (Figure 3.11). They are arranged in *en echelon* fashion with some overlap of their respective fissure swarms. The central volcanoes of these volcanic systems are areas of pronounced focus of volcanic production, with silicic rocks and high-temperature geothermal areas. Krafla and Askja Volcanic Systems have been the most active of these systems. Recent work suggests the existence of an additional central volcano, the Hrúthalsar Central Volcano, north of the Askja Central Volcano (Sæmundsson et al., 2005).

The architecture of volcanic systems lying under the Vatnajökull Icecap, at the junction of the NVZ and the EVZ was evaluated by Björnsson and Einarsson (1990). Mapping by radio echo-sounding has revealed the subglacial topography and

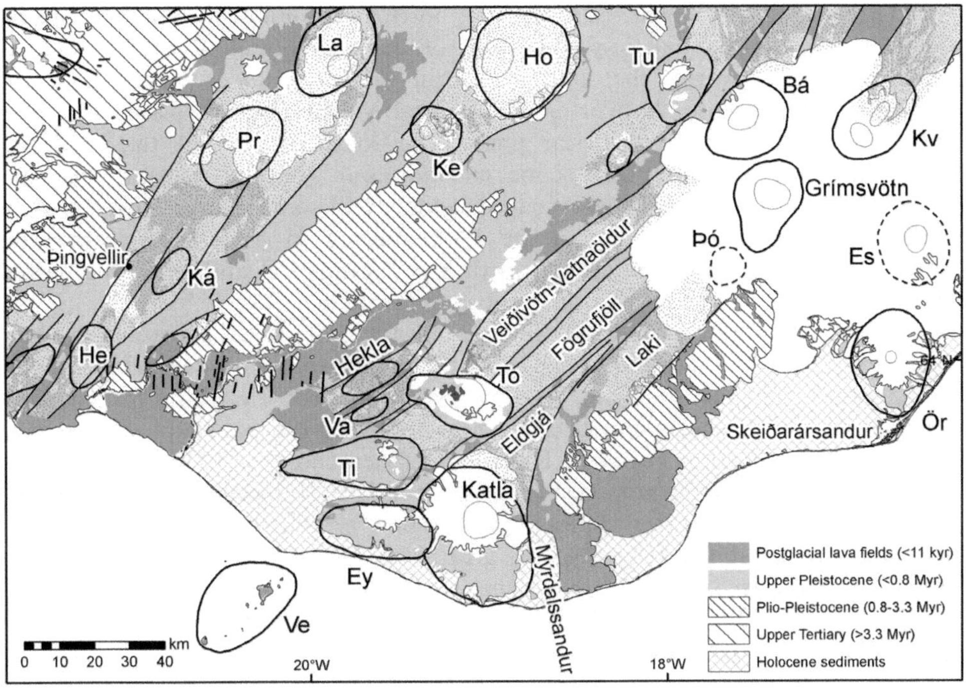

Figure 3.12. Tectonic map of southern Iceland. Volcanic systems as in Figure 3.10 and geologic boundaries as in Figure 3.2.

geothermal areas have been inferred from ice cauldrons on the surface of the icecap. The main volcanic systems are the Kverkfjöll, Bárðarbunga, and Grímsvötn Volcanic Systems (Figure 3.12). The Bárðarbunga Volcanic System consists of the Bárðarbunga Central Volcano and fissure swarms southwest and northeast of it, the Veiðivötn–Vatnaöldur Fissure Swarm and the Dyngjuháls Fissure Swarm. The Laki Fissure Swarm south of Vatnajökull links to the Grímsvötn Volcano and forms the Grímsvötn Volcanic System. Within this system lies also the separate Þórðarhyrna Central Volcano. In historical times, volcanic activity in Vatnajökull has been mainly limited to the Bárðarbunga and the Grímsvötn systems, with the Grímsvötn Volcano having the highest eruption frequency of all volcanoes in Iceland.

A separate volcanic system is suggested (but difficult to define because of complexities) between the Bárðarbunga and Grímsvötn systems, encompassing the Hamarinn Central Volcano, the Loki volcanic ridge extending from Hamarinn towards Grímsvötn, and the Fögrufjöll Fissure Swarm. It has been referred to as the Loki–Fögrufjöll Volcanic System (Björnsson and Einarsson, 1990). The Öræfajökull–Snæfell Volcanic Flank Zone also lies partly under Vatnajökull. It consists of three central volcanoes: Öræfajökull, Esjufjöll, and Snæfell. A fourth central volcano, Breiðabunga, was suggested in this zone (Einarsson and Saemundsson, 1987), but further work has abandoned this idea (Björnsson and Einarsson, 1990).

The volcanic systems in the EVZ differ north and south of its junction with the SISZ. North of that junction and south of the Vatnajökull Icecap, it includes extensive fissure swarms that link to central volcanoes under Vatnajökull, and propagate as well into the Torfajökull Central Volcano. Large fissure eruptions have occurred there in historical times. The southern part of the EVZ, the South Iceland Volcanic Flank Zone includes four volcanic systems: Tindfjöll, Eyjafjallajökull, Katla, and Westman Islands. Hekla and Katla have been the most active. Volcanic activity in the EVZ appears to have been propagating southwards over the last 2–3 Myr, with the Westman Islands volcanic system being the southernmost part of this propagating rift.

The WVZ (Figure 3.12) has been the main focus of crustal spreading in South Iceland for the last 6–7 million years. Its division into volcanic systems is somewhat complicated, but the following systems are suggested: Hengill, Prestahnjúkur, Langjökull, and a volcanic system northeast of Lake Þingvallavatn with a focus on volcanic production near Mt. Kálfstindar (Saemundsson, 1991). The majority of Postglacial lavas in the area were erupted in the early Holocene, and few eruptions have occurred in historical times. Low magma production has led to graben formation in the Lake Þingvallavatn area, with lowest point of the lake being below sea level. At Þingvellir (the former site of the Icelandic parliament from its establishment in 930 AD to 1789) rifting structures and normal faults are particularly pronounced (see cover of book, main figure). The Hengill Volcanic System has been further subdivided into the Hengill proper system and the Hrómundartindur Volcanic System, as these appear to stand out as two separate systems (e.g., Saemundsson, 1992).

Between the northern ends of the WVZ and EVZ lies the Hofsjökull Volcanic Zone as defined here. It includes the Hofsjökull and the Kerlingarfjöll Central Volcanoes, and a fissure swarm extending north and south of Hofsjökull (Figure 3.12).

The RP oblique rift plate boundary in southwestern Iceland (Figure 3.13) is the direct onshore continuation of the MAR. Its structure differs from the rest of the volcanic zones because its overall trend is highly oblique to the plate spreading; a high amount of shearing and relatively little spreading perpendicular to its axis characterizes this volcanic zone. The division into volcanic systems is not particularly clear, and the volcanic centres are not associated with silicic rocks except at the eastern end of the zone at Hengill. Existence of maxima in volcanic production and high-temperature geothermal areas suggests five volcanic systems (Jakobsson et al., 1978): Reykjanes, Svartsengi, Krísuvík, Brennisteinsfjöll, and Hengill. The two westernmost systems, the Reykjanes and Svartsengi, are often classified as a single system (e.g., Einarsson and Sæmundsson, 1987). The Hengill Volcanic System lies at the junction of the RP Plate Boundary with the WVZ, with its fissure swarm north of the Hengill Central Volcano in the WVZ.

The volcanic flank zone in western Iceland, the Snæfellsnes Volcanic Zone is more complex than other zones. Here volcanic products of the last 1–2 million years lie unconformably on top of much older volcanic formations. Volcanism is alkalic in character. The zone is divided into three volcanic systems (e.g., Sigurdsson, 1970;

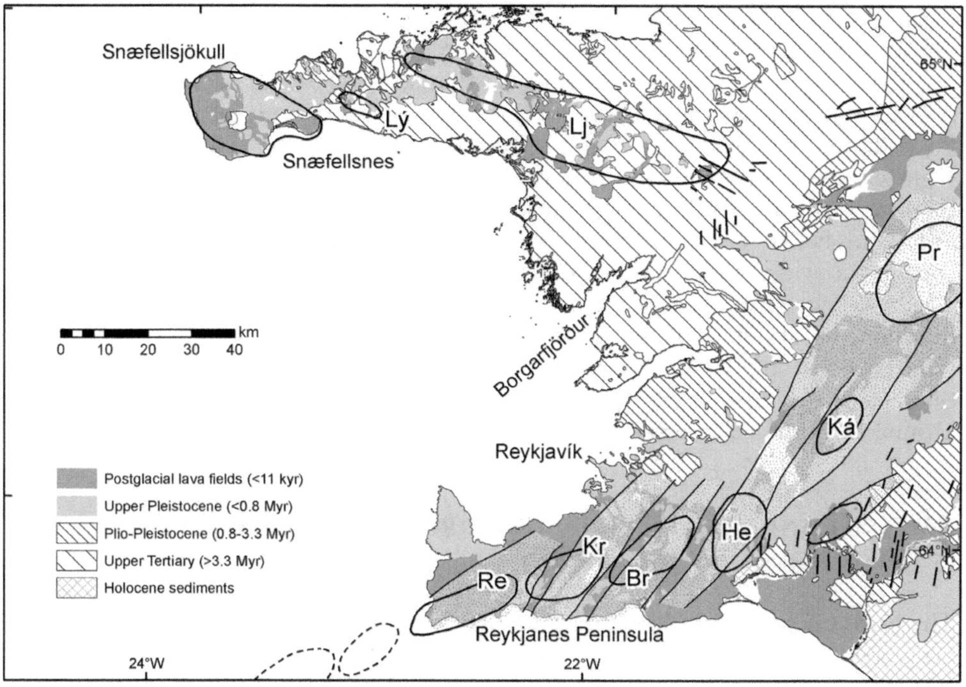

Figure 3.13. Tectonic map of southwestern and western Iceland. Volcanic systems as in Figure 3.10 and geologic boundaries as in Figure 3.2.

Einarsson and Sæmundsson, 1987): Snæfellsjökull, Lýsuskarð, and Ljósufjöll (Figure 3.13).

3.4 RIFT JUMPS AND PAST PLATE BOUNDARIES

The current tectonic configuration of Iceland is the result of a complex interplay between the MAR and the North Atlantic Mantle Plume. This structure is by no means stable, and has been in continuous development throughout the geological history of Iceland. The EVZ is the youngest of volcanic zones in Iceland. Activity in this zone began 2–3 Myr ago, whereas the WVZ has been active for 6–7 Myr (Figure 3.3). Most of the crust in South Iceland has been formed in the WVZ. The NVZ is of similar age to the WVZ. It has been the main zone of spreading in northern Iceland for the last 6–7 Myr. The main pattern of rift jumps was outlined by Jóhannesson (1980).

Prior to establishment of the WVZ and the RP oblique rift, the rift zone at Snæfellsnes, extending northwards to Skagi, was the main locus of spreading from about 15 Myr until about 7 Myr. The Snæfellsnes–Skagi Rift Zone linked directly to the Kolbeinsey Ridge north of Iceland. The centre of the zone is marked by a synclinal structure in the lava pile (Jóhannesson, 1980). Aeromagnetic results are consistent with the existence of this rift zone, and partial overlap in its activity with the WVZ in a transitional period between the two (Kristjánsson and Jónsson, 1998).

Evidence for a still older rift zone is found at the extreme northwest of Iceland. A 14.9-Myr unconformity at the very northwestern extreme of Iceland separates lavas dipping towards an older axis now off the northwest coast of Iceland, and younger lavas dipping towards the younger axis of the Snæfellsnes–Skagi Rift Zone (Hardarson et al., 1997). Furthermore, geochemical differences exist between lavas above and below the unconformity, with basaltic lava flows below the unconformity showing a wider range of incompatible element and radiogenic isotope ratios. The unconformity in the lava pile in northwest Iceland marks a hiatus with a duration of about 200,000 years in the lava succession, around 15 Myr ago (Hardarson et al., 1997). The oldest lavas directly above the hiatus are the oldest ones from the Snæfellsnes–Skagi Rift Zone that is inferred to have initiated at around this time when activity in the older Northwestern Rift Zone died out.

3.5 VOLCANIC ACTIVITY IN HISTORICAL TIMES: WRITTEN RECORDS OF 1,100 YEARS

The Nordic settlers in Iceland arrived in the late 9th century, with Iceland being fully occupied in 874 AD. The Icelandic parliament, Alþing, was established in 930 AD and met every summer thereafter in the Þingvellir area, in the WVZ, with the parliament site located at a major normal fault, the Almannagjá Fault (see main figure on book cover). Oldest written records in Iceland extend back to the 12th century and thereafter traditions of writing prevailed, with the exception of few contemporary written accounts during the 15th–16th century. In the 12th century, oral accounts of activity in the earlier centuries were also written down.

The written records provide information on volcanic and seismic activity in Iceland. Dates of the largest earthquakes in South Iceland are known back to the 12th century, pointing to sequences of large earthquakes in the SISZ at average intervals of 80–100 years (e.g., Einarsson, 1991a). The historical volcanic record shows about 20 eruptions per century, or one eruption about every 5 years on average (e.g., Thorarinsson and Saemundsson, 1979). Extensive literature on volcanic activity in Iceland exists, partly in Icelandic (e.g., Gudmundsson, 2001). Three volcanoes, Hekla, Katla, and Grímsvötn, have by far been the most active in Iceland during historical times, and are responsible for more than half of all eruptions that have occurred in the last 1,100 years. These volcanoes have had profound environmental impact and greatly influenced their surroundings.

The first post-settlement eruption of Mt. Hekla, a volcanic ridge in the EVZ, occurred in 1104 AD. At that time an explosive eruption produced about 2.5 km^3 of rhyolitic tephra which blanketed large parts of Iceland and caused complete destruction of nearby inhabited areas. Through historical times one or two major eruptions occurred each century at Hekla until 1947 (Thorarinson, 1967). Thereafter the eruptive pattern changed to more frequent and smaller eruptions. The initial phase of many Hekla eruptions is explosive and has spread tephra over large parts of Iceland, depending on prevailing wind conditions (Figure 3.14). At Hekla, the length

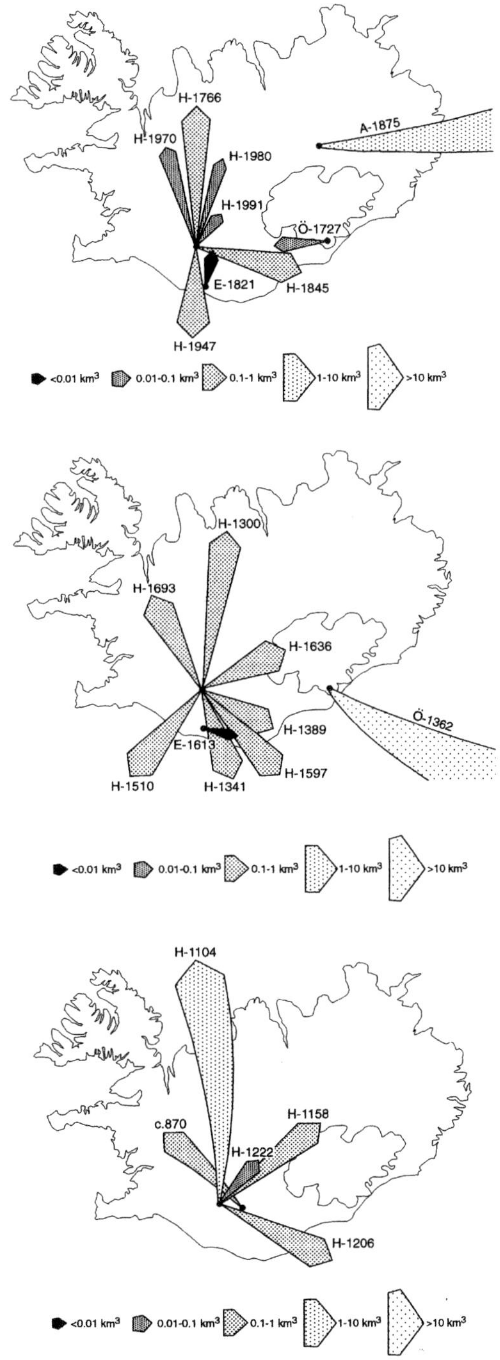

Figure 3.14. Main axes of tephra fallout from historical eruptions of Hekla (H), Askja (A), Eyjafjallajökull (E), Öræfajökull (Ö), and the ~870-AD eruption of Torfajökull.

Reproduced from Larsen et al. (1999).

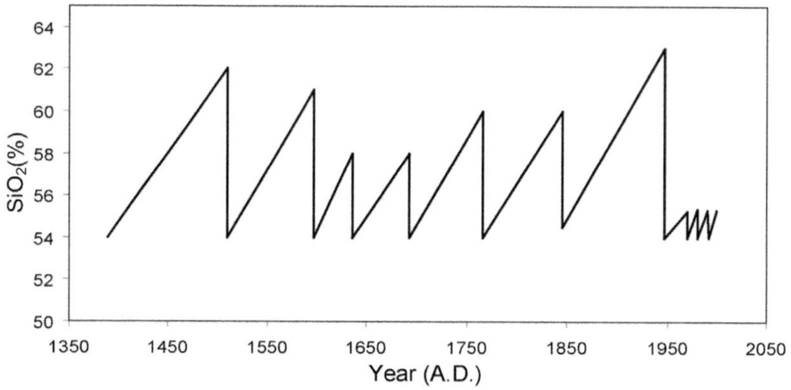

Figure 3.15. SiO_2 content of initial eruptive products during Hekla eruptions versus date of eruptions.

Modified from Thorarinsson (1967). Data for 1970–2000 eruptions from the Nordic Volcanological Centre. Niels Óskarsson (pers. commun., 2005).

of the repose period between eruptions (known from the historical records) scales with the initial silica content of eruptive products (Thorarinson, 1967). The longer the repose period, the higher the silica content of the initial eruptive products (Figure 3.15). In addition to direct effects from tephra, the environmental effects of Hekla eruptions have included effects of soluble fluorine adhering to erupted tephra particles, leading to lethal fluorosis in grazing animals even in areas of minor tephra fallout (Óskarsson, 1980). An interesting feature of Hekla eruptions is that the volume of eruptive products also scales with the preceding repose period, adding up so that about 1 km^3 of magma is erupted each century (Gronvold et al., 1983). This is the only volcano in Iceland with such regular pattern.

The Katla Volcano is a subglacial caldera in the SIFZ. It has also erupted once or twice each century throughout Iceland's history (Larsen, 2000). The eruptions have been phreatomagmatic because the volcano resides under ice up to 500 m thick, which fills the Katla Caldera (Björnsson et al., 2000). In addition to producing large quantities of airborne tephra (Figure 3.16), Katla eruptions cause huge glacial outburst floods with estimated peak flow rates exceeding 100,000 m^3/s. The Grímsvötn Volcano under the Vatnajökull Icecap has also produced a great number of sudden glacial outburst floods. The Icelandic term for these floods, *jökulhlaup*, is used internationally because of good descriptions and early studies of them in Iceland by Sigurður Þórarinsson (e.g., Thorarinsson, 1953). Only a fraction of jökulhlaups originating from Grímsvötn are associated with eruptions. Most of them are due to storage of water and melting of ice by geothermal heat within the subglacial Grímsvötn Caldera. Nowhere else on Earth are jökulhlaups as frequent as in Iceland. Jökulhlaups originating from the Katla and Grímsvötn Volcanoes have produced large outwash plains downstream from the affected glaciers (Figure 3.17). These outwash plains, termed *sandur* in Icelandic, have been greatly augmented in historical times in Iceland, in particular at Mýrdalssandur and

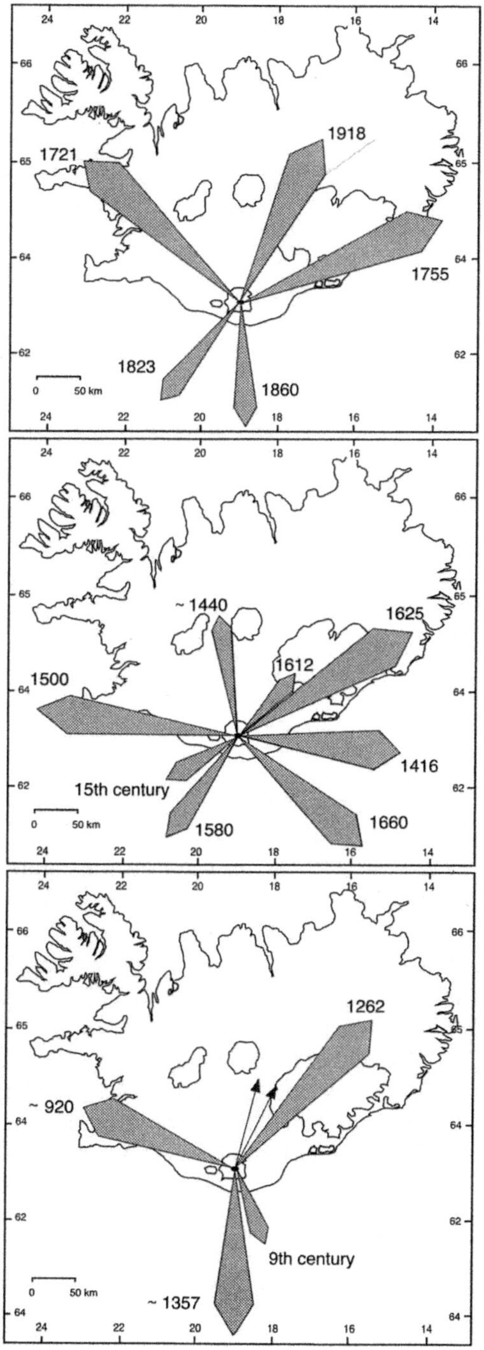

Figure 3.16. Main axes of tephra fallout from historical eruptions of Katla Volcano.
Reproduced from Larsen (2000) with permission of *Jökull*.

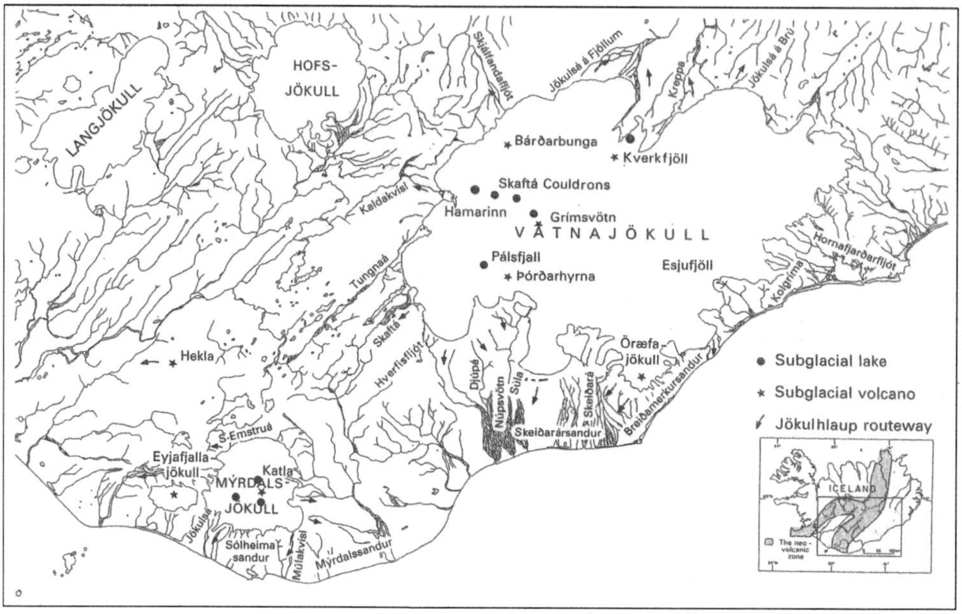

Figure 3.17. Location of subglacial lakes at geothermal areas and site of subglacial volcanic eruptions in Iceland, and rivers affected by jökulhlaups in historical times.

Reproduced from Björnsson (1992).

Skeiðarársandur. At the time of settlement Mýrdalssandur was inhabited, but most of the farmland has since been destroyed and the coast has migrated 4 km southwards (Larsen, 2000). At Katla, large eruptions appear to be followed by longer non-eruptive intervals than after smaller eruptions (Eliasson et al., submitted).

Other volcanoes which produced large explosive eruptions during historical times include the Askja Volcano in the NVZ, which experienced a plinian eruption and an associated caldera collapse in 1875. In 1362, a large explosive eruption of Mt. Öræfajökull (Thorarinson, 1958) devastated large areas in southeast Iceland. Studies of explosive volcanism in Iceland were pioneered by Sigurður Þórarinsson who established the field of tephrochronology through such studies. Origin of tephra layers can often be traced to their source volcano by studies of their chemical composition (e.g., Grönvold et al., 1995). In addition to soil profiles, tephrostratigraphy has been applied to tephra layers in ice within Iceland's icecaps. A study of these layers in the Vatnajökull Icecap has provided a comprehensive record of eruptions within Vatnajökull, revealing periodicity in their frequency (Larsen et al., 1998).

In addition to tephra, large quantities of lava have been erupted in effusive eruptions during historical times in Iceland. Whereas most of the lava-forming eruptions are small in volume (on the order of 0.1 km^3), two exceptionally large-

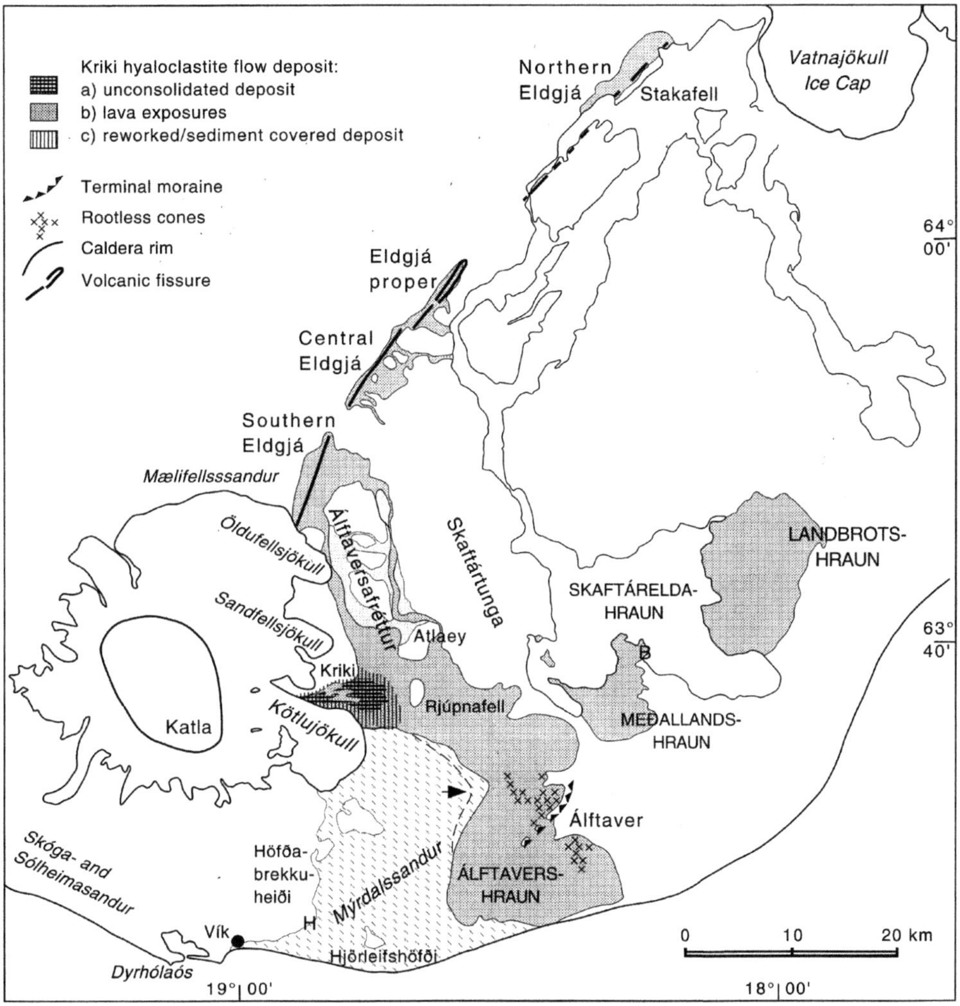

Figure 3.18. The 934 AD Eldgjá Lava Flow (shaded) and the 1783–1784 Laki Lava Flow (Skaftáreldahraun).

Reproduced from Larsen (2000) with permission of *Jökull.*

volume eruptions occurred in the EVZ, including the largest historical lava flow on Earth (witnessed by man). This lava formed in the Eldgjá Eruption in 934 AD (Figure 3.18) and has an estimated volume of 19.6 km^3 (Thordarson et al., 2001). Written descriptions of this event are scarce, consisting of only a few sentences with indirect reference to this eruption in the book "Landnáma" (*Icelandic Book of Settlements*) written in the 12th century. The timing of the eruption is confirmed by an acid peak in the Greenland ice cores (Grönvold et al., 1995). Various historical documents suggest this eruption had a major environmental impact over a large part of Europe, as well as in the Middle East (Stothers, 1998).

Another exceptionally large historical lava flow formed in the 1783–1784 Laki Eruption, when 15 km^3 of lava erupted (Thordarson and Self, 1993). Both the Eldgjá and Laki Eruptions mark major rifting episodes in the EVZ, and consisted of a series of eruptions associated with dike intrusions accommodating spreading across the plate boundary. A good contemporary description exists of the Laki Eruption, written by the Reverend Jón Steingrímsson, an eyewitness to the eruption who recorded his observations in the book "Fullkomið rit um Síðueld" (*A Complete Treatise on the Sida Fires*), recently translated into English (Steingrímsson, 1998). The environmental effects of the Laki and Eldgjá Eruptions were tremendous. Widespread air pollution associated with the Laki Eruption led to the death of livestock by fluoride poisoning and subsequent famine in Iceland. The population of Iceland decreased from about 50,000 before the eruption to about 40,000 in the years after, and the eruption also had an impact on living conditions in Europe (e.g., Thordarson and Self, 1993). As the Laki and Eldgjá Fissure Eruptions both had major environmental impacts, then a similar event in the future could result in widespread fluoride poisoning, air pollution, and disruption of air traffic over large areas (e.g., Stone, 2004).

The historical record in Iceland holds information on occurrence of earthquakes as well as volcanic eruptions. This allows statistical analysis of correlation between the two (Gudmundsson and Saemundsson, 1980). A weak correlation is suggested, with eruptions leading to large earthquakes. In particular, a major earthquake sequence occurred in the SISZ in 1784, eventually triggered by stress change induced by extensive dike formation in association with the Laki Eruption that started a year earlier.

3.6 OVERVIEW OF SEISMICITY OF ICELAND

The SISZ and the TFZ are the main seismic zones in Iceland, each experiencing persistent micro-earthquake activity (Figure 3.19) and earthquakes as large as magnitude 7–7.5 (M_s) occurring in a series typically about once each century (Figure 3.20). The transform zones are associated with a lateral shift in plate spreading, and high stresses build up in response to shearing across the zones (Figure 3.9). The timing of large earthquakes in these zones is partly known from historical records that give an account of the activity extending about 800 years back in time. The record is more complete in the more populated SISZ than in the TFZ which lies mostly offshore north of Iceland.

Only two seismometers were operating in Iceland between 1928 and 1951 (Tryggvason, 1973) and the seismic coverage was poor until the 1970s when a regional network of analogue seismic stations was installed. That network gave results revealing the main seismic characteristics of the plate boundary in Iceland (Einarsson and Saemundsson, 1987; Einarsson, 1991a). In the 1990s a new network of three-component digital seismic stations, the South Iceland Lowland (SIL) network, was installed. The network is run by the Icelandic Meteorological Office (*http://www.vedur.is*). Seismicity in Iceland during 1994–2000 (Figure 3.19) is well

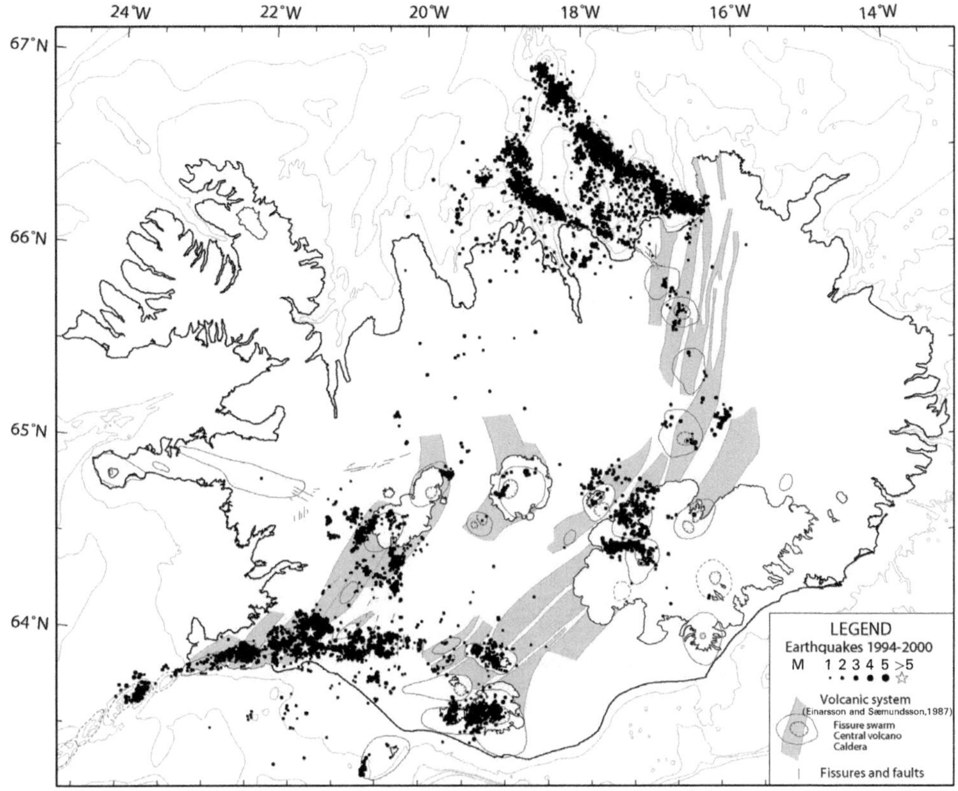

Figure 3.19. Earthquake epicentres 1994–2000 recorded by the South Iceland Lowland (SIL) seismic network of the Icelandic Meteorological Office.

Modified from Jakobsdóttir et al. (2002) with permission of Jökull. Courtesy of Gunnar Guðmundsson, Icelandic Meteorological Office.

captured by the SIL network (Jakobsdóttir et al., 2002), after initial expansion of the network in 1994 to northern Iceland and subsequent additions of stations.

The 1994–2000 epicentral map shows well how seismicity is focused on the transforms zones, and that a large part of the fissure swarms at the spreading plate boundary are devoid of earthquakes. Within the volcanic zones, background seismicity is focused on the central volcanoes. Elevated earthquake activity within the rift zones is often associated with magmatic movements that cause temporarily high local stresses. Such magmatic movements are most frequent at the central volcanoes, but major seismic activity also occurs in the fissure swarms during rifting events. An example is the activity between 1994 and 2000 at the Hengill–Hrómundartindur volcanic area. Over 85,000 earthquakes were recorded in that area 1994–1998 in association with 2 cm/yr of inflation, interpreted to be caused by accumulation of magma at about a 7-km depth (Sigmundsson et al., 1997; Feigl et al., 2000).

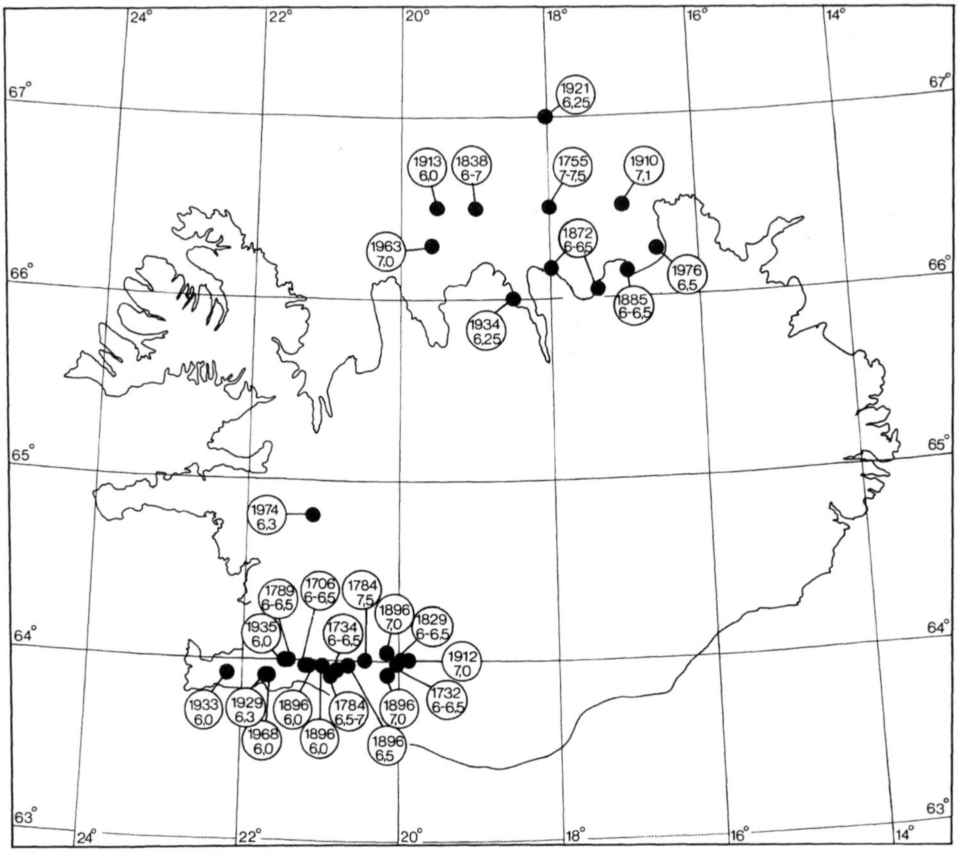

Figure 3.20. Large historical earthquakes in Iceland. For earthquake magnitudes in South Iceland, see also Stefánsson and Halldórsson (1988).

Reproduced from Einarsson (1986) with permission of the Geological Society of America.

The period from 1994 to 2000 includes some significant tectonic and magmatic events, including a major earthquake sequence in South Iceland in 2000, with M_s 6.6 events occurring on June 17 and June 21. Triggered activity followed further to the west, along the RP. Earthquakes associated with subglacial eruptions occurred in 1996 and 1998 at Vatnajökull, and inflow of magma occurred at several central volcanoes, including Hengill–Hrómundartindur, Katla, Eyjafjallajökull, and Grímsvötn. Some earthquakes were associated with the eruption of Hekla in 2000, at this otherwise seismically quiet volcano (Soosalu et al., 2005). Significant earthquake activity immediately precedes most eruptions in Iceland (in association with formation of a feeder dyke), changing at the onset of eruptions to a volcanic tremor that continues throughout the eruption as long as magma flows to the surface (Figure 3.21).

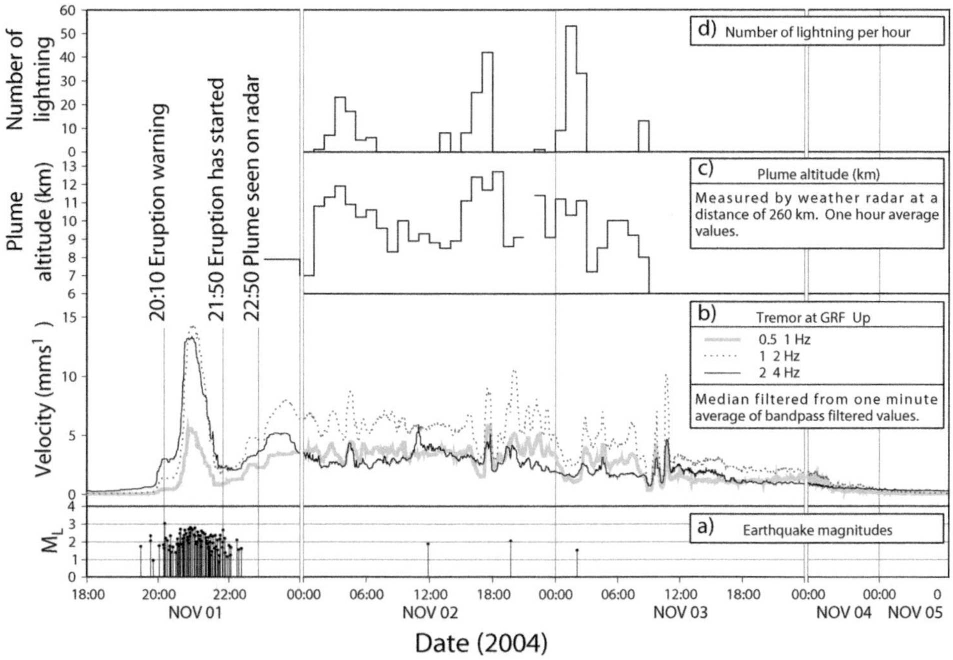

Figure 3.21. Icelandic Meteorological Office observations of (a) earthquakes associated with an eruption at Grímsvötn Volcano in 2004, (b) seismic tremor amplitude in three frequency bands recorded at the GRF seismic station at the volcano, (c) eruption plume altitude, and (d) amount of lightning striking per hour during the eruption. Earthquakes precede the eruption, volcanic tremor continues throughout the eruption.

Reproduced from Vogfjörd et al. (2005). Copyright by the American Geophysical Union.

The historical record shows that major earthquakes have repeatedly caused damage since the settlement in the 9th century. In addition to timing of events, the damage areas are known in some cases. Their mapping has helped reveal fault locations and the tectonic nature of the seismic zones. The damage areas for earthquakes in the South Iceland Seismic Zone are elongated in a north-south direction, providing further evidence that earthquakes in this zone take place along an array of north-south faults in a "bookshelf" faulting mode (Chapter 8).

4

Crustal structure of Iceland

The oceanic crust–mantle boundary marks the transition from peridotitic mantle to gabbroic lower crust, with the bulk of crustal material being formed by material melted and transported from the mantle. For Iceland, the models of crustal structure have changed in recent decades. Results of seismic and magnetotelluric measurements in the 1970s were interpreted in terms of a thin crust, underlain by anomalous mantle with high melt concentrations. A model of 10–15-km-thick relatively hot crust, underlain by anomalous mantle with 10–15% partial melt seemed at that time to be consistent with various types of data, including seismic shear wave profiles collected across Iceland (e.g., Gebrande et al., 1980), extrapolation of near-surface temperature gradients, and results of magnetotelluric measurements indicating a high electrical conductivity zone (e.g., Beblo and Bjornsson, 1978). No seismic reflection from a Mohorovicic discontinuity (Moho) at the crust–mantle boundary was inferred. Extensive seismic surveys in the last decade of the 20th century revealed a different picture. Seismic data strongly argue for a thick cold crust under Iceland, with crustal thickness increasing from ~15 km in the coastal areas towards ~40 km under central Iceland. Clear seismic reflections originate from the Moho. The earlier seismic data can be reconciled with this interpretation (Menke et al., 1996). There is little contrast in density between crust and mantle, and the large crustal thickness in Iceland is consistent with high melt production in a mantle plume under Iceland.

4.1 SEISMIC CONSTRAINTS ON CRUSTAL THICKNESS

The crustal structure of Iceland was initially studied by Båth (1960) and Tryggvason and Båth (1961). Extensive pioneering work was subsequently carried out by Pálmason (1971), who conducted reflection studies on a large number of seismic profiles in Iceland and derived a crustal model for Iceland, in terms of layers 0–4,

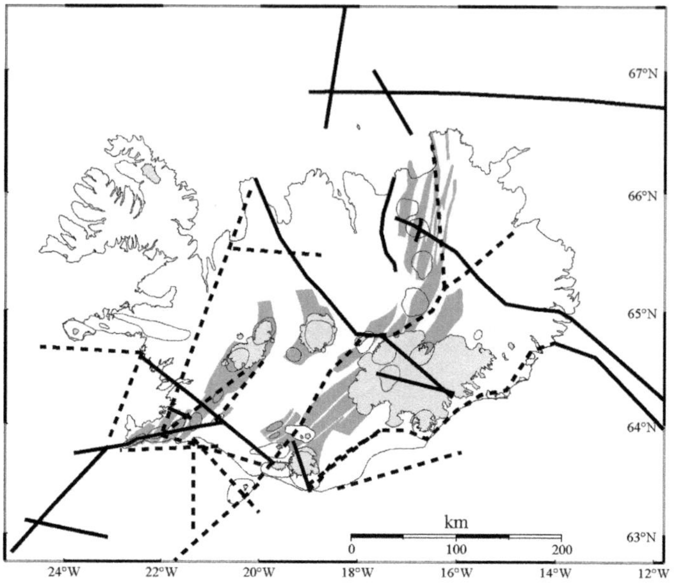

Figure 4.1. Seismic refraction lines 1959–1977 (stippled) and 1991–2000 (solid lines).
Courtesy of Bryndís Brandsdóttir, Institute of Earth Sciences, University of Iceland.

as reviewed by Flóvenz and Gunnarsson (1991). The upper crustal structure has little changed from these earlier models. On the other hand, ideas on the nature of the lower crust have changed in a fundamental way.

A number of important seismic studies of the Icelandic crust were conducted in the last decade of the 20th century which revealed the existence of a thick crust (Figure 4.1). The first of these studies was the South Iceland Seismic Tomography (SIST) project carried out by Bjarnason et al. (1993b). Measurements along a 170-km-long seismic profile consisting of 11 shot points and 210 receiver points, crossing the western volcanic zone and obliquely over the South Iceland Seismic Zone, show large amplitudes of wide-angle reflections and an apparent refractor velocity of 7.7 km/s. These were interpreted to originate from a Moho at a 20–24-km depth outlining the boundary between crust and mantle. The crust was divided into upper crust with P-wave velocities less than 5 km/s, a mid-crust with velocities between 5.0 and 6.5 km/s, and a lower crust with velocities above 6.5 km/s. Along the SIST profile the upper crust varies in thickness from 0.7 to 3.0 km. It is interpreted to be made of subaerial lava flows which become increasingly altered with depth due to secondary mineralization. The transition from upper crust to mid-crust is marked by a change in the secondary mineralization from lighter to heavy minerals like epidote which start to form around 250°C (Bjarnason et al., 1993b). Several drillholes that extend into the mid-crust confirm this change in mineralization. The mid-crust along the SIST profile varies in thickness from 2.0 to 4.5 km. The transition from mid-crust to lower crust is marked by a sharp decrease in P-wave velocity gradients at ~6.5 km/s. In South Iceland, the depth to the lower crust is 3–7 km, and it is 14–20-km thick.

Boundaries in the crust are primarily dependent on the state of alteration of the basaltic crust which leads to a steady increase in velocity with depth as secondary minerals occupy available pore space and cracks in rocks. A higher proportion of intrusions in deeper parts of the crust is also a contributing factor (Flóvenz and Gunnarsson, 1991).

Following the SIST profile, a number of other long seismic profiles were measured in Iceland in the last decade of the 20th century. In North Iceland, the Faroes–Iceland Ridge Experiment (FIRE) conducted in 1994 included an east–west profile across the Northern Volcanic Zone (Staples et al., 1997), and the B96 seismic array measured in 1996 along the western flank of the Northern Volcanic Zone (Menke et al., 1998). The ICEMELT profile measured in 1995 crossed central Iceland above the centre of the inferred Iceland Mantle Plume (Darbyshire et al., 1998). In southwest Iceland, the Reykjanes–Iceland Seismic Experiment (RISE) (Weir et al., 2001) was conducted along the Reykjanes Peninsula. All of these studies have revealed a good reflector interpreted as a Moho. The depth to the Moho is not uniform. It changes in a systematic manner with crustal thickness increasing towards central Iceland. The maximum crustal thickness of over 40 km is inferred to be directly above the centre of the mantle plume (Figure 4.2).

In addition to the explosion seismology profiles, important constraints on crustal structure were derived from the HOTSPOT project, which consisted of a deployment of 30 broadband seismic instruments over a period of 2 years (1996–1998). Data collected during this project formed the basis of a fully three-dimensional study of the crustal structure of Iceland by Allen et al. (2002a). A combination of surface wave and body wave data was used. A crustal S-velocity model and a Moho map were derived (Figure 4.3a, see colour plates). According to this model, the crustal thickness in Iceland varies between 15 and 46 km. A different modelling approach by Darbyshire et al. (2000) using less data (no HOTSPOT data) reveals a similar model, in good agreement with the results of Allen et al. (2002a). Both of these models used gravimetric data in addition to seismic data to constrain the crustal structure (see below).

4.2 GRAVITY AND ISOSTATIC BALANCE OF ICELAND

The main features of the gravity field over Iceland were discovered by the pioneering work of Einarsson (1954) who derived a gravity map of all of Iceland. It revealed a clear Bouguer anomaly low over Iceland. Later work includes the compilation of Eysteinsson and Gunnarsson (1995), who derived a complete Bouguer gravity map of Iceland and the surrounding oceans (Figure 4.4, see colour plates). A still improved Bouguer gravity map based on these data was calculated by Kaban et al. (2002) who demonstrated that the previously used density value for the Bouguer correction should be adjusted. Admittance between topography and gravity was used to conclude that the most appropriate density value for Bouguer correction over Iceland was $2{,}520 \pm 20\,\text{kg/m}^3$. Over the oceans, a standard value of $2{,}670\,\text{kg/m}^3$ was used. The Bouguer anomaly has a value of -40 mGal near central Iceland,

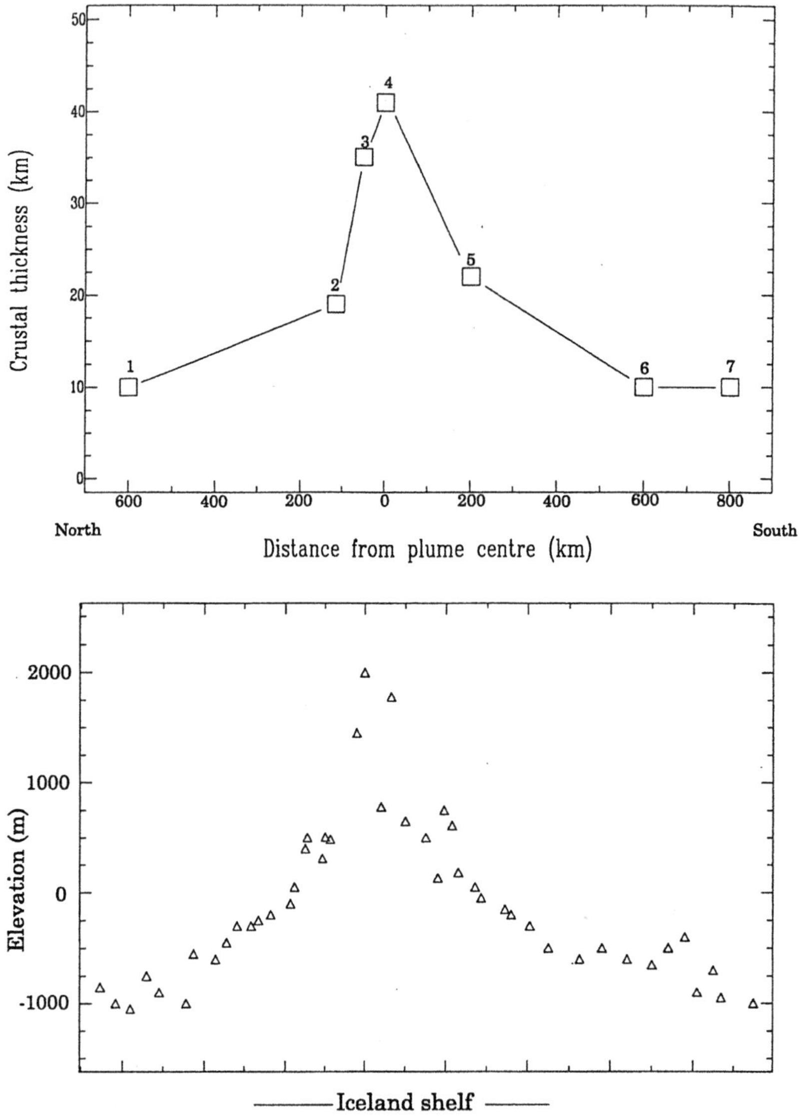

Figure 4.2. Crustal thickness (upper) and topography (bottom) versus distance from the centre of the Iceland Mantle Plume.

Reproduced from Darbyshire et al. (1998).

whereas positive values of over 40 mGal occur at the coast. Kaban et al. (2002) also calculated crustal thickness (Figure 4.3b), revealing similar crust–mantle topography to that derived by Allen et al. (2002a).

The gravity anomaly over Iceland correlates with both the topography (Figure 4.5) and crustal thickness. Comparison of these datasets reveals that the elevation of

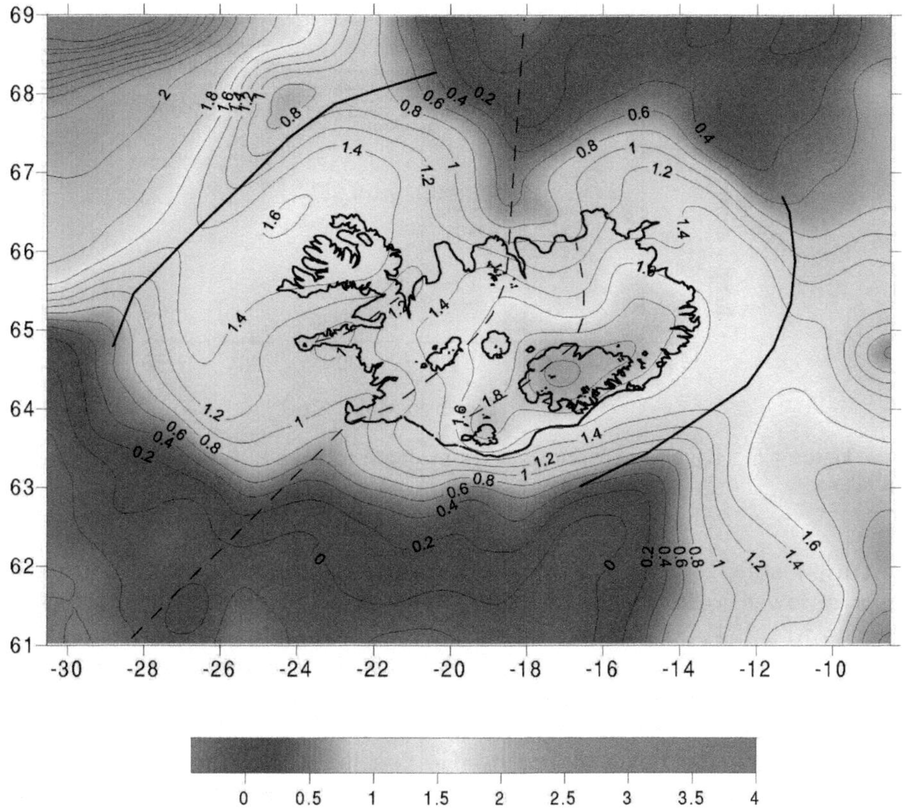

Figure 4.5. Lowpassed residual adjusted topography after removal of the effect of normal oceanic topography and numerical densification of ice, water, and surface rocks to a density value of $2{,}670\,\mathrm{kg\,m^{-3}}$.

Reproduced from Kaban et al. (2002).

Iceland is much lower than expected. Even though Iceland stands up more than 2 km above sea level, the thick crust under Iceland would suggest that it should stand much higher. Menke (1999) inferred an anomalously low density contrast between the crust and mantle from this observation. A regular pattern is revealed when height above sea level is plotted against the inferred depth to the Moho (Figure 4.6). Over Iceland a gradient of 0.030 ± 0.005 is inferred, whereas over the surrounding oceans it is 0.116 ± 0.012 (Gudmundsson, 2003). These gradients provide a direct measure of the density contrast at the crust–mantle boundary. Gudmundsson (2003) derives the relation:

$$\frac{dh}{dz} = \frac{\Delta\rho}{\rho_0} \tag{4.1}$$

where h is elevation above sea level, z is depth to the Moho, ρ_0 is the average upper crustal density equal to $2{,}700\,\mathrm{kg/m^3}$, and $\Delta\rho$ is the density contrast across

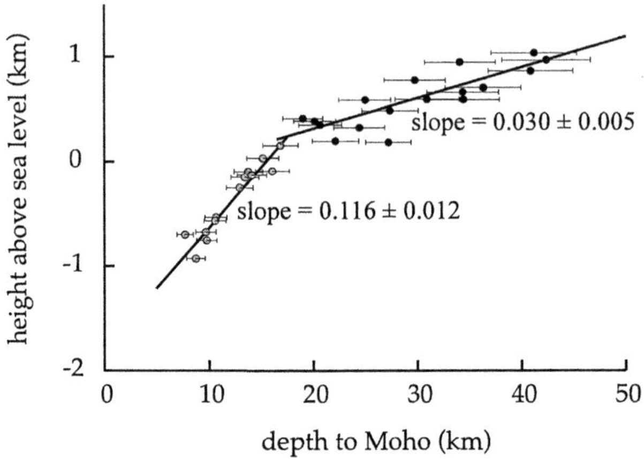

Figure 4.6. Height above sea level versus depth to Moho in the North Atlantic.
Reproduced from Gudmundsson (2003) with permission of Elsevier.

the crust–mantle boundary. In this relation it is assumed that the transition from upper to lower crust is a flat boundary. The observed gradient give values of $\Delta\rho$ equal to $81 \pm 13\,\mathrm{kg/m^3}$ within Iceland and $313 \pm 31\,\mathrm{kg/m^3}$ at the adjacent ridges. Although $\Delta\rho$ is well resolved, it is not easy to derive the absolute densities of crust and mantle as they are inherently difficult to separate. Gudmundsson (2003) conducts, however, further analyses utilizing continuity of h across the change from Iceland to the adjacent oceans. He concludes that the anomalous density contrast is mostly due to a heavy crust, inferring that the lower crust in Iceland is about $200\,\mathrm{kg/m^3}$ denser than the lower crust under the surrounding oceans. Dense lower crust is broadly consistent with the melting models of White and McKenzie (1989). They demonstrate that a high mantle potential temperature at plumes causes the resulting igneous crust above them to be denser than elsewhere.

4.3 THERMAL STRUCTURE OF THE CRUST

A number of observations provide a measure of the thermal state of the crust. Temperature can be measured directly in the uppermost crust in numerous boreholes. Several seismic indicators, including the ratio of P- and S-wave velocities (v_p/v_s ratio), seismic attenuation, and the maximum depth of earthquakes, provide an indirect indication of temperature conditions at deeper levels.

4.3.1 Heat flow

Direct measurements of the temperature gradient in the shallow crust at numerous drillholes in Iceland reveal large variability. At the most active areas of the plate

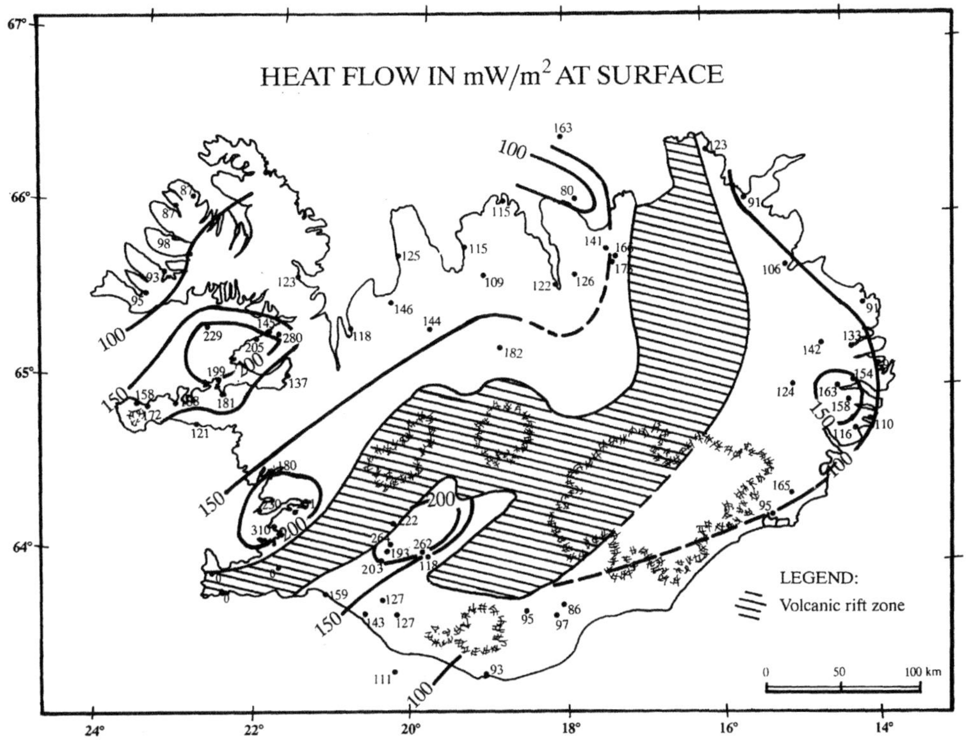

Figure 4.8. Surface heat flow (mW/m^2).

Reproduced from Flóvenz and Saemundsson (1993) with permission of Elsevier.

boundary, convection by hydrothermal circulation dominates over conduction and the temperature gradients are low. Where undisturbed by hydrothermal circulation, the temperature gradients depend on conduction. A study of such data from available drillholes by Flóvenz and Sæmundsson (1993) reveals temperature gradients, $\partial T/\partial z$, ranging from almost 0 to 500°C/km (Figure 4.7, see colour plates). The surface heat flow, Q, can be derived as:

$$Q = A\frac{\partial T}{\partial z} \tag{4.2}$$

where A is the thermal conductivity, inferred to range from $1.6\,\mathrm{Wm^{-1}{}^\circ C^{-1}}$ to $2.0\,\mathrm{Wm^{-1}{}^\circ C^{-1}}$ in the Icelandic crust. The derived surface heat flow is in the range of 0.1 to $0.3\,\mathrm{W/m^2}$ (Flóvenz and Sæmundsson, 1993). The surface heat flow is characterized by a general decrease with distance from the active spreading axis, with local anomalies superimposed (Figure 4.8).

The temperature gradient in the uppermost crust was originally extrapolated linearly to greater depths, suggesting a near-solidus temperature of 1,200°C in the 10–15-km range, in agreement with the thin crustal model and underlying high concentration of partial melt. Later work demonstrates that linear extrapolation of the temperature gradient is not appropriate (see Section 4.3.3). The observed heat flow values are consistent with lower crustal temperatures well below the solidus, and a 1,200°C isotherm at a 30–50-km depth under most of Iceland (Menke and Sparks, 1995; Kaban et al., 2002).

4.3.2 Seismic observations

Seismic waves travelling for long distances in the lower crust in Iceland have clear S-wave arrivals and are little attenuated. Menke and Levin (1994) and Menke et al. (1995) show that this observation requires cold lower crust with temperatures well below the solidus. Seismic attenuation depends on the quality factor, Q, with amplitudes of seismic waves, A, relating to Q in the following way (e.g., Aki and Richards, 1980):

$$A \propto \exp\left(\frac{-\pi f T}{Q}\right) \tag{4.3}$$

where f is the wave frequency and T is its travel time. Q is defined from the above relation. If the attenuation varies along the travel path, the amplitude of the observed seismic wave depends on the path-averaged seismic attenuation. A study by Menke and Levin (1994) of seismic attenuation of S-waves spending most of their travel time in the lower crust of Iceland found high-shear-wave-quality factors in the lower crust of Iceland. Their Q values were found by computing displacement spectra of shear waves using Fourier analysis. The slope of logarithmic displacement amplitude versus frequency gives a direct estimate of the shear wave quality factor. A more extensive study by Menke et al. (1995) concluded that the lowest path-averaged shear wave quality factors, Q_s, for S-waves turning in the mid- to lower crust in southwest Iceland is $Q_s = 250$ with most values being much higher. These values for Q_s are an order of magnitude higher than expected if the lower crust has temperatures close to solidus, according to experimental studies (Kampfmann and Berckhemer, 1985). Assuming a gabbroic lithology is appropriate, Menke et al. (1995) conclude that lower crustal temperatures in Iceland do not exceed 700–775°C.

Crustal temperature also influences wave velocities, with a decrease in S-wave velocity and increase in v_p/v_s ratio if temperatures approach the solidus. In northern Iceland, Menke et al. (1998) find a v_p/v_s ratio of 1.75–1.76 with no significant variation between the mid- to lower crust. They conclude that near-solidus temperatures in the lower crust are ruled out. Such temperatures would cause a v_p/v_s ratio close to 1.9–2.0, whereas for crystalline rocks the ratio is close to $\sqrt{3}$. Experimental data (e.g., Kampfmann and Berckhemer, 1985) suggest a rapid decrease in shear modulus and resulting shear wave velocity for temperatures above 800°C. In central Iceland, Darbyshire et al. (1998) find v_p/v_s ratios similar to those in North Iceland. A number of studies compiled by Allen et al. (2002a) show consistent v_p/v_s ratio in Iceland in the range of 1.75–1.79 for the bulk of the crust, all indicative of relatively cold temperatures. However, Allen et al. (2002a) find a best fit for depth variation of v_p and v_s in an average crustal velocity model for Iceland

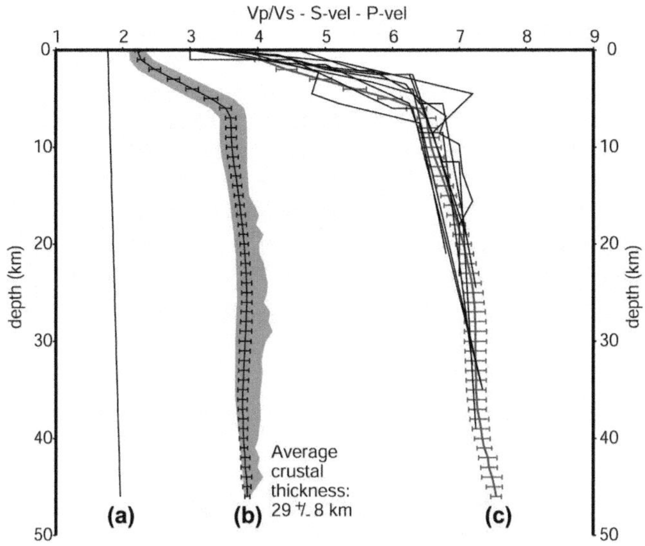

Figure 4.9. Average crustal velocity model for Iceland. (a) v_p/v_s ratio, (b) S velocity, and (c) P velocity.

Reproduced from Allen et al. (2002a). Copyright by the American Geophysical Union.

(Figure 4.9) if the v_p/v_s ratio increases slightly with depth. They find good agreement for available data if:

$$\frac{v_p}{v_s} = 1.78 + (0.004\ \mathrm{km}^{-1})z \tag{4.4}$$

where z is depth in the crust. This suggests temperatures closer to solidus in the lowermost part of the lower crust than in the upper part. Allen et al.'s. (2002a) fully three-dimensional S-wave velocity model for Iceland reveals considerable variation in the velocity structure. In the upper 10–15 km of the crust an elongated low-velocity region extends along some of the volcanic zones, with up to −7% velocity anomalies. At more than the 15-km depth in the crust, they find an indication of a low-velocity region under Iceland that can be represented by a vertical cylinder (Figure 4.10, see colour plates). Allen et al. (2002a) suggest that the low-velocity anomalies in the Icelandic crust reveal the thermal halo of a plume-driven plumbing system under Iceland, where material is fed from the mantle plume vertically up through the lower crust in central Iceland, and then laterally along the upper crustal rift system.

Thickness of the seismogenic crust in Iceland is also in agreement with relatively cold lower crust. Ágústsson and Flóvenz (2005) find depths of earthquakes typically in the range of 10–20 km, varying significantly from one area to another (Figure 4.11). They suggest the base of the seismogenic layer is associated with a temperature of 750° ± 100°C.

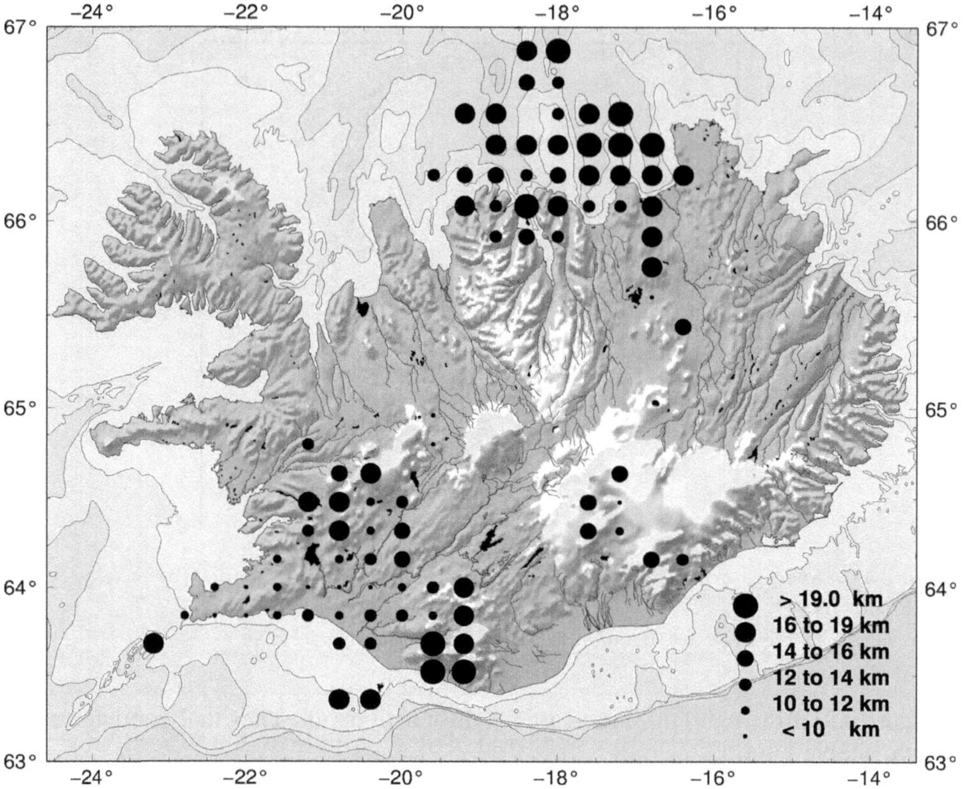

Figure 4.11. Depth of earthquakes in Iceland.
Reproduced from Ágústsson and Flóvenz (2005).

4.3.3 Models of thermal structure

Earlier models of thermal structure extrapolated the geothermal gradient observed in shallow crust, and suggested high temperatures at the bottom of crust in a thin-crust model (e.g., Pálmason, 1986). However, the assumption of linear extrapolation appears to be invalid. A thermal model by Menke and Sparks (1995) incorporating constraints from the seismic data can fit a large number of observations. The model (Figure 4.12) includes mass and heat transfer between upwelling mantle and accreting cooling crust. Melt formed in the mantle is carried rapidly and without heat loss to a 0–4-km-deep crustal accretion zone to form the crust. No extrusive processes are included in the model. The whole of the crust is formed by advection of magma from the shallow accretion zone to depth. The resulting lower crust is relatively cold because it has lost its heat near the surface. A pronounced feature of temperature profiles throughout the crust according to this model is a kink in temperature curves, with much higher gradients in the uppermost part than in the lower crust (Figure 4.13).

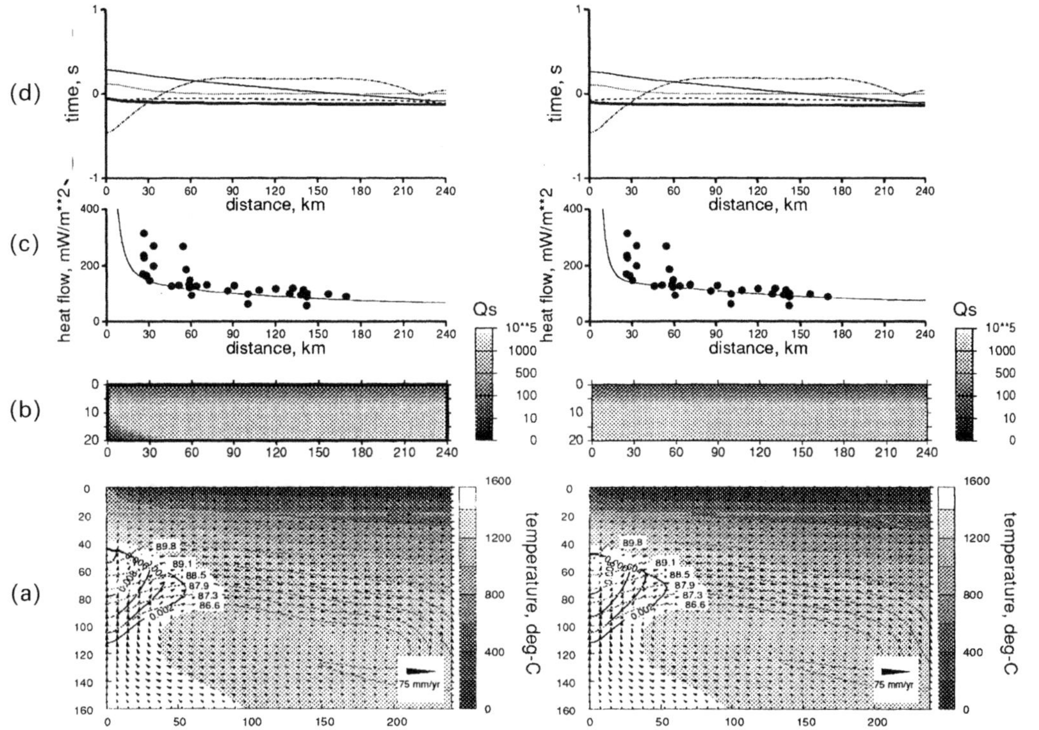

Figure 4.12. Mantle melting and crustal accretion model for Iceland by Menke and Sparks (1995) without (left) and with (right) hydrothermal circulation. Melt formed in the mantle is carried rapidly and without heat loss to a 0–4-km-deep crustal accretion zone to form the crust. No extrusive processes are included in the model. (a) Melting of mantle and accretion of crust, (b) seismic attenuation in crust, (c) heat flow, and (d) teleseismic travel time delays.

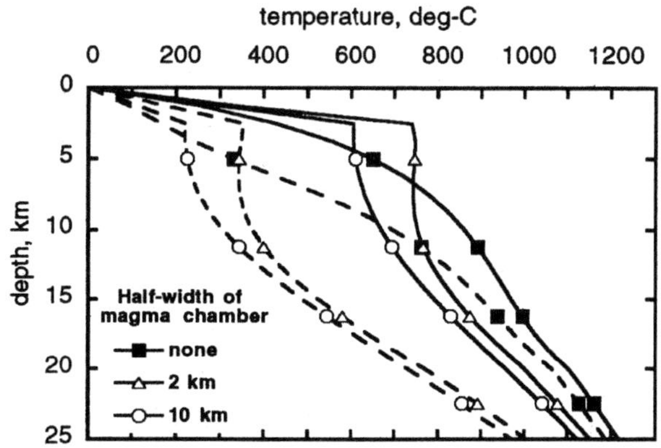

Figure 4.13. Model temperature profiles through the Iceland crust.

Magnetotelluric measurements (e.g., Beblo and Björnsson, 1978; Eysteinsson and Hermance, 1985) indicate the presence of a low-resistivity layer under Iceland. It was originally interpreted as evidence for partial melt under Iceland, but these results have not been fully put into the context of the thick cold crust model strongly favoured by the seismic data.

4.4 THE PÁLMASON MODEL OF CRUSTAL KINEMATICS

The structure of the crust is determined by its mode of accretion. The characteristic regional tilt of the lava pile towards the rift axis (see Figures 3.4 and 3.5) puts constraints on the accretion process. Both horizontal and vertical crustal velocity components need to be considered, as the shallow crust is modified by horizontal strain within the plate boundary zone (from diking and normal faulting), and by vertical loading from lava deposition on the surface. A model describing crustal kinematics considering these factors was developed in the 1970s by Guðmundur Pálmason (Pálmason 1973, 1980, 1981). It has been applied to Iceland as well as other mid-ocean ridges. In addition, an extension of the model has been used to describe the thermal structure of the crust (see Section 4.3.3).

The Pálmason model of crustal kinematics describes the crustal velocity field throughout the crust by considering the time-averaged motion of solid crustal elements. The model is two-dimensional and based on material balance conditions. The main assumptions of the model are that crustal accretion at the plate boundary zone can be described by two input functions, one describing the time-averaged horizontal strain within the plate boundary zone, and another describing the lava deposition rate on the surface of the crust. For a steady-state process, conservation of material requires that lava deposition is balanced by crustal subsidence in the plate boundary zone. Different types of the input functions can be considered. Pálmason (1980) uses input functions having a normal distribution with a certain standard deviation. Such behaviour is consistent with maximum dike density and lava deposition rate near the central axis of a plate boundary, with their intensity gradually decaying away from the rift axis. For a lava deposition rate having a normal distribution with a standard deviation, σ_2, the vertical velocity, v_z, of a crustal material as a function of distance from rift axis, x, is (Pálmason, 1980):

$$v_z(x) = \frac{q}{\sqrt{2\pi}\sigma_2} \exp\left(\frac{-x^2}{2\sigma_2^2}\right) \tag{4.5}$$

The constant q is the integrated rate of lava deposition across the width of the plate boundary zone, along its unit length. The horizontal displacement field depends on the horizontal strain rate at the plate boundary. The horizontal surface velocity, $v_x(x)$, increases from zero at the axis to the plate velocity V in the lithospheric plates on each side of the plate boundary deformation zone (V is half the spreading rate). If the long-term average horizontal strain rate within the plate

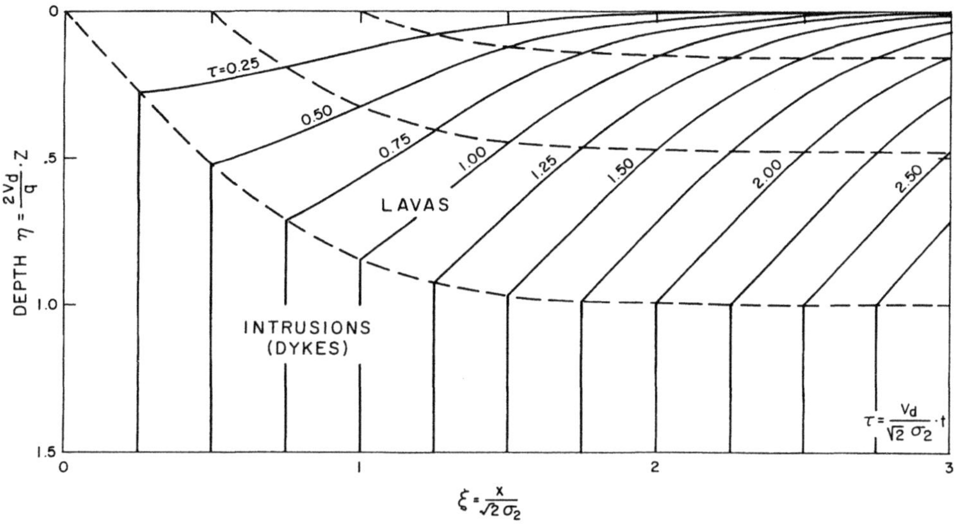

Figure 4.14. The Pálmason model of crustal kinematics. Trajectories (hatched lines) and isochrones (solid lines) of lava mass elements, for the special case of $\sigma_1/\sigma_2 = 0$ (horizontal strain localized at a rift axis).

Reproduced from Pálmason (1980).

boundary deformation zone is assumed to have a normal distribution with standard deviation σ_1, it can be written as:

$$\frac{dv_x(x)}{dx} = \frac{2V}{\sqrt{2\pi}\sigma_1}\exp\left(\frac{-x^2}{2\sigma_1^2}\right) \tag{4.6}$$

The normal distribution density is here scaled such that integration of the strain across the plate boundary zone gives the total relative plate separation, $2V$. Horizontal velocity is found by integration to be:

$$v_x(x) = \int_0^x \frac{dv_x(x)}{dx}dx = V\frac{\sqrt{2}}{\sqrt{\pi}\sigma_1}\int_0^x \exp\left(\frac{-x^2}{2\sigma_1^2}\right)dx = V \cdot \text{erf}\left(\frac{x}{\sqrt{2}\sigma_1}\right) \tag{4.7}$$

In the above presentation four parameters describe the flow field: V, q, σ_1, and σ_2. By assigning values to these parameters, the full crustal flow field can be calculated. Pálmason (1973) identified two sets of curves that are particularly relevant for the understanding of crustal architecture. One set of these is the trajectories, $z(x)$, of individual crustal elements. The gradient of a trajectory with respect to x equals the vertical velocity, divided by the horizontal velocity (Figure 4.14):

$$\frac{\partial z(x)}{\partial x} = \frac{v_z(x)}{v_x(x)} \tag{4.8}$$

Inserting equations (4.5) and (4.7), and considering that the trajectory will depend on the point of origin of the crustal element, it is found that the trajectories are

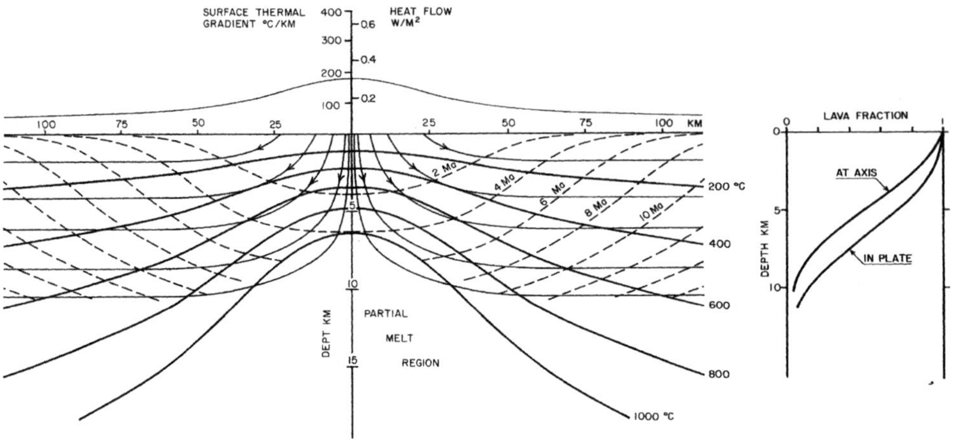

Figure 4.15. The Pálmason model of crustal kinematics. The model parameters are set to: V = half spreading velocity = 1 cm/yr, q = lava production rate per unit length of the rift zone = 1.33×10^{-4} km^2/yr, σ_1 = horizontal strain rate standard deviation = 15 km, and σ_2 = standard deviation of lava deposition rate = 20 km. In addition to the above parameters, the model curves take into consideration the effects of normal faulting in the uppermost crust.

Reproduced from Pálmason (1986) with permission of the Geological Society of America.

given by:

$$z(\xi, \xi_0) = \frac{q}{V} \frac{1}{\sqrt{\pi}} \int_{\xi_0}^{\xi} \frac{\exp(-s^2)\, ds}{\operatorname{erf}\left(\frac{\sigma_2}{\sigma_1} s\right)} \tag{4.9}$$

where ξ_0 is the horizontal coordinate of the point of origin at the surface and s is a variable of integration, equal to $x/(\sqrt{2}\sigma_2)$ (Pálmason, 1973, 1980). Another set of curves of interest are those representing the age of lava in the crust. For the input function given by equations (4.5) and (4.7), these curves are given by (Pálmason, 1981):

$$t(\xi, \xi_0) = \frac{\sqrt{2}\sigma_2}{V} \int_{\xi_0}^{\xi} \frac{ds}{\operatorname{erf}\left(\frac{\sigma_2}{\sigma_1} s\right)} \tag{4.10}$$

where again the integral is along a trajectory originating at the surface position ξ_0. An example of the application of the above equations is shown in Figure 4.15, with parameters appropriate for the Icelandic crust. The Pálmason model provides a good description of layering and tilt in the upper crust, in particular in eastern Iceland (see Figures 3.4 and 3.5).

5

Volcano dynamics

Volcanism in Iceland results from divergent plate movements across the Mid-Atlantic Ridge and excessive production of magma in the North Atlantic Mantle Plume. This excessive production of magma relative to other parts of the Mid-Atlantic Ridge results in thick crust in Iceland, averaging to about 30 km. Generation of crust this thick, over the ~300-km north–south length of Iceland, spreading at 1.9 cm/yr, requires magma generation in the mantle averaging about 0.2 km^3/yr. This magma is injected into the crust as intrusions, deposited on its surface in eruptions, or added to it by underplating. The volume of magma erupted can be directly evaluated, but seismic and geodetic measurements are needed to constrain the style and amount of subsurface magma movements. An extensive geodetic database on crustal deformation in Iceland, in particular on volcano inflation, sheds light on the processes involved. Seismic and geodetic data constrain how magma flows through the deeper ductile crust prior to eruptions or emplacement at shallow levels in the crust.

The volcanic zones in Iceland are divided up into volcanic systems as described in Chapter 3, each volcanic system typically consisting of a central volcano and a transecting fissure swarm. This chapter is mostly about the dynamics of central volcanoes where magma movements are most frequent. The chapter begins with a brief description of the different types of volcanic edifices in Iceland, eruptive styles, and geologic and seismic constraints on volcano interiors. Volcano deformation models are then presented, addressing in particular how volumes of magma moving inside volcanoes can be inferred. Next, separate sections discuss three volcanic systems of special interest, Krafla, Askja, and Hekla. An episode of rifting in the Krafla Volcanic System in 1975–1984 is still the best observed in Iceland. The Askja Volcano exhibits the highest rate of subsidence for any volcano in Iceland during a non-eruptive period. The most recent caldera-forming eruption in Iceland occurred there in 1875. Hekla is one of the most active volcanoes in Iceland, with several well-observed eruptions in recent decades. Hekla is also an important producer of andesitic and silicic rocks. The chapter ends with an overview of present knowledge on magma movements at Icelandic volcanoes.

5.1 VOLCANIC EDIFICES AND STYLES OF MAGMATIC ACTIVITY

The volcanic zones of Iceland are divided into volcanic systems as discussed in Chapter 3, each consisting of a central volcano, often associated with a fissure swarm, a caldera, and geothermal areas. The majority of volcanism in Iceland is basaltic, tholeiitic in the spreading rift zones, and alkaline in the flank zones. Silicic rocks occur at the central volcanoes, both at the active ones, as well as at eroded central volcanoes in the Tertiary sequence. Their volume is about 10–12% of exposed rocks in Iceland, anomalously high compared with other islands built on oceanic crust (Gunnarson et al., 1998). Partial melting of hydrated basalts is considered a likely explanation for the origin of many of the rhyolites, as suggested, for example, by Sigvaldason (1974). Only a few percent of surface rocks are andesitic in composition, most notably at the Hekla Volcano (Jakobsson, 1979a, b).

The morphology of volcanic landforms in Iceland varies widely (Figure 5.1). The largest volcano is the Öræfajökull Stratovolcano, over 2,100 m high. Other stratovolcanoes include the Eyjafjallajökull (1,667 m.a.s.l.) and Snæfellsjökull (1,446 m.a.s.l.) Volcanoes. These high-rising volcanoes are in the volcanic flank zones where little or no crustal spreading occurs. Mt. Hekla is a high-rising (1,491 m.a.s.l.) volcanic ridge, elongated along the strike of its main eruptive fissure. Tholeitic lava shields with gentle slopes are frequent, and monogenic crater rows characterize the fissure swarms of the volcanic systems. An important feature of Icelandic volcanism is the interaction of magma with ice and water. Iceland was mostly ice-covered during the Pleistocene, and today 10% of the country is still covered by ice, including some of the most active volcanoes. Volcanic landforms formed in subglacial eruptions are common outside the currently glaciated regions, resulting from subglacial volcanic activity during previous glaciations. In subglacial eruptions, pillow basalts are formed if the ice/water pressure is sufficiently high. Under lower pressure conditions, magma fragments immediately in response to rapid heat transfer from magma to ice (e.g., Gudmundsson, 2005). Pyroclastic material piles up at the eruptive site, and is later altered to hyaloclastite (e.g., Werner et al., 1996). Subglacial eruptions from a central vent lead to the formation of table mountains or tuyas, whereas subglacial fissure eruptions lead to the formation of elongated hyaloclastite ridges.

The average interval between eruptions in Iceland is 4–5 years and eruptions display a wide range in styles (Figure 5.2). Basaltic fissure eruptions are frequent, producing lava fields ranging in volume from about 0.01 km^3 up to 18 km^3 in historical times (the last 1,100 years). The largest lava flows formed in the Eldgjá (934 AD) and Laki (1783–84 AD) Eruptions (Larsen, 2000; Thordarson et al., 2001). Some eruptions, like those of Mt. Hekla, begin with an explosive initial phase but then the vigour of the eruption decreases and effusive lava production takes over.

Silicic explosive eruptions are infrequent but have caused catastrophic environmental effects. When subglacial volcanic eruptions break through their overlying icecaps, they experience an explosive phase due to magma–water interaction. Basaltic subglacial eruptions of this type at Grímsvötn and Katla Volcanoes are frequent and have resulted in numerous basaltic tephra layers. Subglacial volcanic eruptions are associated with sudden glacial outburst floods (jökulhlaup), one of the main volcanic hazards in Iceland.

Various types of eruptions have occurred in recent decades in Iceland. Mt. Hekla erupted in 1947, 1970, 1980–1981, 1991, and 2000. Nine eruptions occurred at the Krafla Volcano during a rifting episode from 1975 to 1984. A submarine eruption in the Westman Islands Volcanic System off the south coast of Iceland from 1963 to 1967 formed the island of Surtsey. Another eruption in the same volcanic system at the island of Heimaey in 1973 resulted in temporary emergency evacuation of the island and partial destruction of a village. Subglacial volcanic eruptions occurred at the Grímsvötn Volcano in 1983, 1998, and 2004. In 1996, 0.45 km^3 of magma were erupted, under an initially 600-m-thick ice, at the Gjálp eruptive site in Vatnajökull, midway between the Grímsvötn and Bárðarbunga Volcanoes. This was the first large subglacial eruption under thick ice to be monitored in detail, providing new insight into aspects of ice–volcano interaction, such as the rate of ice melting, the efficiency of heat transfer, ice deformation, and subglacial water pressure (Gudmundsson et al., 1997, 2004).

5.2 VOLCANO INTERIORS: GEOLOGIC AND GEOPHYSICAL CONSTRAINTS

The interiors of extinct Icelandic volcanoes can be viewed in the eroded Tertiary formation (see Figure 3.2). Extinct volcanoes are carried away from the rift axis by plate spreading and eroded along with their surroundings (Figure 5.3, see p. 76), down to 1–2 km from the original surface in the far east and west of Iceland (Saemundsson, 1979). Complex intrusive bodies are revealed, representing the uppermost parts of magmatic systems. In favourable cases such as at Breiðdalur, eastern Iceland, a cross section of extinct volcanic structures can be inferred (Figure 5.4, see p. 76). Pioneering studies of such complexes in eastern Iceland were conducted by George Walker (e.g., Walker, 1963). Similar structures are found in western Iceland—e.g., at the eroded Setberg Volcano (Sigurðsson, 1966).

Seismic studies have provided constraints on the internal structure of the currently active volcanoes. The best seismically studied volcanoes include Krafla, Katla, and Hekla. No magma chamber has been seismically detected at Hekla, and Soosalu and Einarsson (2004) argue that if considerable molten volume exists under Hekla, it must be located below the 14-km depth. On the other hand, clear low velocity anomalies and S-wave shadows (Figures 5.5 and 5.6, see colour plates) are found under Krafla and Katla Volcanoes, interpreted in both cases as resulting from the presence of shallow magma chambers.

Figure 5.1. Examples of different volcanic landforms in Iceland. (a) Wintertime view over the Krafla Volcanic System in the Northern Volcanic Zone. Þrengslaborgir Crater Row and Lúdent Tuff Cone in the Krafla Fissure Swarm are in the foreground and the Krafla Central Volcano is in the background. (B) The Öræfajökull Stratovolcano.

Photos courtesy of Oddur Sigurðsson.

Figure 5.1. Examples of different volcanic landforms in Iceland (*cont.*). (c) The Hekla Volcanic Ridge. (d) The Askja Caldera with its nested Lake Öskjuvatn Caldera, with Mt. Herðubreið Table Mountain in the background to the right.

Photos courtesy of Oddur Sigurðsson

Figure 5.2. Examples of styles of volcanic activity. (a) The submarine eruption of Surtsey in 1964, after the island of Surtsey had formed. (b) The subglacial Gjálp Eruption in 1996.
Photos (a) Guðmundur E. Sigvaldason and (b) Freysteinn Sigmundsson.

Figure 5.2. Examples of styles of volcanic activity (*cont.*). (c) Eruption of Krafla in 1980. (d) Eruption of Hekla Volcano in 2000.

Photos (c) courtesy of Halldór Ólafsson and (d) of Sigurjón Sindrason.

Figure 5.3. The Sandfell Laccolith in Fáskrúðsfjörður, eastern Iceland. Intruded magma has lifted up and tilted overlying strata.

Photo courtesy of Ágúst Guðmundsson, Jarðfræðistofan Ltd.

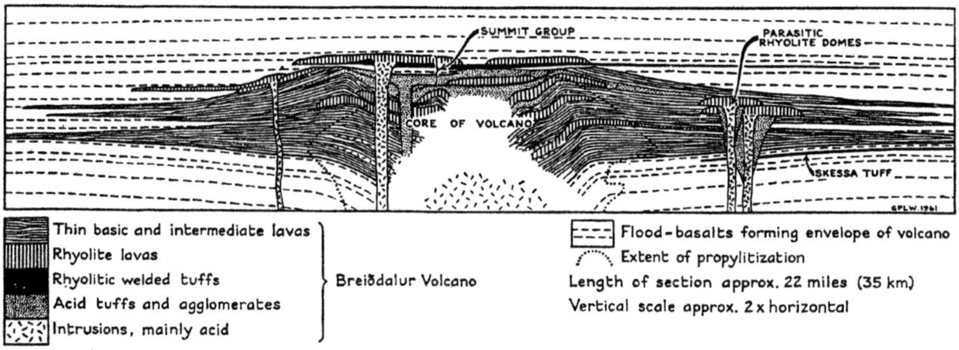

Figure 5.4. Schematic view through the Breiðdalur Tertiary Central Volcano in eastern Iceland.

Reproduced from Walker (1963).

Brandsdóttir et al. (1997) found a low-velocity anomaly under the central part of the Krafla Caldera at about a 3-km depth (Figure 5.5) and previous studies had revealed an S-wave shadow under the caldera at a 3–7-km depth (Einarsson, 1978). Brandsdóttir et al. (1997) interpret their results in terms of a magma chamber with 0.2–0.3-s compressional wave delays and shear wave shadowing. From these observations the thickness of the Krafla Magma Chamber is estimated at 0.7–1.8 km, the north–south length about 2–3 km, and an east–west length of 8–10 km. Its estimated volume is 12–54 km^3. The magma chamber sits at the top of a high-velocity dome. It is concluded that the mid-crust under the shallow magma chamber can neither contain partial melt nor be at near-solidus temperatures (Brandsdóttir et al., 1997).

At Katla Volcano, seismic undershooting shows clear S-wave shadows associated with delays in traveltime due to a shallow body with anomalously slow velocities (Gudmundsson et al., 1994). The seismic results are interpreted in terms of a shallow magma chamber, with a bottom at 3 km below the surface of the volcano (Figure 5.6, see colour plates). The magma chamber is about 5 km across along the seismic profile studied. The chamber is underlain by rocks of average or high velocity for that depth, and fast structures interpreted as crystalline intrusives occur on both sides of the magma chamber. Gudmundsson et al. (1994) estimate the volume of the magma chamber to be about 10 km^3, with about half of that volume being melt, in order to produce the observed low velocities (2.5–3 km/s). Another area where a magma chamber has been inferred from seismic data is at the Torfajökull Volcano. Careful inspection of seismicity by Soosalu and Einarsson (1997, 2004) has revealed a volume with a centre at an 8-km depth and diameter of 4 km that is devoid of earthquakes, surrounded by earthquake hypocentres on all sides. Their interpretation is that this volume is a cooling magma chamber.

Additional constraints on magma chambers are provided by geothermal, gravimetric, and magnetic studies. Example is provided by study of Gudmundsson and Milsom (1997) who show magnetic and gravimetric anomalies at the subglacial Grímsvötn Caldera consistent with magma source at shallow depth under the caldera.

5.3 MODELLING OF VOLCANO DEFORMATION

How does magma accumulate inside volcanoes? The combined use of seismic and geodetic techniques has been particularly useful to provide the answer to this question. Geodetic results from Iceland and elsewhere demonstrate considerable variability in volcano behaviour (e.g., Sigmundsson, 1996; Massonnet and Sigmundsson, 2000; Sturkell et al., 2005). An emerging pattern from the available observations at Icelandic volcanoes reveals that most of them are either non-deforming or subside between eruptions. This quiescent state is interrupted by episodic inflow of magma from depth, continuous for a timespan of only months or years. Such recharging of magmatic systems in Iceland occurs intermittently, and in many cases ends without an eruption.

Table 5.1. Geodetic measurements of inflation at Icelandic volcanoes.

Volcano	Period	Duration (yr)	Source depth (km)	Total uplift (cm)	Average rate (cm/yr)	Uplift volume (10^6 m^3)
Krafla[(1)]	1974–1989	15	2.5	1,360	91	534
Askja[(2)]	1967–1968	1	2.5	14	14	5
Askja[(2)]	1970–1972	2	2.5	40	20	16
Hekla[(3)]	1981–1991	10	6	35	3.5	79
Hrómundartindur[(4)]	1993–1998	5	7	10	2	31
Eyjafjallajökull[(5)]	1994	0.7	4.6	20	29	27
Eyjafjallajökull[(6)]	07/99–05/00	0.8	6.3	20	24	50
Katla[(7)]	2000–2004	4	4.7	16	4	22
Grímsvötn[(8)]	1998–2004	6	2.5	60	10	24
Krafla[(9)]	1993–1999	6	21	8	1.4	219

(1) Cumulative uplift punctuated by series of deflation events—e.g., Tryggvason (1995).
(2) Tryggvason (1989).
(3) Tryggvason (1994); similar behaviour suggested for subsequent inter-eruptive periods.
(4) Sigmundsson et al. (1997); Feigl et al. (2000).
(5) Pedersen and Sigmundsson (2004); Sturkell et al. (2003b).
(6) Pedersen and Sigmundsson (in press); Sturkell et al. (2003b).
(7) Sturkell et al. (2005).
(8) Sturkell et al. (2003, 2005)
(9) Suggested deep magma accumulation at crust–mantle boundary under the Krafla Volcanic System (de Zeeuw-van Dalfsen et al., 2004).

Volcanic unrest associated with magmatic movements is typically accompanied by elevated seismicity as a consequence of stresses induced by intruding magma. Earthquake activity may occur in crustal volumes around magma chambers, next to intrusions, or as a result of fracturing associated with opening of magma conduits or diking events. Magma migration may lead to migrating earthquake activity, and the amount of magma moving can be estimated using geodetic techniques. Crustal deformation associated with volcano inflation can be interpreted in terms of deformation source models which in favourable cases constrain location, volume, and geometry of deformation sources. Inflow of magma to eight volcanic systems has been observed geodetically in Iceland in 1966–2004 (Table 5.1). Magma has travelled upwards from an uncertain depth to 3–7-km levels; one set of observations suggests magma accumulation at much deeper levels, at about 20-km depth (at the crust–mantle boundary) under the Krafla Volcanic System. Deflation of volcanoes associated with pressure decreases in shallow magma chambers at about a 3-km depth has been documented at Krafla, Askja, and Grímsvötn Volcanoes, and co-eruptive subsidence has been documented at the Hekla Volcano as well (Table 5.2).

5.3.1 The Mogi model

Much of the data on volcano deformation in Iceland has been interpreted using the "Mogi model" (Mogi, 1958). Accumulation of magma inside the crust is modelled as

Table 5.2. Geodetic measurements of deflation at Icelandic volcanoes.

Volcano	Period	Source depth (km)	Subsidence (cm)	Description
Krafla[(1)]	1975–1984	2.5	~1,360	About 20 diking events and eruptions
Hekla[(2)]	1991	8?	~20–40	Co-eruptive deflation
Hekla[(3)]	2000	8?	~10–20	Co-eruptive deflation
Grímsvötn[(4)]	1998	~2	>100	Co-eruptive deflation
Grímsvötn[(5)]	2004	~2	>100	Co-eruptive deflation
Krafla[(6)]	1989–present	3	~50	Long-term decaying gradual deflation
Askja[(7)]	1983–present	3	>100	Long-term decaying gradual deflation
Reykjanes[(8)]	1992–present	4.6	~10	Caused by geothermal exploitation

(1) Cumulative subsidence in about 20 abrupt deflation events during the Krafla rifting episode—e.g., Tryggvason (1995).
(2) Tryggvason (1994); Sigmundsson et al. (1992); Linde et al. (1993).
(3) Sigmundsson et al. (2001); Sturkell et al. (2005).
(4) Sturkell et al. (2003a).
(5) Vogfjörd et al. (2005).
(6) Sigmundsson et al. (1997); Sturkell et al. (2005).
(7) Sturkell et al. (in press). A deeper source of subsidence is also suggested.
(8) Vadon and Sigmundsson (1997). Apparent source depth when geothermal exploitation modelled by Mogi source. For a more complete discussion and references see Sturkell et al. (2005).

a point source of pressure in an elastic half-space (Figure 5.7). The equations of elasticity are solved, applying the boundary conditions of zero traction on the half-space surface and a pressure increase, ΔP, applied at a point source (Anderson, 1936; Mogi, 1958; McTigue, 1987). Resulting surface deformation is radially symmetric. In cylindrical polar coordinates, r and ϕ in the horizontal plane, and z along the depth axis, surface displacements are expressed as:

Horizontal radial displacement:
$$u_r = C\frac{r}{(d^2 + r^2)^{3/2}} \tag{5.1}$$

Horizontal tangential displacement:
$$u_\phi = 0 \tag{5.2}$$

Vertical displacement:
$$u_z = C\frac{d}{(d^2 + r^2)^{3/2}} \tag{5.3}$$

where d is the source depth, r is the horizontal displacement away from the source, and C is the source strength parameter. The centre of the coordinate system is set on the surface, directly above the source. These equations are valid for an elastic half-space with Poisson's ratio 0.25—a common assumption when using this model. There are four free parameters in the model, three for location of the source (latitude, longitude, depth) and one for the source strength. The source strength

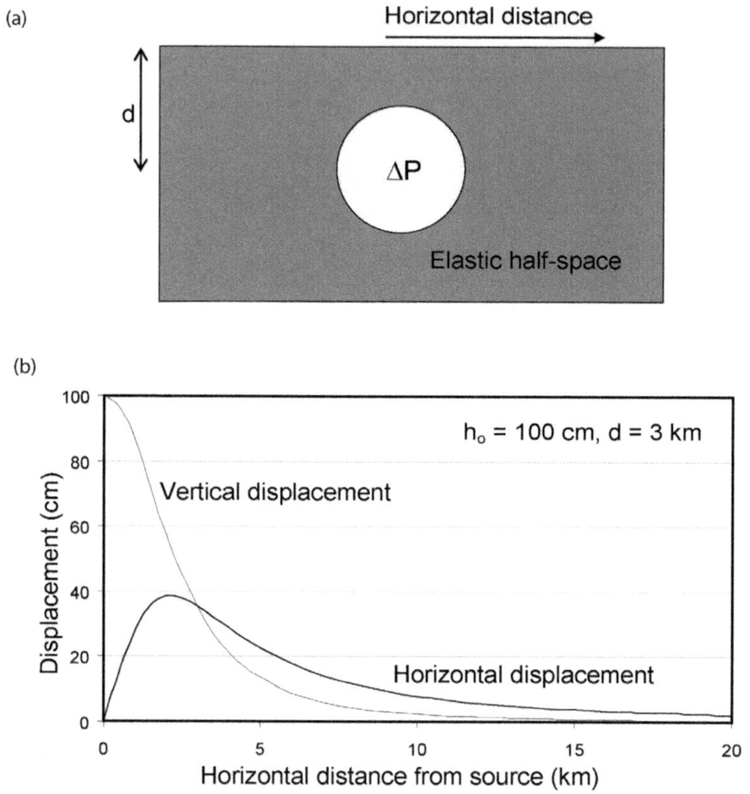

Figure 5.7. The Mogi model. (a) Schematic view. (b) Vertical and horizontal displacements.

parameter, C, is given by:

$$C = \frac{3a^3 \Delta P}{4\mu} \tag{5.4}$$

where ΔP is the change in fluid pressure within the spherical source, a is its radius, and μ is the modulus of rigidity (shear modulus) of the crust surrounding the sphere. It is not possible to separate the contributions from ΔP and a; only the source strength can be derived. Maximum uplift occurs directly above the source. Setting $r = 0$ in equation (5.3) gives:

$$C = h_0 d^2 \tag{5.5}$$

where h_0 is the maximum vertical displacement.

Tilt and strain can be found from expressions (5.1)–(5.3) by taking derivatives. Strain in a cylindrical coordinate system is, for example, given by Laundau and

Lifshitz (1986). For the Mogi model we have:

$$\text{Radial tilt:} \qquad \delta = \frac{\partial u_z}{\partial r} = C\frac{-3dr}{(d^2 + r^2)^{5/2}} \tag{5.6}$$

$$\text{Horizontal radial strain:} \qquad \varepsilon_r = \frac{\partial u_r}{\partial r} = C\frac{d^2 - 2r^2}{(d^2 + r^2)^{5/2}} \tag{5.7}$$

$$\text{Horizontal tangential strain:} \qquad \varepsilon_\phi = \frac{1}{r}\frac{\partial u_\phi}{\partial \phi} + \frac{u_r}{r} = C\frac{1}{(d^2 + r^2)^{3/2}} \tag{5.8}$$

$$\text{Aerial strain:} \qquad \Delta = \varepsilon_r + \varepsilon_\phi = C\frac{2d^2 - r^2}{(d^2 + r^2)^{5/2}} \tag{5.9}$$

Comparison of observations with predicted deformation by the model allows estimation of the location of the magma source and the strength parameter (or alternatively the maximum uplift). Although the Mogi model is originally derived for a point source, it has been shown to be valid also for spherical sources as long as $(a/d)^5 \ll 1$ (McTigue, 1987). The model is therefore valid both for a spherical magma intrusion into a cold structure (no pre-existing magma chamber), or for inflow of magma into a spherical pre-existing magma chamber (pressure change in a finite size pre-existing chamber). Although surface deformation in these two situations is similar, there can be a large difference in the volume of magma needed to cause the same amount of surface deformation in these two cases.

5.3.2 Estimation of magma volumes from the Mogi model

Modelling of volcano deformation can give direct information on the volume of magma, ΔV_{magma}, flowing in or out of a magmatic system. This is difficult with any other technique. The volume estimates from deformation studies do, however, have large uncertainties. There are several steps needed to infer ΔV_{magma}. If a Mogi model is applicable, then the initial and simplest volume to calculate is the integrated volume of surface change. This is also referred to as the edifice volume change, $\Delta V_{edifice}$. It is given by:

$$\text{Integrated ground surface volume change:} \quad \Delta V_{edifice} = \int_{r=0}^{\infty} u_z 2\pi r\, dr = 2\pi C \tag{5.10}$$

The volume change of the Mogi source itself can be found by considering the displacement on the surface of the source at depth (Delaney and McTigue, 1994). It is:

$$\text{Volume change of a Mogi source:} \quad \Delta V_{Mogi} = \tfrac{4}{3}\pi C \tag{5.11}$$

Equations (5.10) and (5.11) show that for a Mogi source the integrated ground volume change is 3/2 times the volume change of the Mogi source, or:

$$\Delta V_{Mogi} = \tfrac{2}{3}\Delta V_{edifice} \tag{5.12}$$

The difference between the two volumes is due to dilation of the crust above the Mogi source (Delaney and McTigue, 1994). If magma is injected into a cold solidified volcanic structure and forms a spherical source, then the volume of magma intruded, ΔV_{magma}, is the same as ΔV_{Mogi}. For the Icelandic cases presented in Table 5.1, this situation may be applicable at the Hrómundartindur and the Eyjafjallajökull Volcanoes, as both of these are characterized by infrequent magmatic activity.

If new magma flows into (or out of) a pre-existing magma chamber, an additional effect has to be considered. In such cases, residing magma in a chamber will compress (expand) as new magma flows in (out). In general, volume change of material due to pressure change depends on the bulk modulus of the material, k. The volume change, ΔV, associated with a change in pressure, ΔP, in volume, V, is:

$$\Delta V = V \frac{1}{k} \Delta P \tag{5.13}$$

Considering this effect, a general equation relating $\Delta V_{edifice}$ and ΔV_{magma} for magma inflow into a spherical source can be derived (Johnson et al., 2000):

$$\Delta V_{magma} = \frac{2}{3}\left(1 + \frac{4\mu}{3k}\right)\Delta V_{edifice} \tag{5.14}$$

where μ is the shear modulus of the host rock and k is the effective magma bulk modulus. For an incompressible fluid, k tends to infinity and equation (5.14) reduces to (5.12). Values for μ and k are needed to infer volume of moving magma, but uncertainties on their values are large. While studying the 1984 eruption of Krafla Volcano, Árnadóttir et al. (1998) used experience from Kilauea Volcano (Johnsen, 1987) to infer ranges for net volume of magma expelled from the shallow Krafla Magma Chamber during that eruption, suggesting $2\Delta V_{edifice}/3 \leq \Delta V_{magma} \leq 2.4\Delta V_{edifice}$. The lower limit comes from equation (5.12). The upper limit comes from the Kilauea results of Johnson (1987), corresponding to $\mu \approx 2k$. If magma flows into a pre-existing magma chamber, utilization of equation (5.14) is required for estimation of the amount of magma flow. This is the situation at Krafla, Askja, Grímsvötn, and Katla Volcanoes, and probably Hekla. Assuming values of $\mu = 2k$ (e.g., $\mu = 30$ GPa and $k = 15$ GPa), an estimate of the amount of new magma flowing into these systems can be derived (Figure 5.8).

5.3.3 Modelling magma sources as sills, dikes, and ellipsoidal sources

Alternative models for magmatic deformation sources include sills, dikes, and ellipsoidal sources. Dikes and sills are frequently modelled as rectangular dislocations with opening parallel to their plane, using formulations given by Okada (1985). A general dislocation model is characterized by ten parameters, whereas a Mogi model is determined by only four parameters. Such planar sheet models may be favoured over a Mogi model because they fit better to geodetic observations, and/or because independent evidence suggests planar geometries. Such evidence may be seismic constraints on magma source geometry or compat-

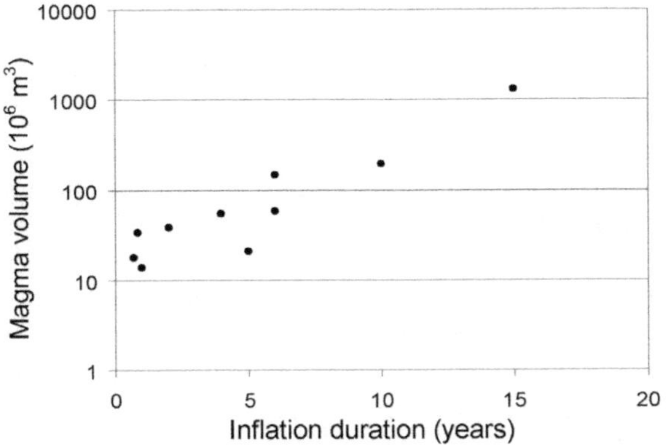

Figure 5.8. Volume of intrusion inferred from geodetic techniques versus duration of inflation episodes recorded in Iceland.

ibility with prevailing stress fields. In Iceland, for instance, magma intrusions in the Eyjafjallajökull Flank Zone Volcano in 1994 and 1999 have been modelled as sills (Pedersen and Sigmundsson, 2004; in press). Prevailing stress fields in volcanoes govern the shape of intrusions, and knowledge about stresses may help to constrain models. If the minimum compressive stress is horizontal, a dike will form, but if it is vertical a sill will form, because a planar sheet intrusion will open along a plane perpendicular to the direction of minimum compressive stress.

The overpressure associated with a spherical magma intrusion depends on its size. If magma flows into a spherical, pre-existing magma chamber that has a volume of 50 km^3 ($a = 2.3$ km), then the associated increase in pressure is 8 MPa if $\Delta V_{Magi} = 0.01$ km^3 and $\mu = 30$ GPa—using equations (5.4) and (5.11). If a similar amount of magma is emplaced in solid rock, with no pre-existing magma chamber, then unrealistically high overpressure is required if the intrusion is to form a spherical source. Equating the Mogi volume in equation (5.11) to the volume of a sphere (equal radii), and inserting the source strength from equation (5.4) one finds for this case that:

$$\Delta P = \tfrac{4}{3}\mu \tag{5.15}$$

For $\mu = 30$ GPa, the required overpressure is 40 GPa. This is four orders of magnitude larger than the tensile strength of the crust (see Section 5.3.5). Long before this stress level is reached, the rock will fail and a planar sheet intrusion will form perpendicular to the direction of minimum compressive stress. If magma is injected into cold structures, a planar sheet model may therefore be a more realistic approximation than a Mogi model. Much lower overpressures are needed for such sheet-like intrusions. The order of magnitude of stresses needed to dilate a magma-filled sheet, ΔP_{sheet}, is (Lister and Kerr, 1991):

$$\Delta P_{sheet} \approx \frac{\mu}{(1-\nu)}\frac{u}{L} \tag{5.15}$$

where μ is the shear modulus and ν the Poisson's ratio of the host rock, u is the typical thickness of the sheet, and L is the shorter of its two dimensions. The uniform opening sill model of Pedersen and Sigmundsson for the 1994 intrusion in Eyjafjallajökull Volcano has $u = 0.22\,\text{m}$ and $L = 4.5\,\text{km}$. The pressure needed to inflate such a sill is on the order of 2 MPa, if $\mu = 30\,\text{GPa}$ and $\nu = 0.25$.

The ratio between intruded magma and integrated surface uplift volume depends on source geometry and elastic properties of the host rock. Assuming Poisson's ratio of 0.25, the Mogi model gives this ratio as 3/2—see equation (5.12). For a sill this ratio is 1, and for a dike the ratio is 3/4 (Delaney and McTigue, 1994). This variability demonstrates the need for careful consideration of deformation source geometry, as it may significantly influence inferred magma volumes.

An alternative model for a pressure change in shallow magma chamber is the ellipsoidal source model. The deformation field due to such a source is described, for example, by Yang et al. (1988). In this model there are eight parameters (semimajor axis a, semiminor axis b, dip angle, orientation angle, three parameters for location, and the pressure change inside the ellipsoid, ΔP). Resulting volume change of the ellipsoidal source can be related to change in pressure within it (Tiampo et al., 2000):

$$\Delta V_{ellipsoid} = \frac{\Delta P}{\mu} \pi a b^2 \tag{5.16}$$

For $a = b$ the ellipsoid becomes a spheroid and relation (5.11) is reproduced. Different types of ellipsoidal models exist, with some simplifying assumptions. One of these approaches has been used by Ewart et al. (1991) to model data from the Krafla Volcano. Their conclusion was, however, that utilization of their model resulted in an unrealistically shallow depth of magma storage. At Askja Volcano, Pagli et al. (in press) successfully used an ellipsoid model to interpret deformation.

5.3.4 Feeder channels for magma chambers and shallow intrusions

How does magma move upwards towards shallow depth during inflation? The analyses in Section 5.3.2 provided an estimate of intrusion volumes (assuming the Mogi model). Dividing this volume by the duration of magma inflow, one finds that the volumetric magma flow rates, Q, average from 0.05 to $5\,\text{m}^3/\text{s}$. These values constrain the dimensions of feeder channels, suggesting they are narrow with a diameter on the order of a few metres or less.

The process can be modelled as fluid flowing through a pipe. The volumetric flow rate for a laminar pipe flow assuming Newtonian behaviour is (Turcotte and Schubert, 1982):

$$Q = -\frac{\pi R^4}{8\eta} \frac{dp}{dx} \tag{5.17}$$

where R is the radius of the pipe, η the viscosity of the fluid, and dp/dx is the pressure gradient along the pipe. If a volcanic pipe is vertical and flow is driven only by overpressure related to density difference between magma and host rock, then $dp/dx = -g\Delta\rho$, where g is the gravitational acceleration, and $\Delta\rho$ is the density difference between magma and host rock. Equation (5.17) becomes:

$$Q = \frac{\pi R^4}{8\eta} g \Delta\rho \tag{5.18}$$

If viscosity and density difference are known or can be estimated, then the radius of the flow channel can be inferred. For an intrusion in 1999 at Eyjafjallajökull Volcano, Pedersen and Sigmundsson (in press) find $Q = 5\,\mathrm{m^3/s}$ and suggest a magma viscosity, η, in the range 10–100 Pa s and $\Delta\rho$ in the range 300–485 kg/m^3. The radius of the feeder channel is then 1 m or less. The order of magnitude for the channel radius is well constrained, as the flow rate scales with the fourth power of the radius of the pipe. Alternative geometries for feeder channels can be considered, but their cross-sectional area has to be similar as for a pipe in order to produce a similar volumetric flow.

A calculation of the Reynolds number for Newtonian pipe flow (Turcotte and Schubert, 1982) demonstrates the validity of the assumption of a laminar flow. Because the width of feeder channels for shallow intrusions or magma chambers is limited, their expected surface deformation is limited. Equations for deformation due to vertical pressurized pipes are given by, for example, Bonaccorso and Davis (1999), but deformation associated with such feeder conduits for shallow intrusions has not been confirmed in Iceland.

5.3.5 Failure criteria for eruptions

Eruptions from inflating magma bodies are associated with tensile failure of the host rock. An alternative triggering mechanism may however occur, such as sudden slip on faults which may in turn rupture magma bodies or cause instabilities. The rupture criterion for tensile failure (e.g., Pinel and Jaupart, 2003) is that the magnitude of the deviatoric stress must exceed a certain threshold value, which is the tensile strength of the crust, T_s. The criteria can be written as:

$$\Delta\sigma_3 < -T_s \tag{5.19}$$

where $\Delta\sigma_3$ is the deviatoric minimum compressive stress (minimum compressive stress minus lithostatic stress). The tensile strength of the crust in Iceland is not well known, but analysis of hydro-fracturing measurements in the uppermost 600 m of a drill hole at Reyðarfjörður, eastern Iceland, have revealed it to be 1–6 MPa (Haimson and Rummel, 1982).

Stresses around a magma body increase with the pressure inside it. For a two-dimensional magma source such as a pressurized pipe in an infinite medium, the hoop stress (tensile stress on the pipe walls) is constant along the walls of the pipe and is equal to the overpressure in the pipe. In this case, and for a pressurized pipe of small radius in an elastic half-space, the eruption criteria will be:

$$\Delta P_c \rightarrow T_s \tag{5.20}$$

The critical overpressure needed for failure, ΔP_c, equals the tensile strength. In general, the geometry of magma sources will cause deviations from this simple relationship. Hoop stresses on the walls of a magma chamber will be modified by the finite dimensions of the source, and also by the presence of a volcanic edifice above the source. Stress concentration around a pressurized pipe in an elastic half-space is well known from the work of Jeffery (1920). This model has been applied to magma chambers, for example, by Gudmundsson (1988), and more recently by Pinel and Jaupart (2003) who analyse the role of volcanic edifices in modifying stress fields at their underlying magma chambers. A general form of the failure criteria includes an amplification factor, k, such that the criterion becomes:

$$k\Delta P_c \rightarrow T_s \tag{5.21}$$

In the two-dimensional case, for a pressurized pipe in an elastic half-space, the stress concentration will vary along the walls of the pipe if its radius, a, is not small compared with its depth, d. The maximum value of k is (e.g., Pinel and Jaupart, 2003):

$$k = \frac{1 + (a/d)^2}{1 - (a/d)^2} \tag{5.22}$$

For $a/d = 0.5$, the amplification factor will be 5/3. Less overpressure is needed to cause failure than for a pipe with smaller radius. The stress concentration is different in three dimensions. For example, the hoop stress at the boundary of a sphere in an infinite elastic half-space is equal to half the overpressure in the source, corresponding to $k = 0.5$ (McTigue, 1987). Numerical models are, in general, needed to evaluate the stress concentration.

Stresses around magma chambers are also influenced by plate movements. Extension across the spreading plate boundary in Iceland causes gradual reduction in the minimum compressive stress (horizontal in the rift zones). This causes a second contribution to deviatoric stresses, in addition to those of magmatic origin, and can modify the conditions for eruptions (Gudmundsson, 1988).

5.4 THE KRAFLA VOLCANIC SYSTEM AND ITS 1975–1984 RIFTING EPISODE

On December 20, 1975 a minor eruption began at Krafla Central Volcano that lasted 2 days. Seismicity began 15 minutes before the eruption, and continued until March 1976 (Einarsson, 1991b). Harmonic tremor was recorded. The ground within the Krafla Caldera subsided by 2 m, and a major diking event occurred along the plate boundary north of Krafla. Maximum widening during this diking event, about 2 m, appears to have occurred near the north coast of Iceland, 50 km north of the caldera (Björnsson et al, 1977, 1979). In a 9-yr period $\sim$1 km^3 of magma was transported from depth, feeding 9 eruptions and $\sim$20 diking events. Cumulative widening of the Krafla Volcanic System amounted to up to 9 m, with average widening of 5 m along an 80-km segment of the Krafla Fissure Swarm (Tryggvason, 1984).

The Krafla Central Volcano is a shield built by repeated eruptions with its oldest surface formations around 200,000 years old (Saemundsson, 1991). It has a caldera

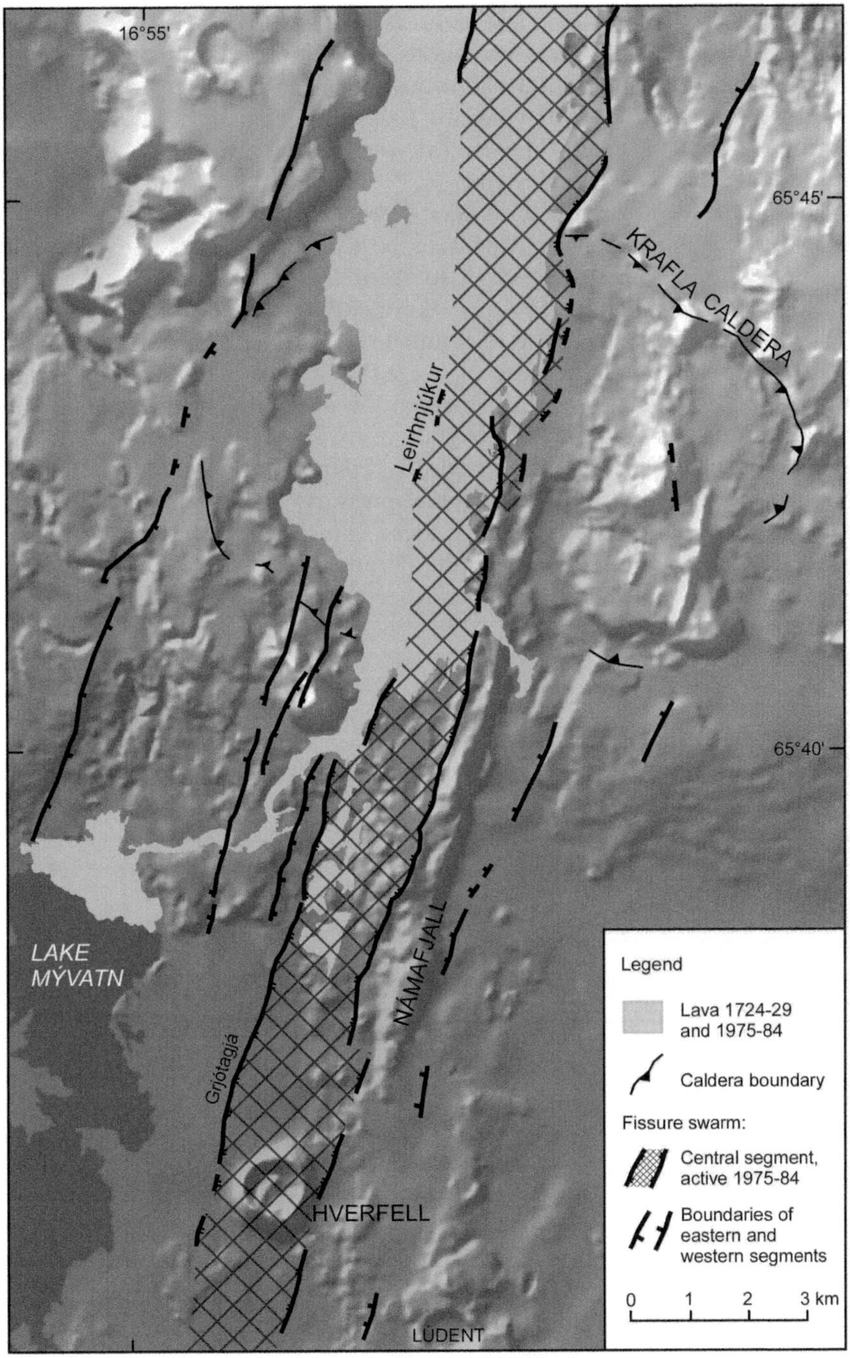

Figure 5.9. Simplified geologic map of the Krafla area.
Modified from Sæmundsson (1991).

that formed during interglacial time about 100,000 years ago (Figure 5.9) and is now mostly filled with younger rock formations. Rhyolite formations border the caldera and extensive geothermal activity is in the area. The Krafla area has all the characteristics of a typical central volcano. It is transected by a 100-km-long fissure swarm that is 5–10 km wide. The central volcano and the fissure swarm constitute together the Krafla Volcanic System.

The subsurface structure of Krafla is known from extensive geothermal exploration and drilling, and has been mapped by seismic, gravimetric, and geodetic techniques. Good seismic evidence exists for a shallow magma chamber under the central part of the caldera at about a 3-km depth (see Section 5.2).

Extensive geodetic studies were conducted at Krafla during the rifting episode. The results show a consistent pattern, where the shallow magma chamber at a 3-km depth, in the same location as the observed low-seismic-velocity anomaly, played a key role. Continuous inflow of magma towards the shallow magma chamber occurred throughout the rifting episode, punctuated by diking events and eruptions that temporarily lowered the pressure in the magma chamber and caused subsidence. Deflation of the caldera coincided with activity in the fissure swarm outside the boundaries of the Krafla Central Volcano (Figure 5.10).

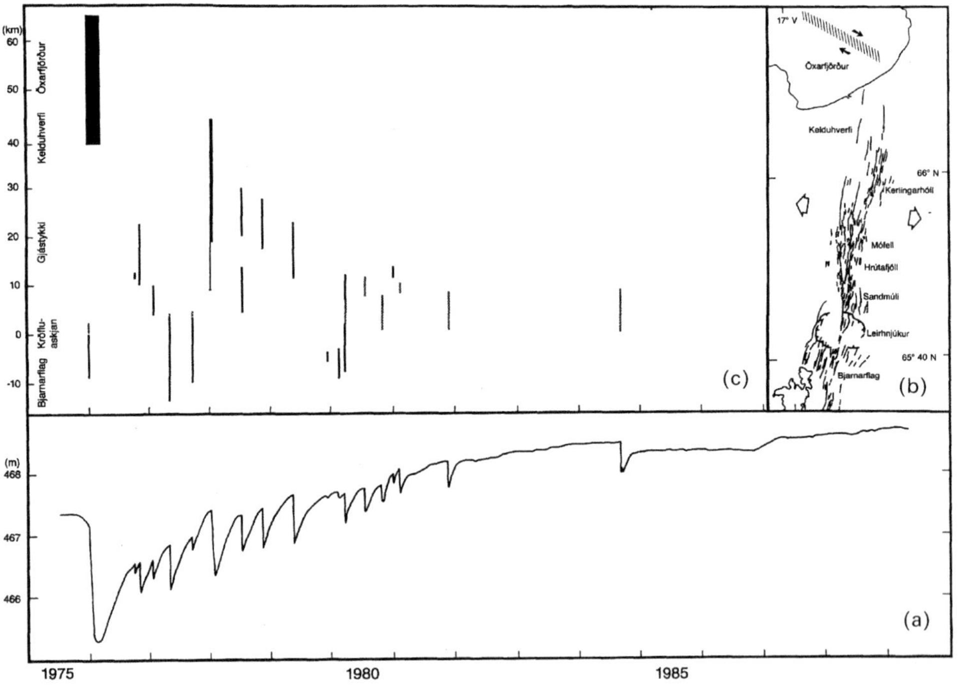

Figure 5.10. The Krafla Rifting Episode. (a) Land elevation at the Krafla Power Station 3 km south of the inferred shallow Krafla Magma Chamber. (b) Overview of the Krafla Fissure Swarm. (c) Areas of the Krafla Fissure Swarm affected in each rifting event.

Modified from Einarsson (1991b).

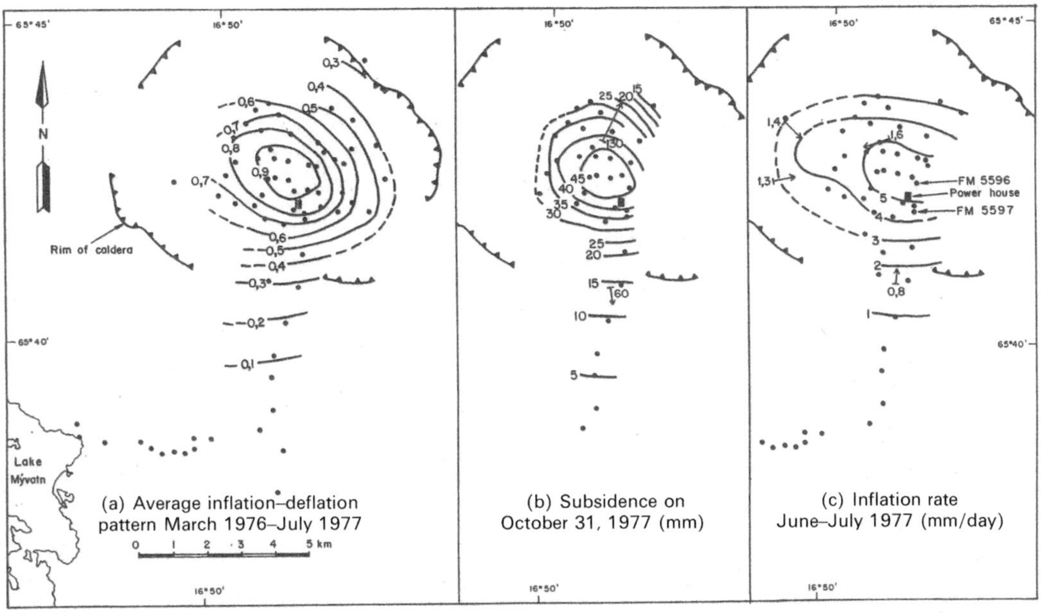

Figure 5.11. Pattern of uplift and subsidence in the Krafla area. (left) Average total ground movement March 1976–July 1977 in fractions of maximum movement. (middle) Total subsidence in cm during deflation event October 31–November 1, 1976. (right) Rate of uplifting in millimetres per day during June–July 1977. Arrows show tilt changes in microradians at four sttions.

Reproduced from Björnsson et al. (1979). Copyright by the American Geophysical Union.

The continued inflow of magma to the shallow magma chamber then caused renewed uplift and increased pressure until the breaking limit of the crust was again attained.

A number of geodetic techniques were used to measure crustal deformation during the rifting episode, including levelling, optical levelling tilt, automated electronic tiltmeters, electronic distance measurements, and strainmeters (e.g., Tryggvason, 1980, 1984, 1986, 1987; Hauksson, 1983; Ewart et al., 1991; Björnsson and Eysteinsson, 1998). Geodetic observation of deflation events and inflation show surface deformation that fit well with a Mogi model (Figure 5.11). Strainmeters also showed a consistent pattern, with areal expansion above the shallow magma chamber during inflation periods being taken up along nearby fissures, and reversal of this process during deflation episodes (Hauksson, 1983). Out in the fissure swarm, the diking events were associated with major surface faulting (discussed in Chapter 6). Gravity observations are consistent with magma leaving the shallow magma chamber during deflation events (Johnsen et al., 1980).

The joint interpretation of seismic and geodetic results at Krafla revealed a coherent pattern. After each rifting event, seismicity was low. As magma flowed into the shallow reservoir, seismicity gradually increased as stresses in the roof above the magma chamber increased. A cyclic pattern was observed (Figure 5.12).

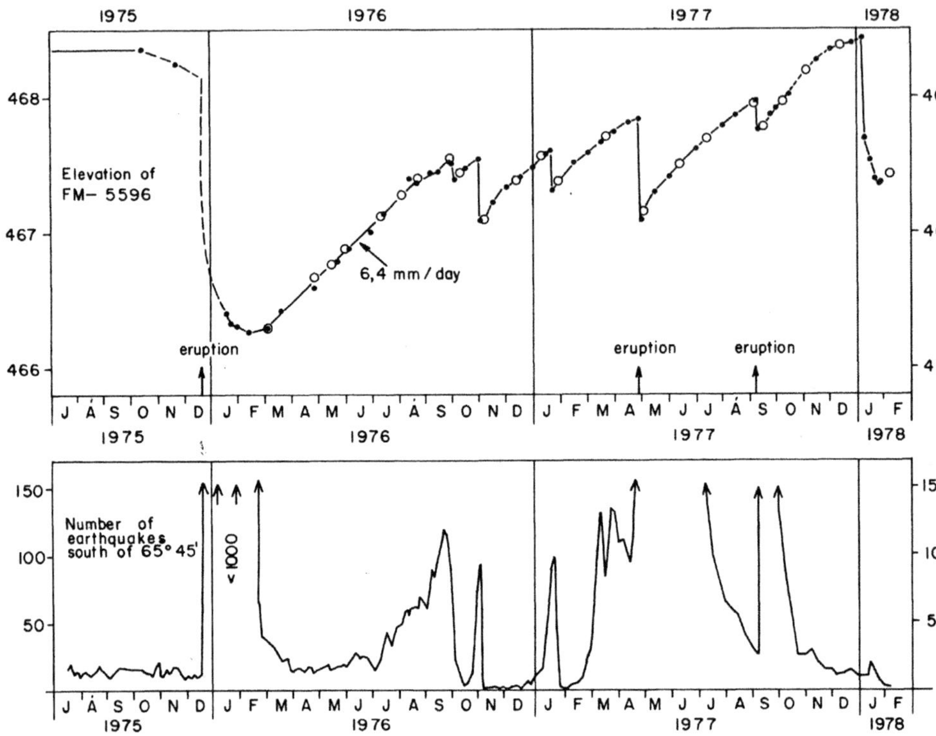

Figure 5.12. Elevation change (upper panel) and number of earthquakes within and south of the Krafla Caldera (lower panel) during the initial years of the Krafla Rifting Episode. Reproduced from Björnsson et al. (1979). Copyright by the American Geophysical Union.

When the magma chamber failed, a dike propagated away from it. The gradient of the uplift curve demonstrates that after each rifting event, flow was most rapid to the shallow chamber, and then it declined. This pattern is consistent with fluid flow between two reservoirs, where pressure in the deeper one is always higher and does not drop significantly each time the shallow reservoir fails in diking events or eruptions. The pressure in the upper reservoir was always at a minimum immediately after a diking event had occurred. Pipe flow from the lower to the upper reservoir between eruptions, with eruptions then reducing pressure in the upper one, is consistent with this behaviour. Exponential decay in flow rate can be expected between eruptions:

$$\text{Flow rate} \propto \exp(-\alpha t) \tag{5.23}$$

where t is time and α is a constant that depends on the magmatic system. Continuous measurements of ground tilt provided insights into the details of the process (Figure 5.13), as well as fissure strain measurements (Hauksson, 1983).

The pattern of vertical movements and gradients of the uplift curve at Krafla supports the above model of two magma sources, although the location of the deeper source remains uncertain. Geodetic data led Tryggvason (1986) to suggest a series of

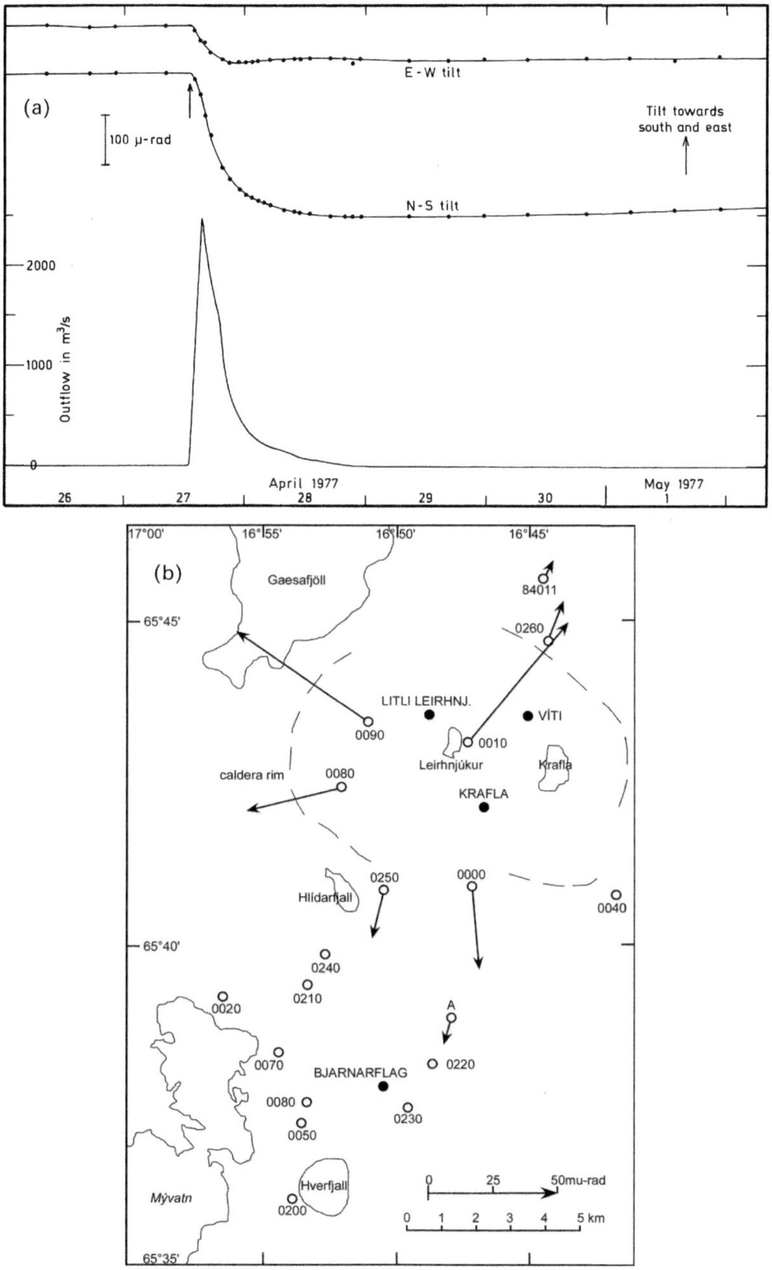

Figure 5.13. Ground tilt changes during eruptions of Krafla Volcano. (a) 1977 eruption; deflation April 27–28. Tilt measured by water tube tiltmeter and inferred rate of magma outflow from the Krafla Magma Chamber. (b) Tilt vectors associated with the 1984 eruption of Krafla and deflation measured by repeated optical levelling.

Reproduced from Tryggvason (1980, 1986).

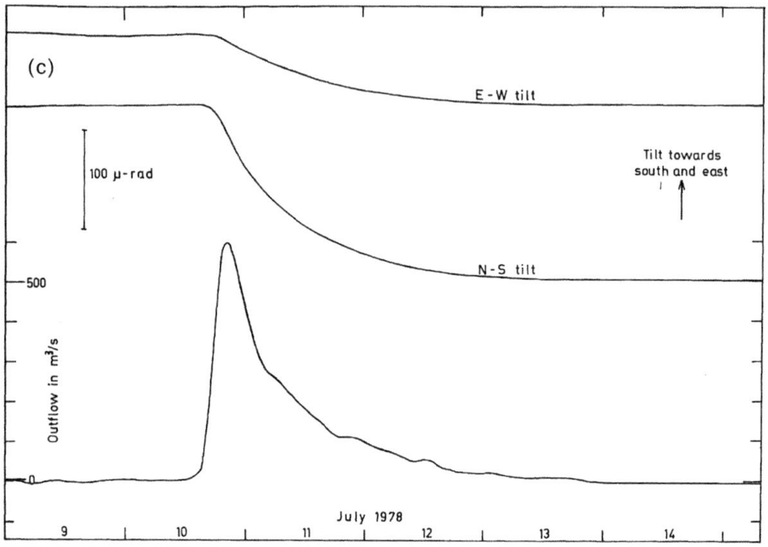

Figure 5.13. Ground tilt changes during eruptions of Krafla Volcano (*cont.*). (c) Tilt measured by electronic tiltmeter during an eruption in July 1978.

Reproduced from Tryggvason (1980).

several stacked magma chambers under Krafla, but the geodetic evidence was scarce. Geodetic data spanning the 1984 eruption have been interpreted by Árnadóttir et al. (1998) who conclude that a second magma reservoir was active during that eruption. Its depth was inferred to be more than 5 km, but it is difficult to constrain due to the limited aperture of the geodetic network. New observations may shed some light on the location of this deeper reservoir. An SAR interferometry (InSAR) study of the deformation at Krafla 1993–1999 shows a wide inflating area with uplift rates up to 10 mm/yr, over an ~50-km-wide area (de Zeeuw-van Dalfsen et al., 2004). The favoured explanation for these observations is renewed magma accumulation at an ~21-km depth, near the crust–mantle boundary, that can explain a large part of the observed deformation (Figures 5.14 and 5.15, see colour plates for the latter).

5.5 CALDERAS: THE 1875 CALDERA-FORMING ERUPTION AT ASKJA AND CURRENT UNREST

Many of the central volcanoes in Iceland have calderas (Figure 3.10) formed in explosive eruptions associated with partial destruction of volcanic edifices. The most recent caldera-forming eruption in Iceland occurred in 1875 AD, at the Askja Volcano in the Northern Volcanic Zone (Figure 5.16). An explosive eruption on March 28–29, 1875 caused fallout of rhyolitic tephra over large parts

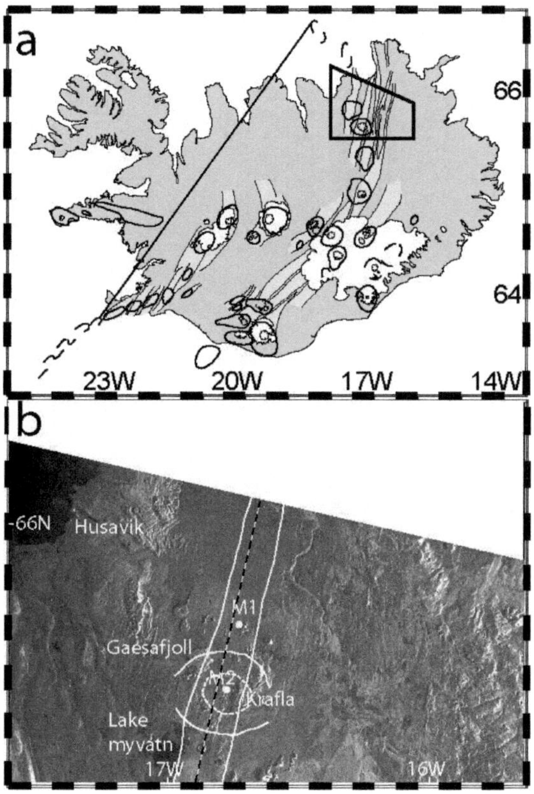

Figure 5.14. Location of a SAR interferometry study of Krafla Volcano.
Reproduced from de Zeeuw-van Dalfsen et al. (2004). Copyright by the American Geophysical Union.

of eastern Iceland. An estimated 0.21 km^3 of rhyolitic tephra (dense rock equivalent) were erupted (Thorarinsson, 1944; Sparks et al., 1981). The eruption was related to the formation of the Lake Öskjuvatn Caldera but not in a simple way. The caldera appears to have formed gradually over a period of decades following the eruption (Sigurdsson and Sparks, 1978). Today, the Öskjuvatn Caldera is 5 km wide, mostly filled with water up to 220 m deep, and has an estimated total subsidence volume of 2–2.5 km^3 (Sigvaldason, 1979). Lake Öskjuvatn is the deepest lake in Iceland. The Lake Öskjuvatn Caldera is nested within an older caldera-like structure called Askja. It has been suggested (Sigvaldason, 2002) to be a caldera-like basin between tectonically well-defined uplifted crustal blocks, with a rhyolitic eruption near the beginning of the Holocene forming an embayment into it. An alternative suggestion is that repeated subglacial ring fracture eruptions have been important in Askja's structural formation (Sigurdsson and Sparks, 1978). Still older calderas than Askja are found at the volcano, demonstrating that it is in fact a complex of several calderas. A similar complex of several calderas is also found at the subglacial Grímsvötn Volcano. The largest caldera within the active volcanic zones of

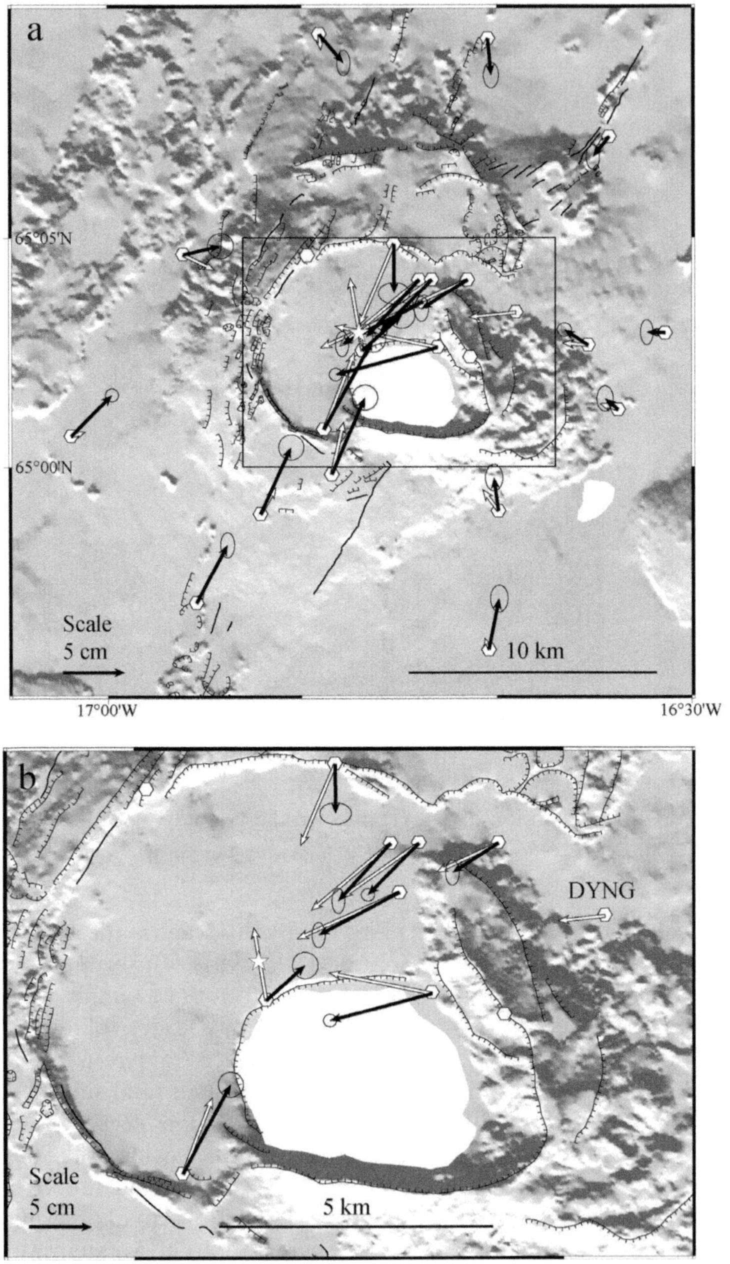

Figure 5.16. (a) Shaded topography of the Askja Caldera and surroundings. Black arrows show measured horizontal displacements 1993–1998 by GPS, white arrows show model displacement. Lake Öskjuvatn Caldera shown in white. (b) Close-up of the Askja Caldera. See Figure 5.1(d) for a photo of the area.

Iceland is at Torfajökull, an 18-km-wide composite caldera. The estimated age of initial caldera collapse is uncertain, but estimated to be around 0.6 Myr (Sæmundsson and Fridleifsson, 2001). Large historical explosive eruption in addition to the Askja 1875 eruptions, include the 1362 eruption of Öræfajökull that devastated a large part of its surrounding farmland when $\sim 10\,km^3$ of rhyolitic tephra may have erupted (Thorarinsson, 1958), and the 1104 AD eruption of Hekla Volcano (Thorarinsson, 1967).

5.5.1 The 1874–1875 rifting episode at Askja

The explosive eruption on March 28–29, 1875 associated with the formation of Öskjuvatn Caldera was part of a rifting episode at the Askja Volcanic System in 1874–1875 (Sigurdsson and Sparks, 1978). An ~100-km-long segment of the Askja Fissure Swarm was activated, with a series of basaltic fissure eruptions occurring 60 km north of Askja at the Sveinagjá Graben, producing $0.3\,km^3$ of lava. Extensive dike formation is likely to have taken place; major earthquake activity occurred during the rifting episode, particularly during the initial months (Brandsdóttir, 1992).

The initial sign of the rifting episode was increased steam emission in the Askja area during February 1874, a year before the plinian rhyolitic eruption. Direct observations were, however, very limited, as Askja is in a remote part of Iceland's central highlands. Movement of magma towards shallow depth appears to have already started by this time. Lateral flow of magma from the Askja Central Volcano north into the Askja Fissure Swarm is suggested as a cause of earthquakes and the series of fissure eruptions at Sveinagjá (Sigurdsson and Sparks, 1978). Petrological arguments suggest, however, that lavas that erupted within the caldera are from a different reservoir than those erupted in the fissure swarm at Sveinagjá; those from within the caldera are aphyric, but those erupted at Sveinagjá have ~13% phenocryst content (Gudmundur Sigvaldason, pers. commun., 2004).

Volcanic activity in the Askja Volcanic System in 1874–1875 was both basaltic and rhyolitic. It has been argued that magma mixing might have been the cause of the explosive eruption. Basaltic magma may have intruded into acidic magma, triggering the plinian rhyolitic eruption (Sparks et al., 1977). Lateral flow of magma may be a common process in Iceland, contributing to caldera formation at the spreading plate boundary in Iceland, although other explanations have been suggested as well. Activity at Askja since the 1874–1875 episode has included a series of basaltic fissure eruptions in 1921–1929, and an eruption in 1961 on the rim of the main Askja Caldera which produced $\sim 0.09\,km^3$ of lava (Sigvaldason et al., 1992).

5.5.2 Current unrest at Askja Volcano

Askja Volcano has the longest time series of deformation measurements in Iceland, going back to 1966 when the initial levelling line was established in the area (Tryggvason, 1989).

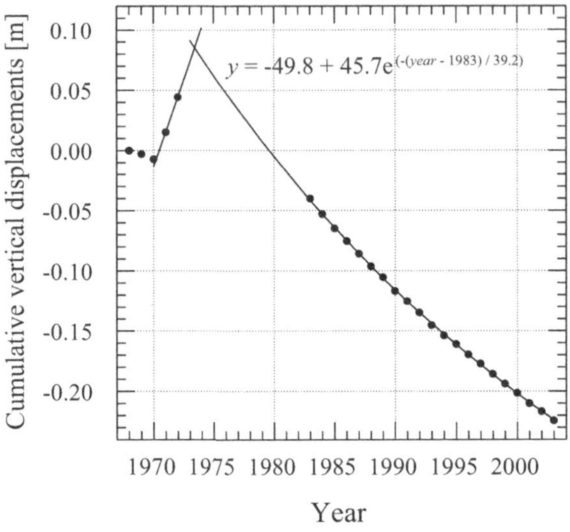

Figure 5.17. Subsidence of the Askja Volcano. Cumulative vertical displacement between two benchmarks spaced about 1 km apart on a levelling profile at Askja are shown versus time. The relative movement of these stations is 18.4 cm from 1983 to 2003, decaying slightly with time in an exponential manner. The inferred subsidence at the caldera centre in 1983–2004 is over 1 m.

Reproduced from Sturkell et al. (2005) with permission of Elsevier.

Levelling in the area was conducted yearly in the area between 1966 and 1971, and revealed alternating periods of uplift and subsidence (Table 5.1 and 5.2). No levelling measurements were conducted from 1972 to 1983, but yearly measurements have been conducted since. The geodetic work has been expanded and now includes a network of Global Positioning System (GPS) stations (Camitz et al., 1995; Sturkell and Sigmundsson, 2000; Sturkell et al., in press). Measurements have revealed high rates of deformation, the highest rate of subsidence observed for any volcano in Iceland during a non-eruptive period. The central area of the Askja Caldera subsided by about 1 m from 1983 to 2004 without any magmatic activity. InSAR observations have also revealed the subsidence of Askja (Pagli et al., 2003b; Pagli et al., submitted). The deflation causes horizontal contraction towards Askja (Figure 5.16), at the same time as the caldera subsides (Figure 5.17). The main source of deformation at Askja is pressure decrease in a shallow magma chamber, as a Mogi source at an ~3-km depth can explain a large part of the observed deformation. However, the deformation field is wider than expected if it were only due to a shallow source. A revised interpretation of available data suggests that, in addition to the shallow source, a much deeper source, at about a 16-km depth, is deflating as well (Sturkell et al., submitted).

Why does Askja deflate at this high rate? The long time series available at Askja shows that deflation is a long-term process. The subsidence appears to decrease exponentially with time, scaling with $\exp(-t/\tau)$, where t is time in years.

The relaxation time, τ, found for the decay in vertical deformation at Askja is about 40 years (Sturkell et al., submitted). Additional constraints on the responsible process are provided by results from gravity measurements. Interpretation of a time series of repeated gravity measurements suggests that mass is leaving the shallow magma chamber at Askja (de Zeeuw-van Dalfsen et al., 2005). During the time interval 1988–2003, a net gravity decrease of 115 μGal is unaccounted for after correcting for height changes using the free air gradient. Magma is likely flowing out of the shallow chamber and down to deeper levels, as there is no seismic indication of dike injections during this period. Accommodation of plate spreading by ductile deformation in the lower crust might lower pressure more in the deeper parts of the magmatic system, causing pressure to reduce in an upper chamber if an open channel links the upper and lower parts. Earlier, Tryggvason (1989) had suggested that the vertical deformation at Askja reflects pressure fluctuations in the Iceland Mantle Plume.

5.6 HEKLA: ONE OF ICELAND'S MOST ACTIVE VOLCANOES

Hekla Volcano (Figure 5.1(c)), one of the most active in Iceland, is located at the intersection of the South Iceland Seismic Zone and the Eastern Volcanic Zone. It is a young volcanic ridge built up mostly during Postglacial time through repeated eruptions on the same fissure that splits the Hekla Ridge open during the initial phase of most Hekla eruptions. Fissures which radiate from the summit area are also active, and in later stages of Hekla eruptions, activity is often localized on the lower flanks of the volcano.

Eruption products from fissures on the Hekla Edifice range from rhyolites to basaltic andesites, whereas basalts are produced in the surrounding area. The first Hekla eruption after the settlement of Iceland (in 874 AD) occurred in 1104, and produced a large amount of silicic tephra. An eruption in 1158 also produced silicic tephra, but in much reduced quantities. After that, the volcano erupted once or twice per century until 1947 (see also Section 3.5). These eruptions were characterized by an initial explosive phase, followed by effusive production of andesites or basaltic andesites in later stages of the eruptions (Thorarinson, 1967). The behaviour of the volcano has in recent decades changed to more frequent, smaller volume eruptions. These occurred in 1970, 1980–1981, 1991, and 2000. A unique feature of Hekla eruptions is that silica content of initial eruptive products is proportional to the length of the preceding repose period (Figure 3.20). The initial phase of the eruptions since 1947 has been less explosive than for the earlier eruptions which were preceded by longer repose periods. The lower explosivity may be a consequence of the lower silica content in the initial eruptive products, or relate to lower magma flow rate. The evolved rocks at Hekla appear to be formed by partial melting of hydrated basalts, as suggested by Sigvaldason (1974). Isotope studies show that the Th/U isotope ratio in the basalts and basaltic andesites at Hekla are different from that in silicic rocks, dacites, and rhyolites, demonstrating that fractional crystallization is not the main differentiation process at Hekla (Sigmarsson et al., 1992).

Magma production at Hekla has been fairly constant since its 1104 AD eruption; close to 1 km^3 of magma each century (Thorarinsson, 1967). Furthermore, there is a correlation between the volume of eruptive products in each eruption and the length of the preceding repose period. Eruptions in 1980–1981, 1991, and 2000, each preceded by about a 10-yr repose period, resulted in eruptive products amounting to about 0.1–0.15 km^3, whereas the 1947–1948 eruption, preceded by a 101-yr repose period, resulted in eruptive products of about 0.9 km^3 (Gronvold et al., 1983; Gudmundsson et al., 1992; Ólafsdóttir et al., 2002). An eruption in 1970 preceded by a 22-yr repose period produced about 0.2 km^3.

Various studies have addressed the existence of a magma chamber under Hekla. Sigmarsson et al. (1992) argue for a chemically zoned bell-shaped reservoir, 5 km wide and 7 km high, with its top 8 km deep. The model has basalts at the base of the chamber, and rhyolites near its top, with a compositional range inbetween. The bell shape of the reservoir is suggested to explain the distribution of eruptive products on the surface, under the assumption that magma rises vertically during eruptions at Hekla. On the other hand, Soosalu and Einarsson (2004) examine seismic waves to conclude that if a substantial magma chamber (dimensions larger than 800 m) exists under Hekla it has to be located either in the uppermost 4–5 km or it has to be below about a 14-km depth, as nearly all seismic records with raypaths under Hekla show no signs of attenuated S-waves. The first possibility is not supported by other geophysical measurements. It may also be difficult to use the distribution of eruptive products on the surface to infer the shape of a magma reservoir. Volcanic edifices can modify stress fields at volcanoes, generating a compressive stress field in the upper crust which affects magma transport and may cause lateral transport. Pinel and Jaupart (2004) consider these effects and reproduce in their model observed distribution of eruptive products at many volcanic fields, such that, with increasing distance from the volcano centre, magma compositions are less and less evolved. For these volcanic fields, denser more primitive magmas are unable to erupt through topographic highs of volcanic edifices.

Crustal deformation studies have been conducted at Hekla with various techniques, including electronic distance measurements (Kjartansson and Gronvold, 1983), optical levelling tilt (Tryggvason, 1994), GPS (Sigmundsson et al., 1992), borehole strain (Linde et al., 1993), and InSAR (Sigmundsson et al., 2001). The results have revealed more irregular deformation patterns than at other well-studied volcanoes in Iceland, and several deformation processes appear to be active. The studies suggest eruptions are associated with a pressure decrease in a source whose centroid depth is uncertain. Estimates vary from 5–9 km. Further joint interpretation of the available geodetic data is needed to constrain these estimates better. In particular, a model that can explain all the available observations remains to be found.

A peculiarity of Hekla is the almost aseismic behaviour of the volcano, except immediately prior to eruptions, and during eruptions. Numerous small earthquakes are associated with movement of magma from depth towards the surface prior to eruptions, but only in the last 1–2 hours or less before an eruption starts. After the onset of eruptions, the most prominent seismic signal is a volcanic tremor, which

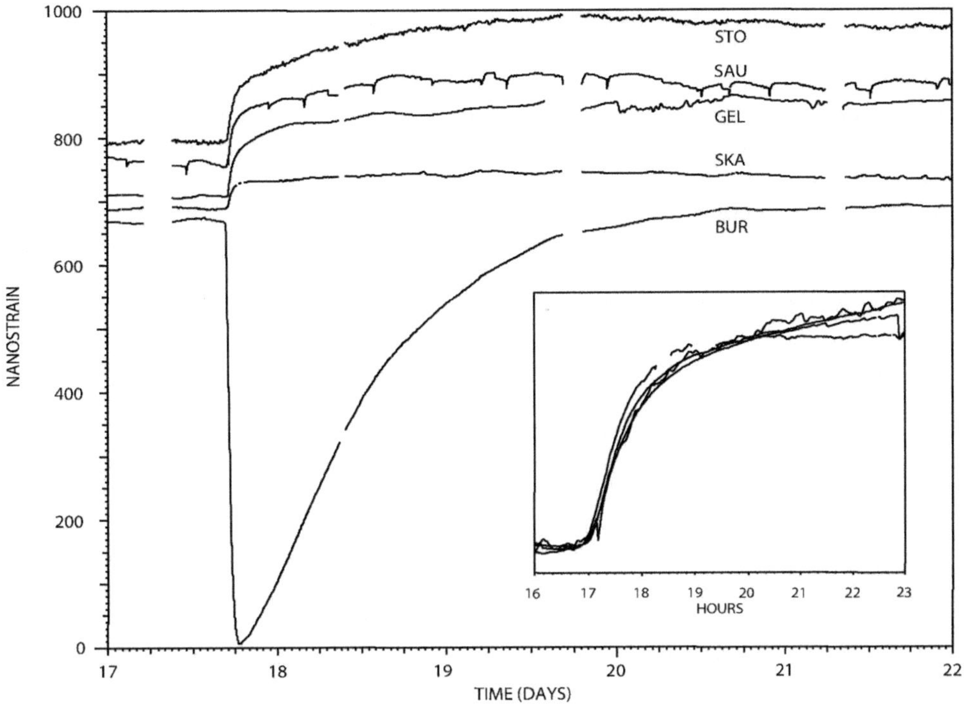

Figure 5.18. Strain changes associated with the 1991 eruption of Hekla Volcano. Station closest to the volcano (BUR) goes into compression during the opening of the eruption feeder dyke, stations farther away go into expansion as magma is withdrawn from beneath the volcano. Strain recovery at the BUR site is a site effect.

Reproduced from Linde et al. (1993) with permission of *Nature*, London.

starts when the magma reaches the surface and then continues at a diminishing magnitude until the end of eruptions (e.g., Soosalu and Einarsson, 2002). In addition to the monitoring of small earthquakes preceding eruptions, a network of continuous strainmeters has been important for detecting the onset of the propagation of a feeder dyke prior to the eruptions (Figure 5.18). Detection of both earthquakes and strain changes prior to the Hekla eruption of 2000 resulted in a prior warning being issued to the public before the onset of the eruption (Agustsson et al., 2000).

5.7 ADDITIONAL EXAMPLES OF VOLCANO UNREST: GRÍMSVÖTN, KATLA, HENGILL, AND EYJAFJALLAJÖKULL VOLCANOES

The Krafla, Askja, and Hekla Volcanoes are among the best studied in Iceland. Other volcanoes where magma movements have been detected by geodetic techniques are listed in Tables 5.1 and 5.2. An overview is also given by Sturkell et al. (2005). Two of the additional volcanoes, Grímsvötn and Katla, are among the most active in Iceland. Both have calderas, shallow magma chambers, and are actively deforming. Their geodetic study is difficult because they are subglacial, but nevertheless, interesting time series of deformation have been obtained. At Grímsvötn there is only one stable nunatak (mountain sticking out of the ice) where precise geodetic measurements can be conducted. Intermittent GPS measurements at this site did reveal co-eruptive deflation associated with an eruption in 1998, and subsequent uplift and horizontal displacement away from the caldera, indicative of recharging of a shallow magma chamber (Sturkell et al., 2003a) until a new eruption occurred in 2004. At Katla Volcano, GPS measurements show inflation beginning in 1999 at the same time as a small jökulhlaup occurred (Sturkell et al., 2003b). The inflation was associated with elevated seismic activity and increase in geothermal activity, demonstrating recharging of this volcano as well. A stratovolcano, Mt. Eyjafjallajökull, is located west of Katla. It has had increased activity levels since 1992, when earthquake activity began to occur after an almost total quiescence since the beginning of seismic observations (Sturkell et al., 2003b), with an intrusion forming in 1994 (Pedersen and Sigmundsson, 2004) (Figures 5.19 and 5.20, see colour plates for the latter). Magmatic activity was simultaneous in both of these neighbouring volcanoes in 1999. At that time, another intrusion formed under Mt. Eyjafjallajökull.

Numerous periods of elevated earthquake activity have been documented at Icelandic volcanoes during non-eruptive periods. In some cases geodetic data exist revealing deformation associated with the earthquake activity. One of the main areas of elevated earthquake activity in the last decade of the 20th century was the Hengill area, at the Hrómundartindur Volcanic System. It is located in the Western Volcanic Zone at its junction with the South Iceland Seismic Zone. The system is drifting out of the active rift zone and has had only one eruption in the last 10,000 years. However, from 1993 to 1998, about 85,000 earthquakes exceeding $M_l = 1$ occurred in this area (Jakobsdóttir et al., 2002). GPS, levelling, and InSAR show that the seismicity was associated with inflation at a steady rate of 19 mm/yr during the period of elevated seismicity, related to a pressure increase at a 7-km depth (Sigmundsson et al., 1997; Feigl et al., 2000). The activity culminated with two $M > 5$ earthquakes in 1998 (Clifton et al., 2002). The high level of seismicity, despite the small amount of inflation, was interpreted as a result of injection of small batches of magma into a highly stressed shear zone. The amount of seismicity relative to the amount of uplift varies greatly from one volcano to another, and will depend on ambient stress levels. Intrusions in the Eyjafjallajökull Volcano in 1994 and 1999 represent an example of the opposite character, where intrusions were associated with relatively little earthquake activity.

5.8 OVERVIEW AND IMPLICATIONS

In Iceland, the flow of magma through the lower crust towards shallow levels is highly episodic. It results in episodic inflation periods and measurable ground

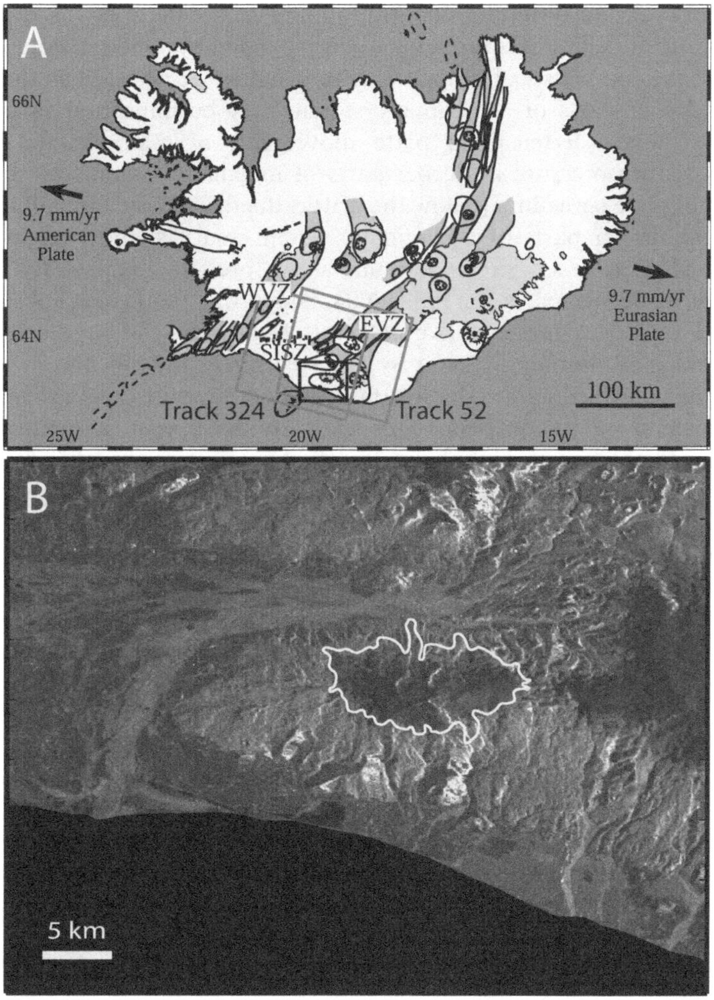

Figure 5.19. Location of an InSAR study of Eyjafjallajökull Volcano.
Reproduced from Pedersen and Sigmundsson (2004). Copyright by the American Geophysical Union.

deformation on the surface of the Earth. Recorded inflation episodes range in time from several months up to 15 years, with cumulative magma volumes ranging from ~0.001 to 1 km^3 (Figure 5.8). Only few of these episodes result in eruptions; often magma is emplaced at depth in the crust without an eruption at the surface.

Between the relatively short periods of inflation, Icelandic volcanoes subside or show no signs of deformation. In the absence of renewed magma inflow, the rate of volcano deflation generally decreases with time from the last eruption. The processes responsible for deflation are magma cooling and solidification, pressure reduction, and outflow of magma. The volumetric contraction associated with magma solidification and cooling to ambient temperatures is $\sim$10%. In some cases, the rate of deflation is too large to be explained by solidification, such as at the Askja Volcano (unrealistically large magma volumes are needed). In these situations, a link to deeper parts of a magmatic system may be important in producing the observed changes. Extensional plate movements across volcanic systems may effectively reduce pressure in deeper parts of magmatic systems by ductile accommodation of plate spreading (below the brittle–ductile transition). A fluid connection between the deeper parts of a magmatic system and a shallow reservoir will then cause the shallow reservoir to respond as a "pressure gauge" for reduction of pressure in its deeper parts. In 1983–2004, Askja was the fastest subsiding volcano in Iceland during a non-eruptive interval. Following the arguments above, an extensive magma plumbing system is suggested under the volcano.

The channels feeding the shallow magma bodies in the crust are narrow (on the order of meters) and are only active for relatively short periods, separated by periods of no magma transport. The channels solidify unless reactivated by new magma batches. The rate of solidification will depend on heat transfer away from these channels, influenced by the ambient temperature. A shallow magma chamber will be sustained only if the rate of inflow of magma is sufficiently high. In many cases, intrusions are the heat source for geothermal systems, as solidification and cooling of intrusions can provide extensive heat (e.g., Björnsson and Gudmundsson, 1993)

Because of limited magma inflow, shallow crustal magma chambers at a 3–7-km depth are only found at the most active volcanoes in Iceland. The seismic and geodetic evidence suggest shallow magma chambers at least at Krafla, Askja, Grímsvötn, Katla, and Torfajökull Volcanoes. If a magma chamber exists under Hekla, it is likely to reside at a significantly deeper level than at the other volcanoes. Examples of volcanoes with no signs of recent magma inflow include the volcanic centres on the Reykjanes Peninsula west of Hengill. The most recent eruption occurred there about 700 years ago. Ongoing subsidence on the peninsula can be attributed to pressure reduction in geothermal areas, as well as to subsidence along the plate boundary due to lack of inflow of magma. In the Northern Volcanic Zone, volcanic systems other than Krafla and Askja show no signs of significant local deformation.

In summary, magma movements in Iceland are focused on time and space at a few volcanic centres which deliver the bulk of magma, creating the uppermost part of the Icelandic crust. Deformation of Icelandic volcanoes during non-eruptive periods is characterized by deflation or absence of deformation. The rate of deflation decreases with time since last recharging of the system, and non-deforming volcanoes dormant for $\sim$700 years may have no magma at shallow depth. Magma inflow is sufficiently high to sustain shallow magma chambers at only few of the volcanoes in Iceland, the ones that are the most active.

6

The plate-spreading deformation cycle

The deformation cycle along the divergent plate boundary in Iceland can be observed in some detail, with different parts of the plate boundary being at different stages in the associated deformation cycle. Co-rifting deformation occurs in episodic diking events that contribute to the plate spreading. The best observed rifting events to date took place in the Krafla Volcanic System in North Iceland from 1975 to 1984 (the Krafla Rifting Episode; Krafla Fires). After a rifting episode, a post-rifting style of deformation will dominate for years or decades. Such deformation was measured in North Iceland by Global Positioning System (GPS) after the Krafla Fires. In South Iceland, current extension across the Eastern Volcanic Zone (EVZ) is most representative of gradual stretching across a plate boundary deformation zone, characteristic of inter-rifting deformation. This gradual continuous stretching builds up stresses that will be released in a future series of diking events along the plate boundary. The cyclic co-rifting, post-rifting, and inter-rifting deformation stages form the plate boundary deformation cycle. Furthermore, more local deformation associated with magma accumulation at shallow depth at central volcanoes may precede rifting episodes, leading to characteristic pre-rifting deformation fields which differ from deformation during inter-rifting periods.

6.1 CONTINUOUS GPS MEASUREMENTS

A network of continuous GPS stations in Iceland (Figure 6.1) reveals well some of the characteristics of the plate spreading in Iceland (Geirsson, 2003; Geirsson et al., submitted). The overall rate of opening across Iceland inferred from GPS was discussed in Section 2.2.3, based on data from the REYK station in southwestern Iceland and the HOFN station in southeastern Iceland. Other stations in the network reveal the spatial and temporal variation in the velocity field. Stations

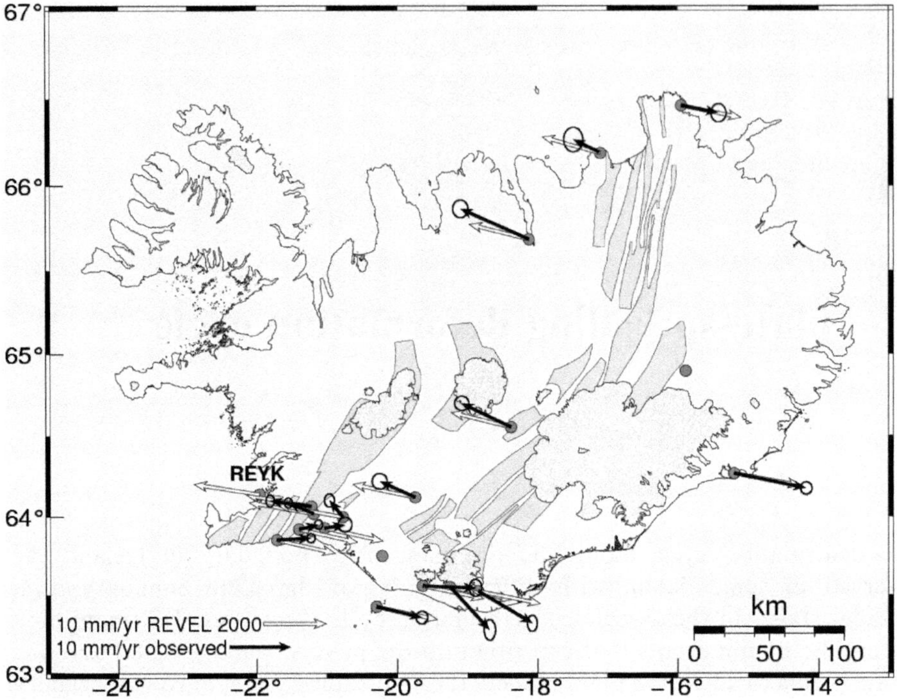

Figure 6.1. Velocities of continuous GPS stations in Iceland 1999–2004 (black arrows), assuming the REYK station moves at 10.5 mm/yr towards east and 1.6 mm/yr towards north. Confidence limits at the 2σ level are shown. The white arrows are velocities from the REVEL plate motion model (Sella et al., 2002), assuming stations on the North American Plate move with a velocity equal to half of the inferred spreading across Iceland, and stations on the Eurasian plate move equally but in opposite direction (movements relative to the central axis of the plate boundary).

Courtesy of Halldór Geirsson, Icelandic Meteorological Office (see also *http://www.vedur.is*).

between the EVZ and Western Volcanic Zone (WVZ) in southern Iceland show the partitioning of spreading between these overlapping rift zones; their movement is similar to the movement of the REYK station (on the North American Plate) demonstrating that the majority of crustal spreading in southern Iceland is currently accommodated by the EVZ. Shearing is observed across the South Iceland Seismic Zone and the Reykjanes Peninsula. Here some of the continuous GPS stations are within the plate boundary deformation zone and show only a fraction of the faraway plate movements. Stations at southernmost coast of Iceland reveal displacements towards the east along with the Eurasian Plate and show that this area moves with that plate. However, stations next to the southern edge of the Mýrdalsjökull Icecap have velocities that deviate significantly from the plate-spreading direction. This is due to a local component of deformation, originating from magma accumulation under the icecap that causes an outward component of movement away from the icecap.

6.2 INTER-RIFTING DEFORMATION

6.2.1 Measurements in North Iceland prior to the Krafla Rifting Episode

Initial attempts to measure plate movements in Iceland were conducted by German geodesists who in 1938 installed the first geodetic network specifically to measure the extension across the rift in North Iceland (Niemczyk, 1943). The work consisted of over 20 geodetic benchmarks in a 100-km-wide area centred on the plate boundary. The work was inspired by Wegener's ideas of continental drift. Resurveying of this network was first conducted in 1965, but comparison with the earlier measurements was difficult because of the evolution of observational techniques, large uncertainties in the measurements, and partial vandalism of some benchmarks during the Second World War. Both the 1938 and 1965 surveys were based on triangulation, requiring the precise measurement of a scale in the network to transfer angles into distances.

Electronic distance measurements and trilateration were first used to measure the geodetic network in North Iceland in 1971 and partly in 1975. Comparison of the 1965 and 1971 measurements (Ritter, 1982; Wendt et al., 1985) did not reveal the expected spreading across the plate boundary. On the contrary, the data suggested contraction across the plate boundary (Figure 6.2). Different observational procedures were used in the two observation periods. Although the 1965 measurements relied on triangulation, Wendt et al. (1985) argue that the unexpected contraction across the network is not due to a scale error. They argue that apparent rotation of sets of points east and west of the rift axis could be caused by non-detected

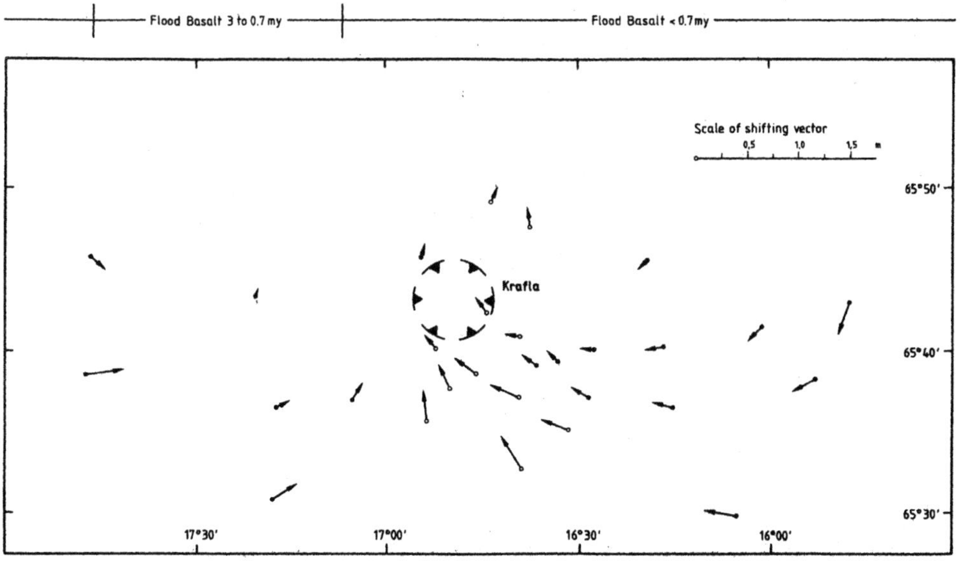

Figure 6.2. Geodetic work in North Iceland by German scientists. Arrows show inferred horizontal displacements 1965–1971, based on triangulation. See text for discussion.

Modified from Wendt et al. (1985). Copyright by the American Geophysical Union.

observational errors, and conclude that the low signal-to-noise ratio in these measurements is too small to warrant an interpretation. The 1975 measurements were conducted on only a few central stations of the network. A comparison of the 1971 and 1975 observations does not reveal a clear spreading signal, but the spatial and temporal coverage of the data may have been too limited to detect the signal. Furthermore, measurements in the area in this time period may be influenced by magma accumulation at Krafla preceding the rifting episode that began in December in 1975, a few months after the completion of the 1975 geodetic measurements. Geodetic data collected in North Iceland between 1975 and 1984 are then influenced by the rifting episode in that period, and after 1984 deformation fields are influenced by post-rifting deformation.

6.2.2 Inter-rifting deformation at overlapping rift zones in South Iceland

Attempts to measure plate spreading in South Iceland began in the 1960s, when Electronic Distance Measurements (EDMs) were conducted on two profiles crossing the EVZ and WVZ, respectively (Decker et al., 1971, 1976), and at several arrays, including one at the Reykjanes Peninsula (Brander et al., 1976). The profiles established by Bob Decker and co-workers were about 50 km long, with average station spacing of 2.7 km. Initial resurveying of these profiles did reveal complicated changes. Measurements across a part of the EVZ north of Hekla revealed widening of the rift by 6–7 cm, associated with the 1970 eruption of Hekla. A zone of apparent extension inferred from 1967–1970 data did, however, change to a zone of apparent contraction in 1970–1973. Local processes, such as magmatic movements at Hekla Volcano, were suggested as a cause of the observed irregular changes. A consistent inter-rifting deformation signal was not well resolved. The EDM measurements have now been replaced by GPS.

GPS measurements were the first to conclusively reveal spreading across the plate boundary in Iceland. Initial GPS measurements in North Iceland did actually reveal spreading rates higher than the long-term average. They were influenced by post-rifting deformation after the Krafla Fires, with displacements higher than average during the inter-rifting period (see Section 6.4). The results of measurements across the rift zones in South Iceland are more representative of stretching across the divergent plate boundary during inter-rifting periods. Spreading here is partitioned between the overlapping EVZ and WVZ (Figure 6.3), with the eastern one currently taking up most of the spreading across South Iceland (Jónsson, 1996; Jónsson et al., 1997; La Femina et al., in press). Spreading across each of the rifts is accommodated in zones over 50 km wide, with horizontal displacements increasing gradually away from a central axis (Figures 6.3–6.5). Stretching across the plate boundary causes relatively uniform strain accumulation within these zones at a rate of up to 0.3 μstrain/yr. The displacements are parallel to the far-field spreading vector, despite the obliqueness of the plate boundary. The summed spreading rates across the EVZ and WVZ is about 18–20 mm/yr. Spreading rates in the EVZ decrease from 19.0 ± 2.0 mm/yr in the northeast to 11.0 ± 0.8 mm/yr in the southwest, whereas the spreading rates across

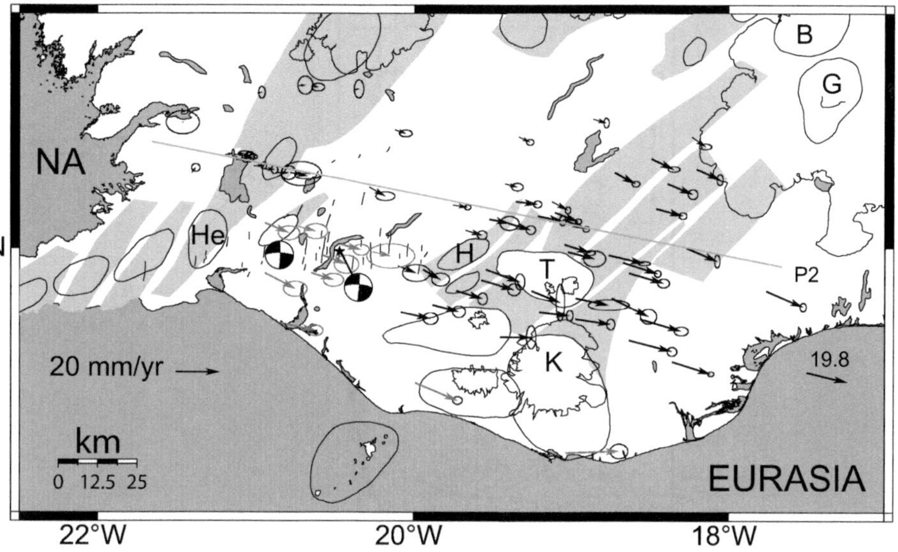

Figure 6.3. The secular displacement field in South Iceland, 1994–2003, relative to stable North America. The displacement field is corrected for co-seismic offsets and magmatic sources. Inter-rifting deformation is partitioned between the Eastern and Western Volcanic Zones. Line P2 shows the location of the profile in Figure 6.5. Grey arrows show pre-2000 displacements in the South Iceland Seismic Zone. GPS data processing performed at the University of Miami Geodesy Laboratory.

After La Femina et al. (in press). Copyright by American Geophysical Union.

the WVZ increase from 2.6 ± 0.9 mm/yr in the northeast to 7.0 ± 0.4 mm/yr in the southwest (La Femina et al., in press). On a 1,000-yr timescale, the history of rifting events in southern Iceland (see Section 6.3) suggests similarily larger amount of spreading there than in the WVZ. The EVZ has been much more active in historical times. On a still longer timescale, the rifting may be more equally divided between the Eastern and Western rift zones in South Iceland. Extensive normal faulting and fissuring in the WVZ at Þingvellir suggests widening there of about 100 m in Postglacial times, averaging about 10 mm/yr. Focusing of spreading may accordingly shift between overlapping rifts on short timescales, eventually depending on magma availability in each of the rifts.

Stretching across a rift zone will lead to buildup of tectonic stress. If displacements only take place perpendicular to the rift zone axis, then we have the conditions of plane strain. Furthermore, if we assume the brittle upper crust that is being stretched behaves as an incompressible elastic plate, the stretching will be balanced by thinning of the plate. Conservation of volume requires the horizontal and vertical strain to be equal. Under these assumptions, the relation between horizontal strain, ε_{xx}, and tectonic stress, $\Delta\sigma_{xx}$, perpendicular to the rift is (e.g., Turcotte and Schubert, 1982):

$$\Delta\sigma_{xx} \approx 2(\lambda + \mu)\varepsilon_{xx} \tag{6.1}$$

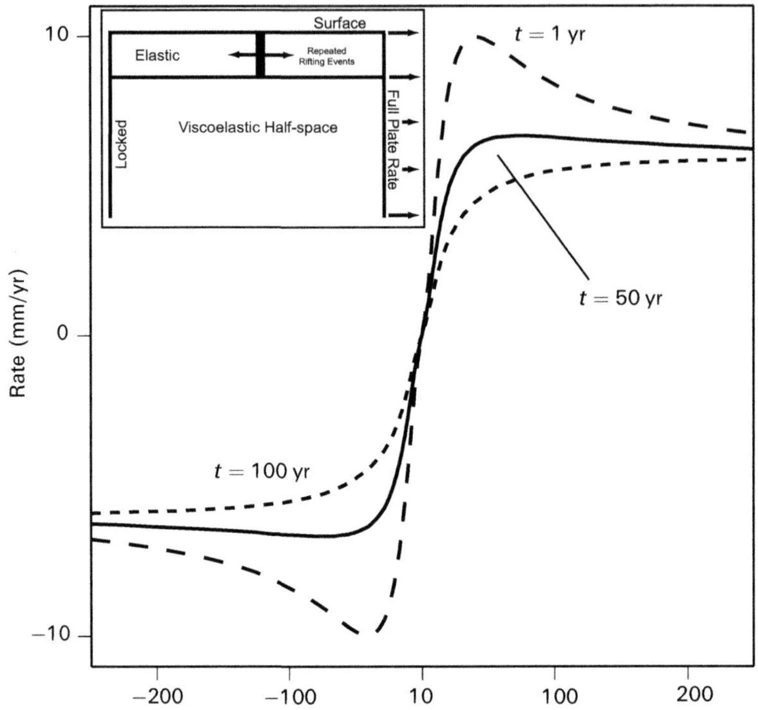

Figure 6.4. A viscoelastic plate boundary deformation model. The curves show predicted surface velocities 1, 5, and 100 years after a rifting event, consisting of a single dike 3 m thick cutting a 3-km-thick elastic layer. The dike opens periodically every 250 years, to give average spreading across a rift zone of 12 mm/yr.

After La Femina et al. (in press). Copyright by American Geophysical Union.

where λ and μ are the Lamé moduli for the elastic plate. If we take $\lambda = \mu = 30$ GPa then we have $\Delta\sigma_{xx} \approx (120\ \text{GPa})\varepsilon_{xx}$. Strain accumulation of 0.1–0.3 μstrain/yr will cause tectonic stress buildup of about 0.01–0.04 MPa/yr.

When strain accumulation has reached a critical limit the plate boundary will fail and rifting occurs. The critical limit is highly variable and depends strongly on availability of magma. If no magma is present at shallow depth along the plate boundary, then normal faulting will relieve the stresses (e.g., Sigmundsson, 1992b). In that case, the critical deviatoric stress is the one needed to cause normal faulting. For example, to initiate slip of a normal fault at a 5-km depth may require deviatoric stresses on the order of 65 MPa, according to the Anderson theory of faulting (e.g., Turcotte and Schubert, 1982). If magma is in contact with stretched brittle crust, then diking events will relieve the stress and accommodate the spreading. The condition for rifting is then that the deviatoric stress exceeds the tensile strength of the crust (see Section 5.3.5). The tensile strength in Iceland has been inferred to be less than 10 MPa, much smaller than the stress needed to cause normal faulting. As a consequence, inflow of magma towards shallow depths may be a precursor to

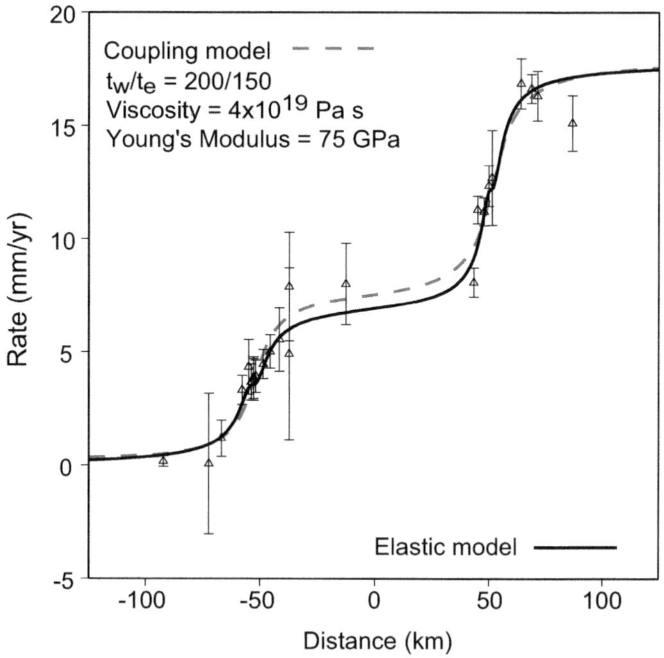

Figure 6.5. Profile (P2 on Figure 6.3) across the Eastern and Western Volcanic Zones in South Iceland showing GPS-derived site velocities (triangles), predictions from a cross-sectional viscoelastic model from Figure 6.4 (shaded hatched line) and an elastic deformation model (solid line). The velocity field in the viscoelastic model is modelled 200 and 150 years after the last rifting event in the WVZ and EVZ, respectively. The elastic model consists of uniform opening of 7 mm/yr under 4-km locking depth in the WVZ, and opening of 11 mm/yr under 3-km locking depth in the EVZ.

After La Femina et al. (in press). Copyright by the American Geophysical Union.

many rifting events where tensional stress may have previously built up to high levels.

6.2.3 Models of inter-rifting deformation

Different types of models have been applied to model stretching across rift zones, including ones with elastic and viscoelastic behaviour. Surface displacements during inter-rifting periods can be mimicked by gradual opening of a dike in an elastic half-space, extending from infinite depth towards a certain locking depth below the surface. The width of the deformation zone depends on the depth to the dike top. Application of ductile deformation processes may, however, be more appropriate in rift zones, and viscoelastic deformation models are consequently more realistic. One such model consists of repeated dike injections into an elastic layer overlaying a viscoelastic half-space (Figure 6.4). Repeated rifting events will cause a spatially and temporally variable deformation field that depends on the rheological parameters of the model, and the style of diking. Rate of displacement is high in periods immediately after a rifting event, and low in the later half of the inter-rifting periods. Such a model can explain the main features of the inter-rifting strain field observed in South Iceland (Figure 6.5).

6.2.4 Vertical rift zone deformation during inter-rifting periods

As the plate boundary deformation zone is stretched, it also subsides. Observations in Iceland, including optical levelling since the 1960s, suggest that subsidence at the divergent plate boundary may be focused on individual fissure swarms. Subsidence is, for example, well observed by levelling across the Þingvellir Graben in the western rift zone (Figure 6.6). Subsidence in the centre of the Þingvellir Graben, relative to stations west of the main boundary faults of the graben amounted to about 1 mm/yr in the 1966–1971 period (Tryggvason, 1974). Another example is the subsidence at the Askja Volcanic System in northern Iceland. Subsidence occurs there above a shallow magma chamber, but SAR interferometry (InSAR) observations show as well subsidence along the Askja Fissure Swarm (Figure 6.7, see colour plates) (Pagli et al., in press). Tilt towards the central axis of the plate boundary has also been

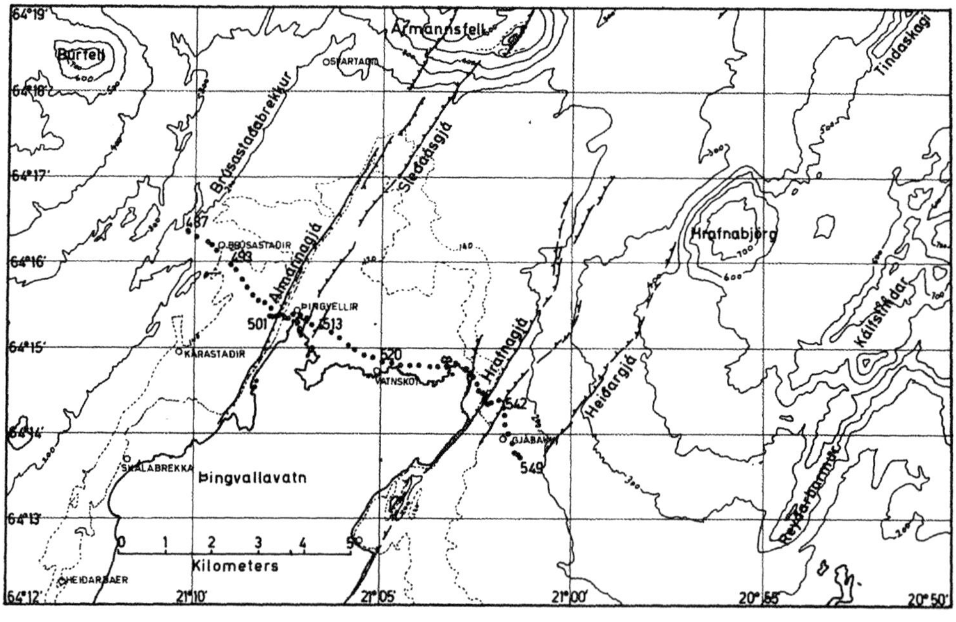

Figure 6.6. Inter-rifting subsidence at Þingvellir (the fissure swarm north of the Hengill Central Volcano). (a) Map of the Þingvellir area showing the location of benchmarks on a levelling profile crossing the Þingvellir Graben, principal active faults, and Lake Þingvallavatn.

Reproduced from Tryggvason (1974).

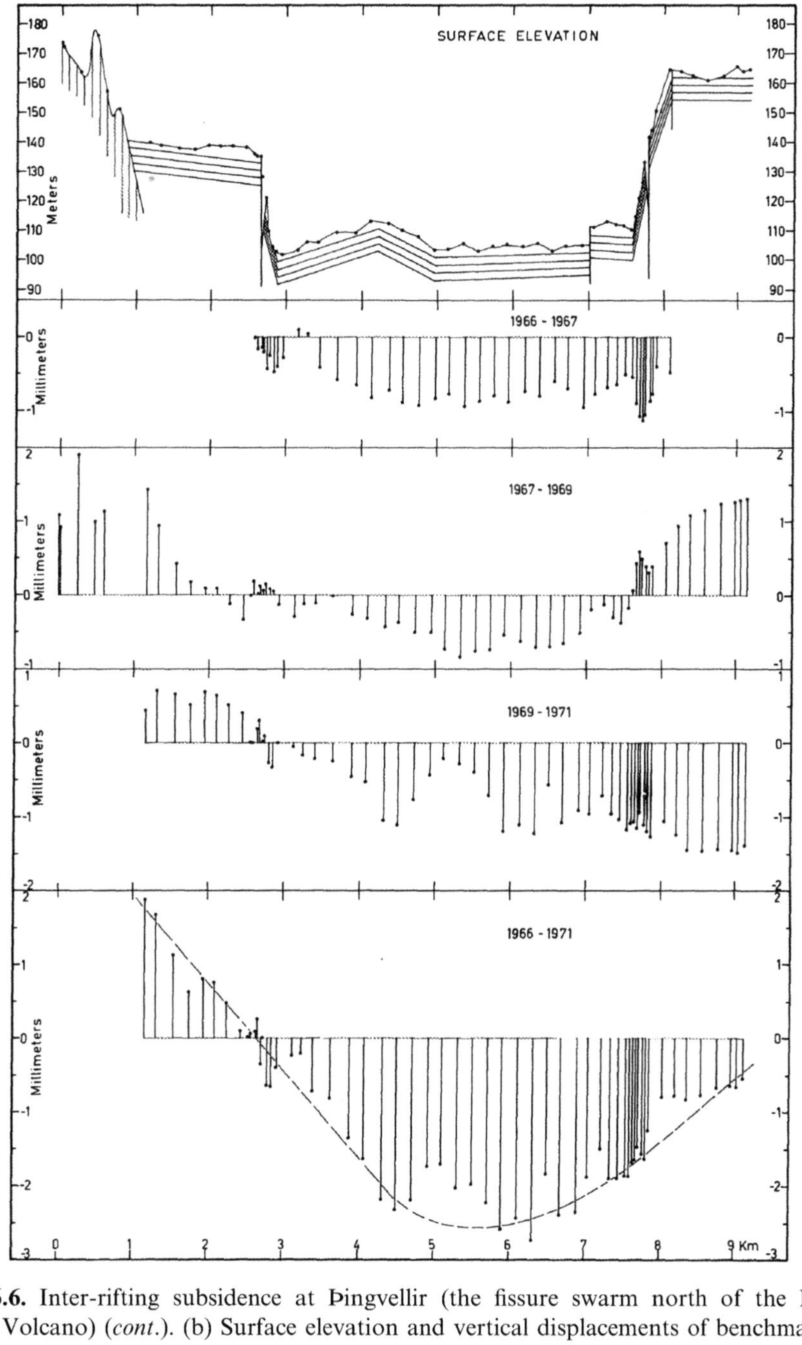

Figure 6.6. Inter-rifting subsidence at Þingvellir (the fissure swarm north of the Hengill Central Volcano) (*cont.*). (b) Surface elevation and vertical displacements of benchmarks on the Þingvellir profile. Benchmark movements are arbitrarily referenced to a benchmark near the west end of the profile.

Reproduced from Tryggvason (1974).

observed in northern Iceland adjacent to the Krafla Fissure Swarm. Along the Reykjanes Peninsula in southwestern Iceland subsidence has been documented by levelling (Tryggvason, 1974), GPS (e.g., Hreinsdóttir et al., 2001), and InSAR (Vadon and Sigmundsson, 1997). It has been modelled as being due to a line source of pressure decrease within an elastic half-space, causing about 6.5 mm/yr of subsidence along the plate boundary. The model considers loss of material below a "locking depth" at the plate boundary. Extension within a ductile layer below it may be more local than in the elastic crust above, causing subsidence if flow of material from below does not occur to replace the laterally displaced material.

6.3 RIFTING EVENTS

Diking events along the spreading plate boundary relieve stresses built up during inter-rifting periods. Individual diking events may lead to extension on the order of a metre, and rifting episodes associated with multiple diking events may cause cumulative widening amounting to more than several metres over long distances along fissure swarms. Major rifting episodes are known to have occurred about once every century in Iceland (Table 6.1). In the EVZ in southern Iceland, the most recent rifting episode occurred 1862–1864 in a remote area at Tröllagígar near the southwestern edge of Vatnajökull (Thorarinsson and Sigvaldason, 1972). The best documented historical rifting episodes in the EVZ was the catastrophic Laki Rifting Episode in 1783–1784 (see Section 3.5). In the Northern Volcanic Zone, a

Table 6.1. Major rifting episodes documented in historical times in the Northern, Eastern and Western Volcanic Zones.[1]

Location	Year
Northern Volcanic Zone	
Krafla: Krafla Fires	1975–1984
Krafla: Mývatn Fires	1724–1729
Askja	1874–1875
Eastern Volcanic Zone	
Vatnaöldur	871 ± 2
Eldgjá	~934
Veiðivötn	~1480
Laki	1783–1784
Tröllagígar	1862–1864
Western Volcanic Zone	
Þingvellir	1789

[1] Plate-spreading contributions of these are, for example, discussed by Jónsson et al. (1997), Björnsson et al. (1977), and Sigmundsson et al. (1995).

major rifting episode occurred at the Askja Volcanic System in 1874–1875, and at the Krafla Volcanic System 1975–1984. Deformation in the centre of these systems during these rifting episodes is discussed in Sections 5.4 and 5.5, but more details on diking events follow here. Diking events during the Krafla Rifting Episode were the first instrumentally recorded, and are still the most important examples of diking events associated with large-scale widening across the rifts in Iceland.

The diking events at Krafla are a topic of continued discussion, in particular the role of lateral versus vertical flow of magma during these events. The seismic and geodetic results fit well with models where magma flows laterally from a shallow magma chamber under the centre of the Krafla System, and is emplaced as dikes in the shallow crust out in the Krafla Fissure Swarm during diking events. Such a model is well consistent with seismic and geodetic observations (e.g., Einarsson, 1991b). A difference in magma chemistry of lavas erupted within and outside the caldera has on the other hand been taken as a strong argument for vertical flow of magma.

An important constraint on the diking events at Krafla is provided by seismic observations. Seismicity associated with diking events propagated away from the Krafla Centre out along the fissure swarm during each of these events, demonstrating lateral growth of dikes (Figure 6.8). The observations indicate that the rate of dike lengthening decreased as the dike length increased. A fluid-dynamical model of lateral flow can explain this pattern. Einarsson and Brandsdóttir (1980) model dike formation at Krafla as flow of viscous fluid through a rectangular box, and derive the following approximate relation for dike lengthening versus time:

$$\frac{dL}{dt} = \frac{b^2 \Delta P}{12 \eta L} \tag{6.2}$$

where L is the dike length, b is the dike width, η is the magma viscosity, and ΔP is the difference in pressure between the reservoir and the tip of the dike. If all parameters are constant except the dike length, then the differential equation can be solved to reveal that L is proportional to $\sqrt{t}$, where t is the time from the beginning of the intrusion. Observed seismic propagation during a rifting event in July 1978 demonstrates this behaviour (Figure 6.8). The fit of equation (6.2) is, however, not particularly good, possibly due to the many simplifying assumptions (Einarsson and Brandsdóttir, 1980).

Geodetic measurements provided complementary data to seismic observations on the Krafla diking events. The geodetic data include repeated EDMs of networks on different scales, including local observations next to the Krafla Fissure Swarm (e.g., Tryggvason, 1984) in addition to more regional observations (e.g., Wendt et al., 1985). Cumulative widening up to 9 m is inferred to have occurred (Figures 6.9 and 6.10) in the complete series of events. Average widening over an ~80-km-long segment of the Krafla Fissure Swarm is inferred to have been about 4–6 m (Tryggvason, 1984, 1986). In addition to horizontal movements, extensive vertical movements were associated with the rifting events. Areas on each side of the Krafla

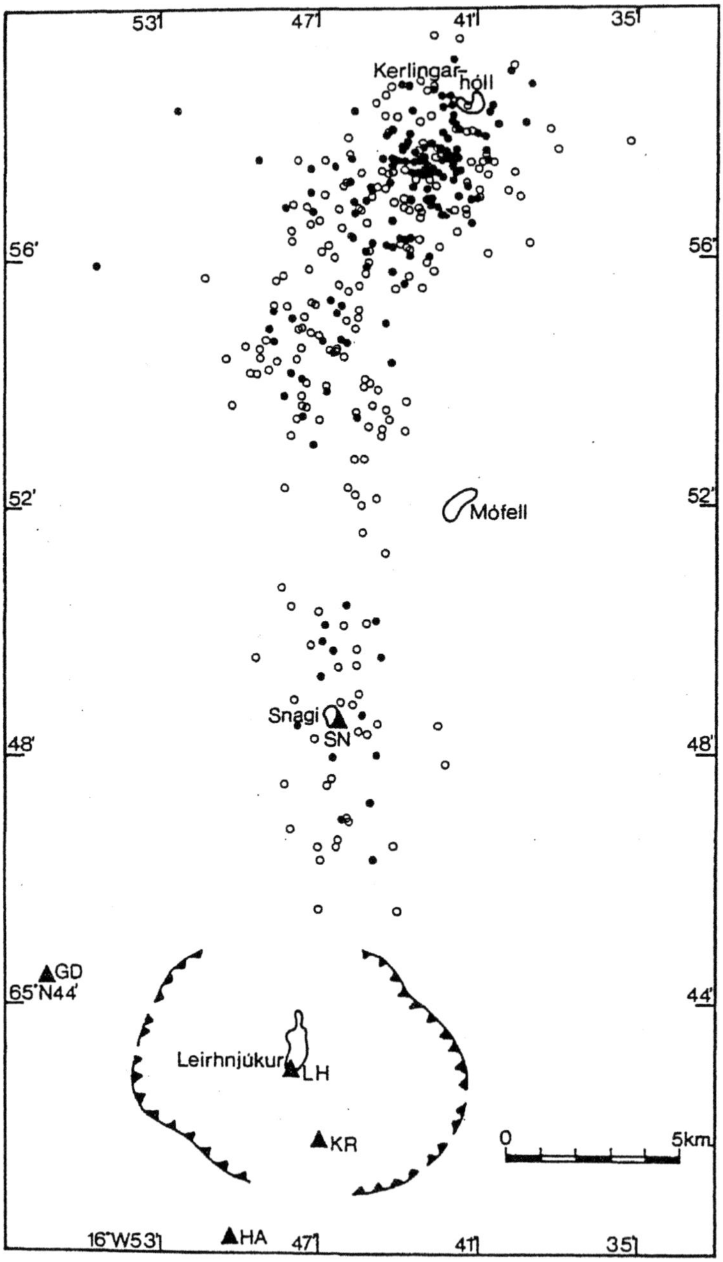

Figure 6.8. (a) The July 1978 earthquake swarm associated with a rifting event in the Krafla Fissure Swarm. Dots mark epicentres located with horizontal standard error of 1 km and less, circles denote epicentres with errors between 1 and 2 km. Triangles denote seismic stations (more stations are located outside the map).

Reproduced from Einarsson and Brandsdóttir (1980).

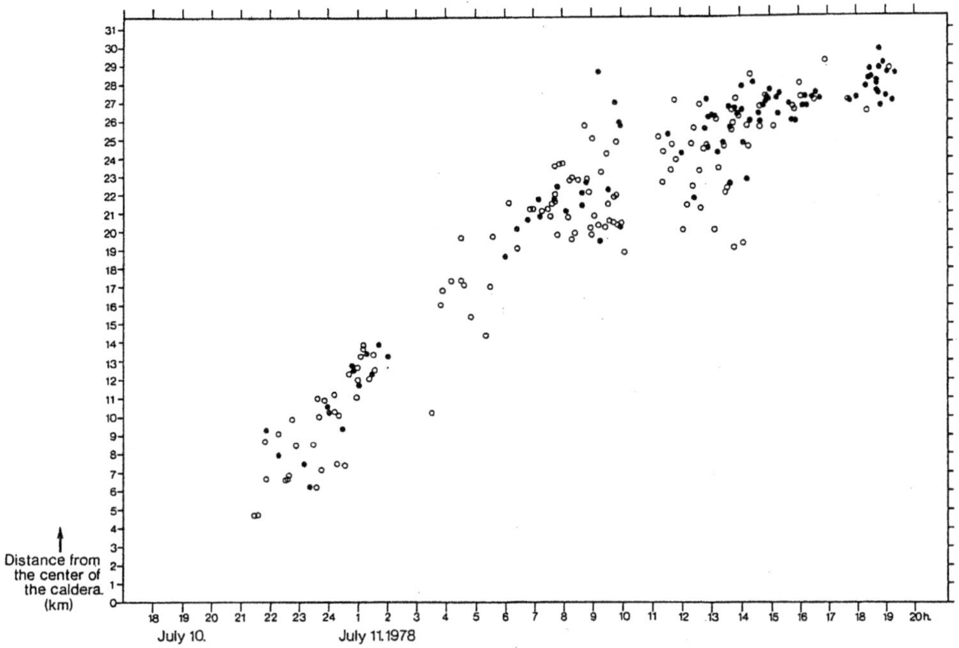

Figure 6.8 (*cont.*). (b) The migration of seismic activity. The distance of epicentres from the centre of the Krafla Caldera plotted as a function of time. The apparent gap in activity between 10 h and 11 h on July 11 is caused by a time signal failure. Continuous tremor and rapid deflation of a shallow magma chamber under Krafla started at 17 h on July 10, associated with onset of the rifting event.

Reproduced from Einarsson and Brandsdóttir (1980).

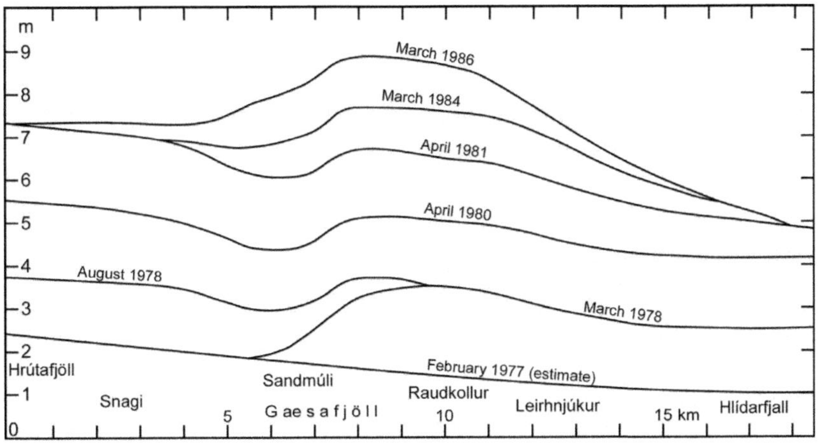

Figure 6.9. Cumulative opening across the Krafla Fissure Swarm during the Krafla Rifting Episode. Contributions of individual events are shown versus distance from the centre of the Krafla Caldera.

Modified from Tryggvason (1984) and Tryggvason (2005, pers. commun.).

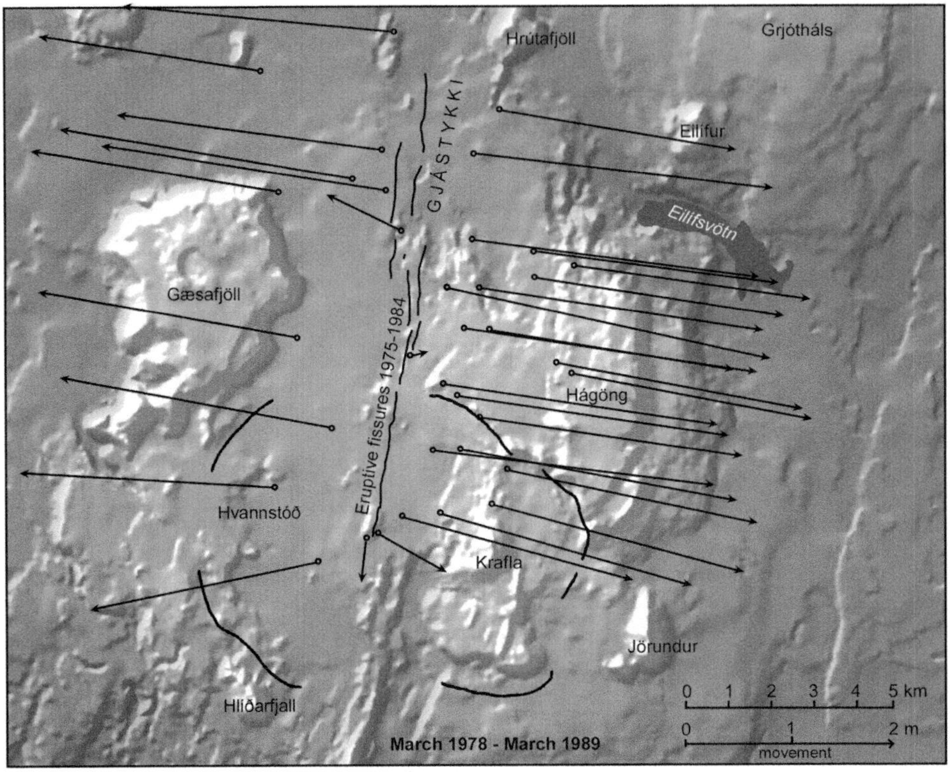

Figure 6.10. Map view of horizontal displacements of benchmarks in an EDM network in the Krafla area between observations of March 1978 and March 1989. Displacements before March 1978 are less well documented.

Reproduced from Tryggvason (1991).

Fissure Swarm were compressed and contracted when the dikes intruded, and consequently uplifted (Figure 6.11).

6.3.1 Models of rifting events

Rifting episodes are a series of rifting events. Each individual rifting event can be modelled in general as a combination of dike intrusion, earthquake faulting, and pressure changes in magmatic sources (e.g., deflation of a shallow magma chamber). Dislocations in an elastic half-space have been used to mimic dikes and faults, and Mogi sources have been used for magma chambers as outlined in Chapter 5. One approach to the modelling of rifting episodes is therefore to sum up deformation from all individual events. As an example, the last of the Krafla events has been modelled by Árnadóttir et al. (1998), combining EDM, levelling, and optical levelling tilt spanning the 1984 eruption of Krafla (Figure 6.12).

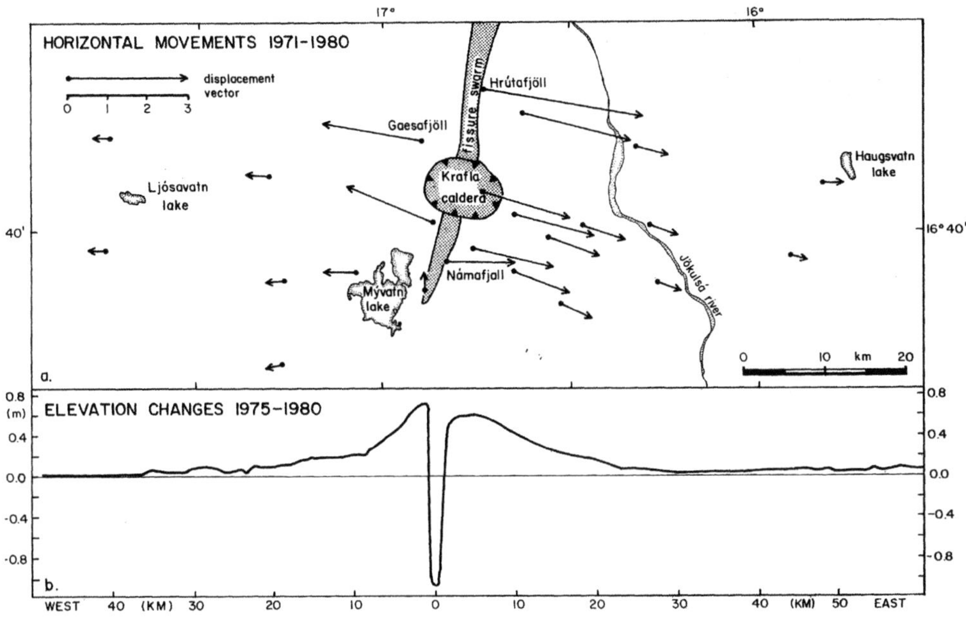

Figure 6.11. Horizontal displacement 1971–1980 inferred from regional EDM measurements (upper panel), and elevation changes 1975–1980 associated with the Krafla Rifting Episode. Reproduced from Björnsson (1985). Copyright by the American Geophysical Union.

6.4 POST-RIFTING ADJUSTMENT

GPS measurements in North Iceland in the years following the Krafla Fires revealed a higher than average extension rate across the plate boundary. Initial measurements in the period 1987–1990 revealed extension rates across the Northern Volcanic Zone as high as 5.6 cm/yr (Figure 6.13), three times the average spreading rate (Foulger et al., 1992; Jahn, 1992). Horizontal displacements increased away from the rift axis and reached a maximum at a distance of about 25–50 km from the spreading axis (Figure 6.14). At larger distances, the displacement rates decreased again. Measurements in 1992, 1993, and 1995 revealed decaying extension rates compared with the 1987–1990 period (e.g., Völksen and Seeber, 1998; Völksen, 2000), with rates approaching the long-term average. The observed rate in 1993–1995 was 2.1 cm/yr.

The GPS observations in North Iceland reveal significant spatial and temporal variability in deformation at the plate boundary in the 1987–1995 period. No significant tectonic events happened along the plate boundary during this time, suggesting that variations may be due to prior events. Modelling shows that observed displacements can be explained as a response to transient post-rifting stress relaxation following the 1975–1984 Krafla Rifting Episode. Higher extension rates across the boundary in the period after the rifting events originate from interaction of a ductile lower crust and an elastic brittle uppermost crust.

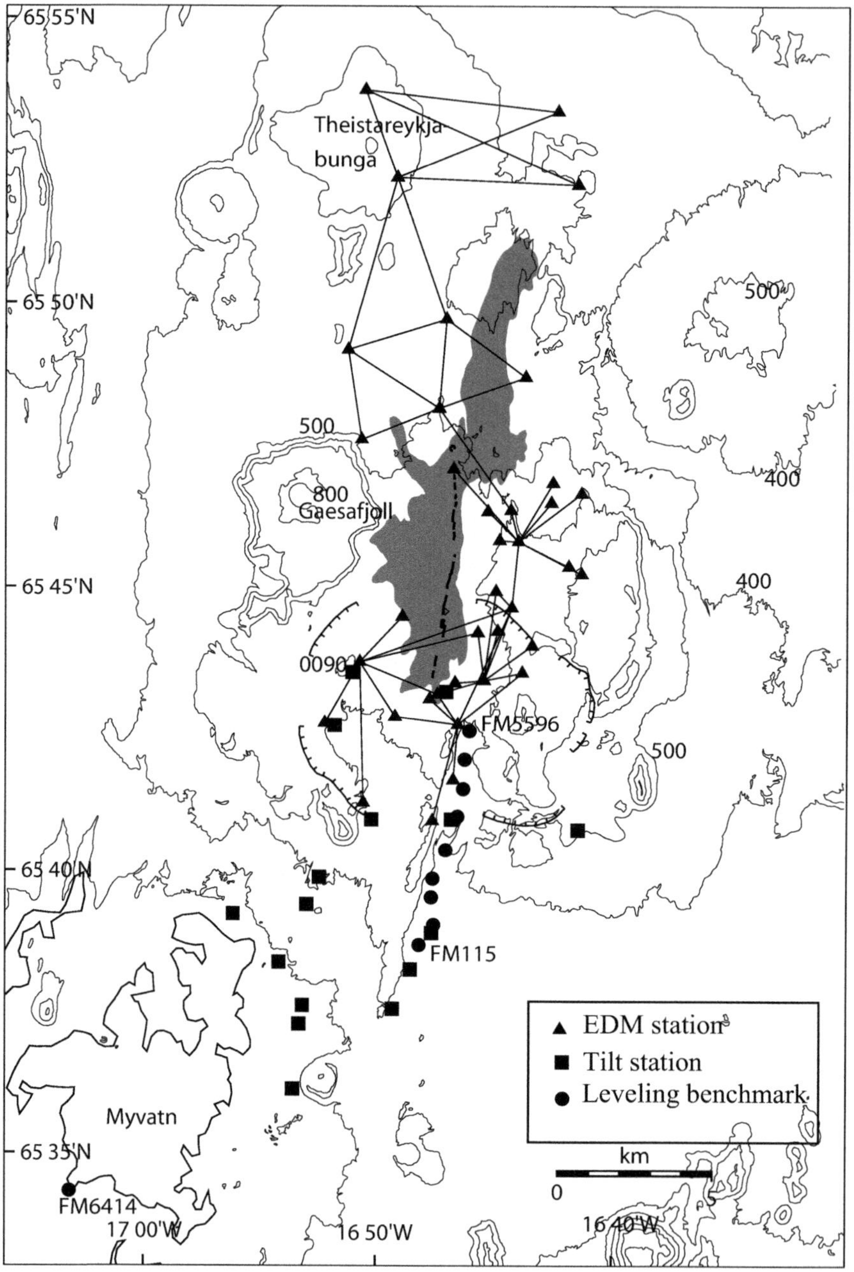

Figure 6.12. (a) Geodetic network used to constrain deformation during the 1984 eruption of the Krafla Volcano (EDM stations, optical levelling tilt stations, and levelling benchmarks). Shading shows the extent of a lava flow formed in 1984, with the broken line on top outlining the eruptive fissure.

After Árnadóttir et al. (1998). Copyright by the American Geophysical Union.

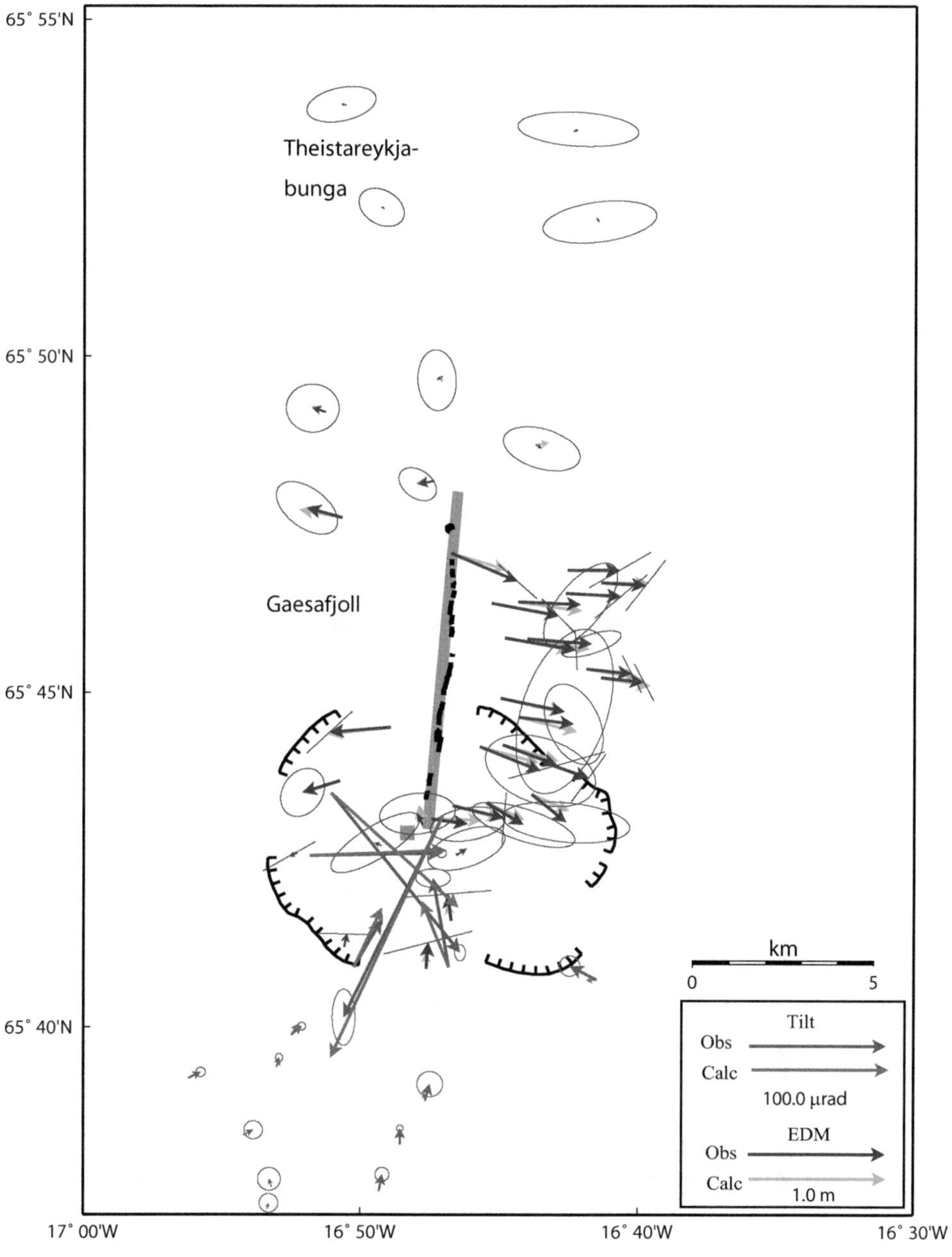

Figure 6.12 (*cont.*). (b) Observed and modelled tilt and horizontal displacements inferred from EDM. Eruptive fissure and extent of the modelled dike are indicated by broken and thick shaded line, respectively. The dike extends to a 7-km depth. Grey rectangle denotes the location of a Mogi pressure source, located at a 3-km depth.

After Árnadóttir et al. (1998). Copyright by the American Geophysical Union.

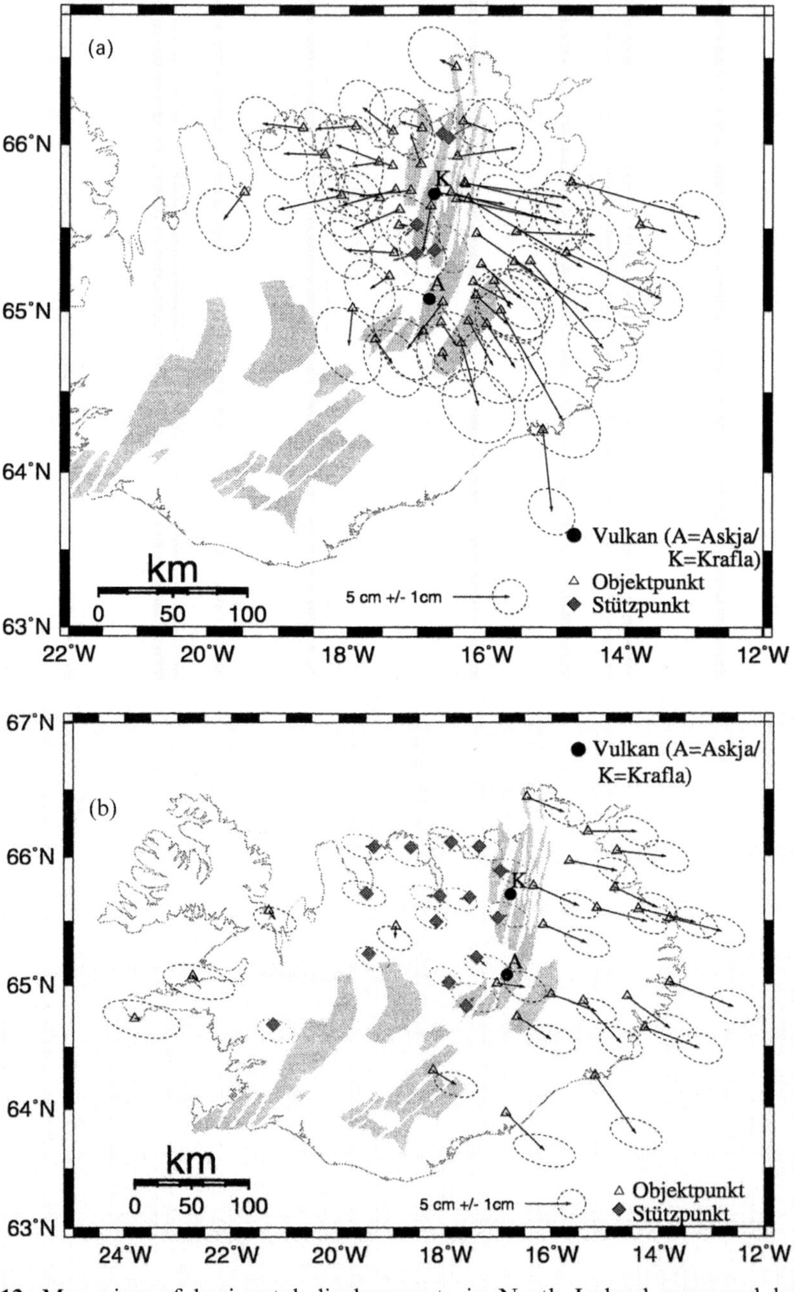

Figure 6.13. Map view of horizontal displacements in North Iceland measured by GPS. (a) 1987–1990 and (b) 1992–1995. Refernce stations are near the central axis of the plate boundary in (a), but on the North American Plate in (b). Confidence limits at the 1σ level are shown.

Modified from Völksen (2000).

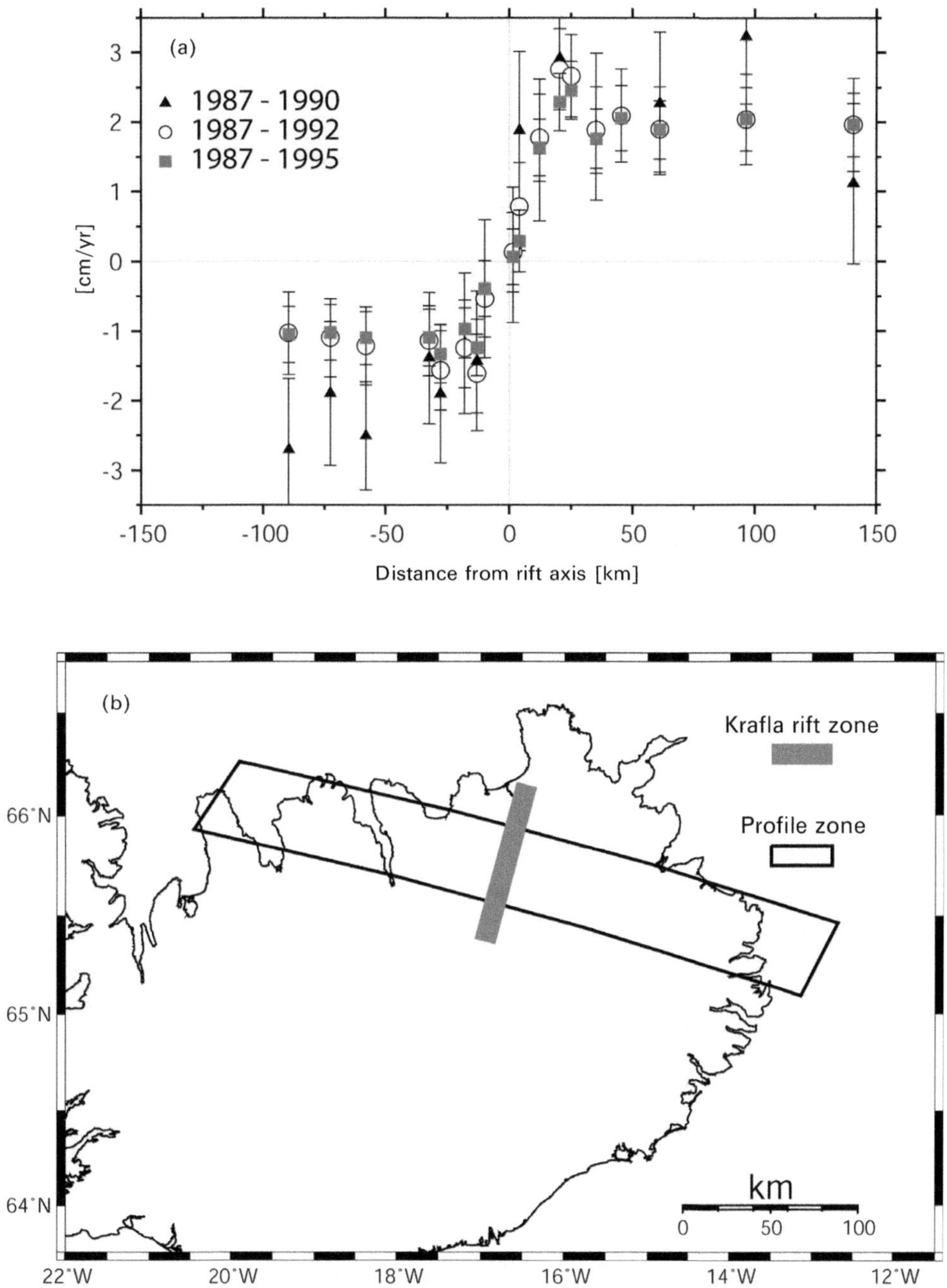

Figure 6.14. Displacement profiles across North Iceland. (a) The highest rate is observed in the initial period, 1987–1990, with lower average rates in the longer periods. (b) The location of the profiles; GPS data from stations within the outlined area are displayed on the profiles. Modified from Völksen (2000).

During rifting, stress builds up in the ductile layer. At later times it is released, driving excessive displacement away from the plate boundary in the post-rifting period.

6.4.1 Newtonian viscosity models of post-rifting deformation

A simple cross-sectional model, consisting of a dike injected into a thin elastic layer overlying a Newtonian viscous layer (Figure 6.15) can mimic the pattern of deformation observed in North Iceland after the Krafla Fires. A dike, injected at time $t = 0$, extends through the thickness of the elastic layer. The model assumes the elastic layer is thin and conditions of plane strain exist within it. The horizontal displacement within the plate, $u(x, t)$, is in that case only a function of time and distance, x, from the dike. The horizontal velocity is $\partial u/\partial t$. The underlying viscous layer, with thickness b and dynamic viscosity η, rests on a rigid surface. Velocity gradients within the layer amount to approximately $(1/b)\ \partial u/\partial t$. Following from the basic definition of dynamic viscosity, the traction exerted by the viscous layer on the base of the elastic layer is $-(\eta/b)\ \partial u/\partial t$. This traction is balanced by elastic forces within the plate. Consideration of force balance on an infinitesimal element within the plate gives (e.g., Foulger et al., 1992; Heki et al., 1993):

$$\frac{\eta}{b}\frac{\partial u}{\partial t} = h\frac{\partial \sigma_{xx}}{\partial x} \tag{6.3}$$

where h is the thickness of the elastic layer and σ_{xx} is the normal stress in the elastic layer in the x-direction. In the elastic layer, stress is related to strain by:

$$\sigma_{xx} = M\frac{\partial u}{\partial x} \tag{6.4}$$

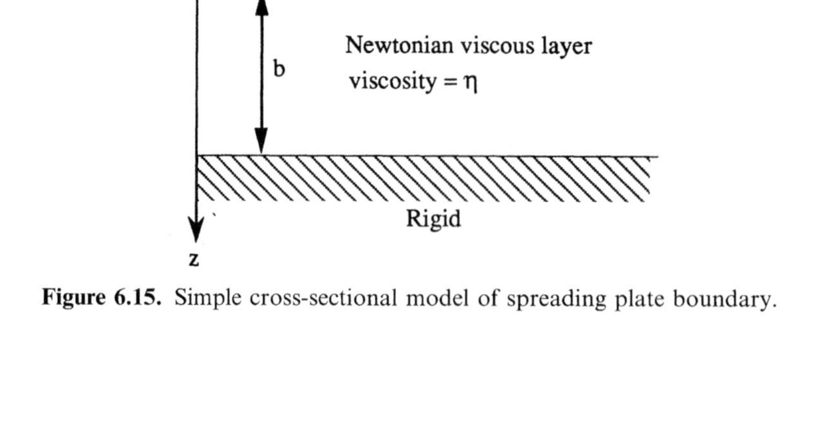

Figure 6.15. Simple cross-sectional model of spreading plate boundary.

where M is an elastic modulus. For plane stress conditions the elastic modulus equals (e.g., Heki et al., 1993):

$$M = \frac{4\mu(\lambda + \mu)}{(\lambda + 2\mu)} \tag{6.5}$$

where λ and μ are the Lamé moduli for the elastic plate. Substituting (6.4) into (6.3), the equation of motion is found to be:

$$\frac{\partial u}{\partial t} = \kappa \frac{\partial^2 u}{\partial x^2} \tag{6.6}$$

with κ equal to:

$$\kappa = \frac{Mhb}{\eta} \tag{6.7}$$

Equation (6.6) is the diffusion equation and in this context the diffusivity term, κ, has been termed stress diffusivity. For a dike of half-width U_0 intruded into the elastic layer at time $t = 0$, the resulting horizontal displacement field is:

$$u(x, t) = U_0 \operatorname{erfc} \frac{x}{2\sqrt{\kappa t}} \tag{6.8}$$

and the horizontal velocity field is:

$$\frac{\partial u(x, t)}{\partial t} = \frac{U_0}{t\sqrt{\pi}} \frac{x}{2\sqrt{\kappa t}} e^{-x^2/4\kappa t} \tag{6.9}$$

The above model predicts displacements and velocities (Figure 6.16) of a similar type as observed in North Iceland after the Krafla Rifting Episode. High rates of displacements dominate during a period after dike injection because of stress interaction; stresses built up in the viscous layer during the diking event relax and drive excess displacements away from the dike axis. A velocity pulse diffuses away from the plate boundary, in a characteristic pattern determined by stress diffusivity. The fit of GPS-derived displacements in 1987–1990 with this one-dimensional model suggest a stress diffusivity of 1.1 m^2/s (Foulger et al., 1992). Interpretation of GPS data collected in 1992 with the same model provides a similar conclusion for the stress diffusivity (Foulger et al., 1994). The stress diffusivity can be used to infer the viscosity, provided b, h, and M are known.

The previous analysis is instructive and demonstrates well the nature of post-rifting displacement and its temporal variations. However, it depends on a number of simplifying assumptions. In reality, the finite length of dikes and associated lack of opening off their ends will limit the amount of deformation. An extension of the above model, considering the finite length of dikes, gives a map view of the horizontal deformation field. Such a model by Heki et al. (1993) provides an improved fit to the observations and gives a stress diffusivity of 10 m^2/s, an order of magnitude higher than application of the cross-sectional model. Using equation (6.7) with $h = 8$–30 km, $b = 5$–10 km, and $\lambda = \mu = 28$ GPa, Heki et al. (1993) derive a viscosity of 0.3–2×10^{18} Pa s.

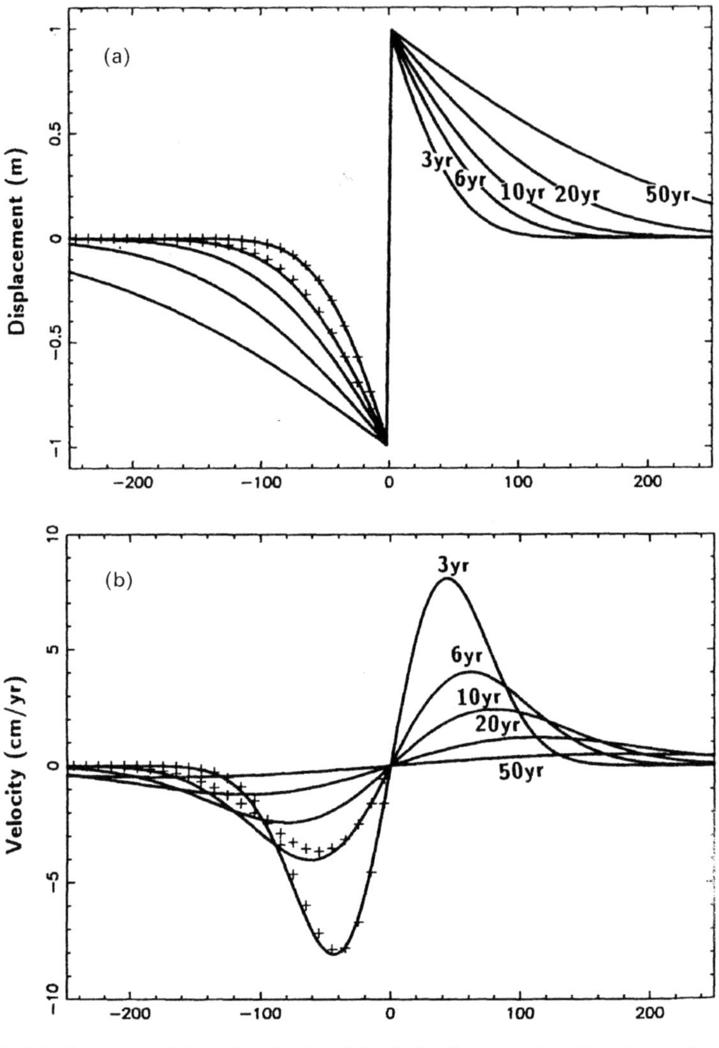

Figure 6.16. Displacement (a) and velocity (b) of the flanks of a plate boundary 3–50 years after a dike intrusion, according to equations (6.8) and (6.9). The curves are for a dike half-width of 1 m and stress diffusivity of $10\,m^2/s$. Pluses show predictions from finite difference simulation study for comparision.

6.4.2 Viscoelastic models of post-rifting deformation

The post-rifting models presented in Section 6.4.1 assumed Newtonian viscosity, whereas a viscoelastic behaviour is a better representation of the Earth response. Several studies of the post-rifting deformation in Iceland have incorporated viscoelastic models. Hofton and Foulger (1996a, b) invoke a model consisting of

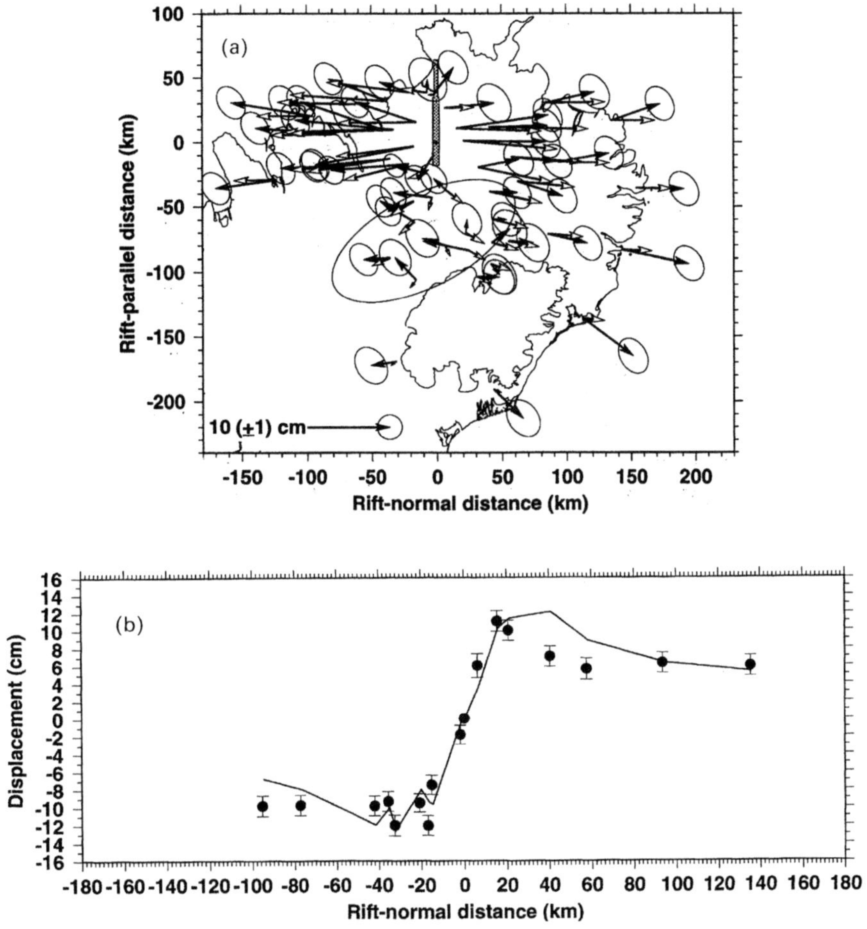

Figure 6.17. Map view (a) and displacement profile (b) showing observed horizontal displacements 1987–1992 measured by GPS, and best-fit-simulated displacements from a viscoelastic model.

Reproduced from Hofton and Foulger (1996a). Copyright by the American Geophysical Union.

an elastic layer over a viscoelastic half-space, with dike injection into the elastic layer. By fitting the prediction of this model to GPS observations 1987–1992 (Figure 6.17), they conclude the elastic layer thickness is 10 km in North Iceland and the viscosity below it is 1.1×10^{18} Pa s.

An alternative interpretation is given by Pollitz and Sacks (1996), who consider an Earth structure consisting of an elastic upper crust and a viscoelastic lower crust, underlain by a viscoelastic half-space with different viscosity than the lower crust. Their favoured lower crustal viscosity is 3×10^{19} Pa s, and the inferred upper mantle viscosity is about 3×10^{18} Pa s. These values yield the closest agreement between model predictions and data. Further viscoelastic modelling efforts have been conducted by Völksen (2000), who considered a more extensive GPS dataset than the other studies, including a GPS survey in 1995.

6.4.3 Elastic dike-opening models of post-rifting deformation

A third type of model that has been applied to post-rifting deformation is based on gradual dike opening at depth in an elastic half-space. During rifting events, dikes form in the uppermost part of the elastic half-space. After the rifting events, continued opening takes place on the downward continuation of the dike plane. Although such models can fit observations (e.g., Foulger et al., 1994; Hofton and Foulger, 1996a), models incorporating viscoelastic behaviour are considered physically more realistic.

A model consisting of continued dike opening at depth was nevertheless used by de Zeeuw-van Dalfsen et al. (2004) to mimic the combined effects of post-rifting deformation and plate spreading across the Krafla Volcanic System, when interpreting InSAR observations covering the area. The elastic model is attractive for computational convenience, as it can be easily implemented along with other sources in an elastic half-space. De Zeeuw-van Dalfsen et al. (2004) find a decreasing rate of dike opening over the 1993–1999 period, averaging 3.4 cm/yr in 1993–1999, but 2.7 cm/yr in 1996–1999. Decay in post-rifting deformation is thus suggested from the InSAR observations as well as from the GPS data.

6.5 OBLIQUE SPREADING: THE REYKJANES PENINSULA

Oblique rifting occurs when the spreading vector is not perpendicular to a rift axis. This is the case for most of the plate boundary in Iceland, but the obliquity is most pronounced on the Reykjanes Peninsula in southwestern Iceland. There the plate-spreading direction and the trend of the plate boundary axis deviate only by about 30°. Shearing and extension across the plate boundary zone contribute both to the style of deformation and overall structure in the area. Evidence for such movements was already found in a pioneering crustal deformation study of Brander et al. (1976), who used precise distance measurements 1968–1972 to conclude that a combination of left-lateral and extensional movement takes place on the Reykjanes Peninsula. Detailed geodetic observations have been conducted in the area with space-geodetic techniques. Horizontal displacements inferred from continuous GPS (Figure 6.18) and campaign GPS (Figure 6.19) show that shearing is the dominant movement across the peninsula; the area south of it being part of the Eurasian Plate and the area north of it belonging to the North American Plate (e.g., Sturkell et al., 1994; Hreinsdottir, 1999; Hreinsdottir et al., 2001). InSAR observations have also been interpreted in terms of shearing across, as well as subsidence of, the plate boundary (Vadon and Sigmundsson, 1997). Furthermore, GPS and InSAR data in the area have been combined to derive three-dimensional motion maps for the area (S. Gudmundsson et al., 2002).

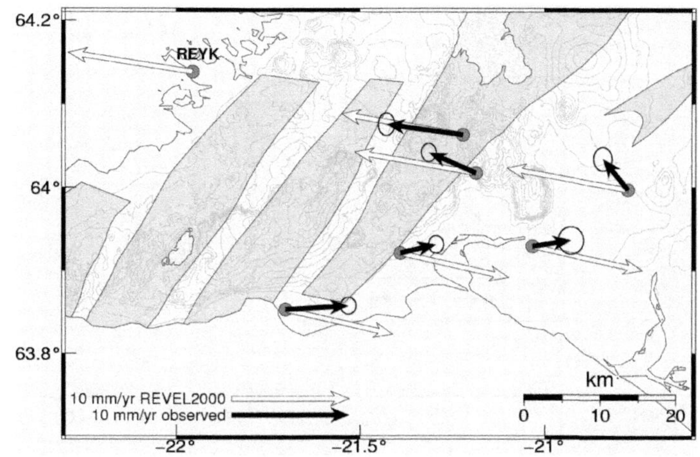

Figure 6.18. Velocities of continuous GPS stations on the Reykjanes Peninsula 1999–2004 (black arrows), assuming the REYK station moves at 10.5 mm/yr towards east and 1.6 mm/yr towards north. Confidence limits at the 2σ level are shown. White arrows show reference velocities from the REVEL plate motion model (Sella et al., 2002), assuming stations north of the central axis of the plate boundary move with a velocity equal to half of the inferred spreading across Iceland, and stations south of it move equally but in the opposite direction. The observed velocities are only a fraction of the REVEL velocities, as the continuous GPS stations are located within the plate boundary deformation zone.

Courtesy of Halldór Geirsson, Icelandic Meteorological Office.

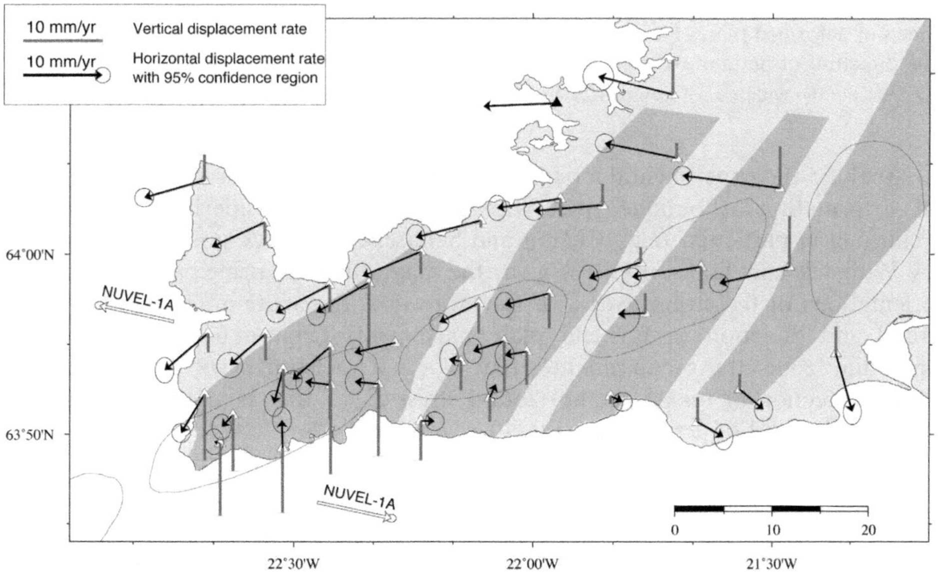

Figure 6.19. GPS-derived velocities (horizontal and vertical) on the Reykjanes Peninsula inferred from campaign GPS measurements 1993–1998.

Modified from Hreinsdóttir et al. (2001). Copyright by the American Geophysical Union.

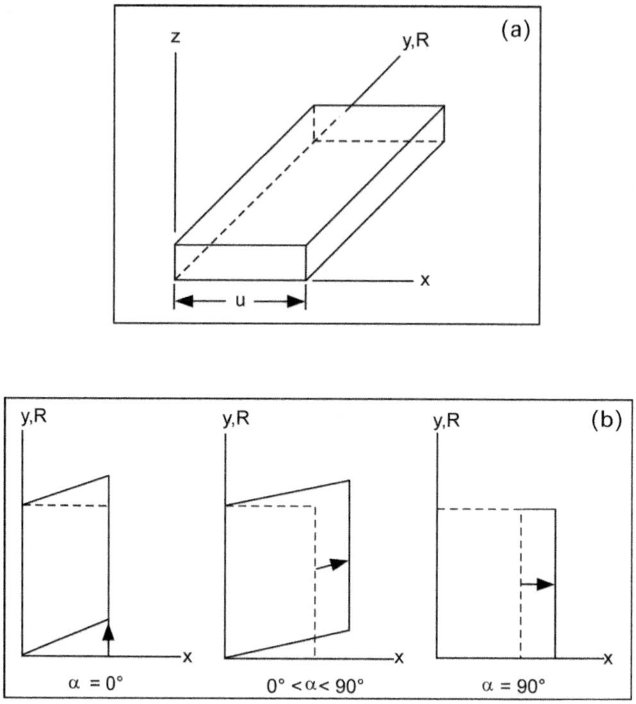

Figure 6.20. Oblique spreading. (a) Undeformed rectangular plate of width u, representing a rift in the upper brittle lithosphere. The long axis of the plate parallels the rift trend, R. (b) Map views of deformed plates for several values of α, the acute angle between the rift trend, R, and the direction of plate movements (arrows).

Reproduced from Withjack and Jamison (1986) with permission from Elsevier.

Analytic and experimental models have been used to study fault patterns and style of deformation at oblique rifts, with some studies considering the Reykjanes Peninsula in particular (e.g., Clifton and Schlische, 2003). A simple analytic model by Withjack and Jamison (1986) gives the magnitudes of principal strains and the orientations of the strain axes as a function of α, the acute angle between the rift trend and direction of far-field plate movements (Figure 6.20). In an oblique spreading zone, the predominant strike of eruptive fissures will be perpendicular to the direction of maximum horizontal stress E_{hmax} that is given with respect to the overall trend of a rift zone as (Clifton and Schlische, 2003):

$$E_{hmax} = 90^\circ - \tfrac{1}{2}\tan^{-1}(\cot\alpha) \tag{6.10}$$

The resulting angle between eruptive fissures and a rift trend is zero only if spreading is fully perpendicular to the rift zone ($\alpha = 90^\circ$). The deviation between direction of eruptive fissures and rift trend will increase with decreasing α, up to a value of 45° for a zone of simple shear ($\alpha = 0^\circ$). For the Reykjanes Peninsula, α is about 30°, giving E_{hmax} as 60° clockwise from the rift trend. The average trend of eruptive fissures on the westernmost part of the Reykjanes Peninsula suggest that the direction of E_{hmax} is 55° clockwise from the rift trend, in good agreement with the analytic model (Clifton and Schlische, 2003).

6.6 THE RIFTING CYCLE

Some main aspects of the plate-spreading deformation cycle have been outlined in the previous chapters. In general, displacement field at spreading plate boundaries can be regarded as the sum of deformation associated with the latest rifting episode, superimposed on background movement, being the summed contributions of all prior episodes. A simplified model of the cyclic behaviour can, for example, be constructed from the model presented in Section 6.4.1. By summing up deformation from rifting events, Heki et al. (1993) derive the following relation:

$$U(x,t) = u(x,t) + \sum_{n=1}^{\infty} u(x, t + nT) \tag{6.11}$$

where $U(x,t)$ is the current displacement field across the plate boundary, t is the time since last rifting event, T is the recurrence interval of diking events, and $u(x,t)$ is given by equation (6.8). Predictions from this kind of plate boundary behaviour are shown in Figure 6.21, demonstrating the temporal and spatial variations in displacement fields next to spreading plate boundaries.

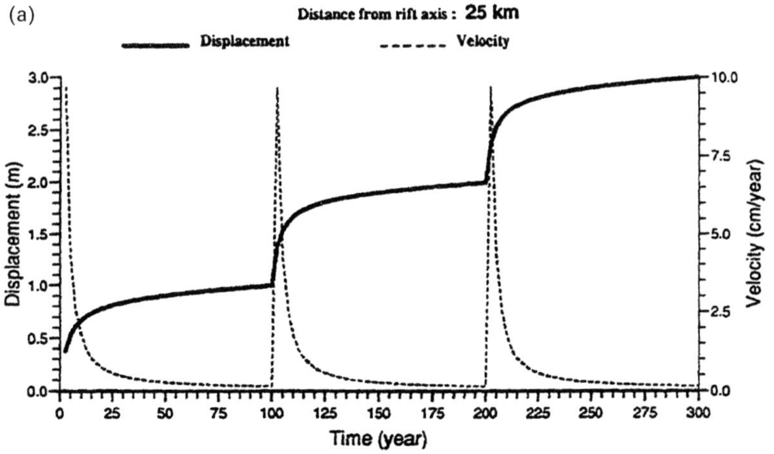

Figure 6.21. Behaviour of a spreading plate boundary. Displacement (solid lines) and velocity (dotted lines) versus time, shown for different distances from a central axis of a plate boundary – (a) 25, (b) 50, (c) 100, and (d) 200 km – according to equation (6.11). A 2-m-wide dike is intruded every 100 years at the central axis of the plate boundary. Stress diffusivity is set to $10\,m^2/s$.

Reproduced from Jahn (1992).

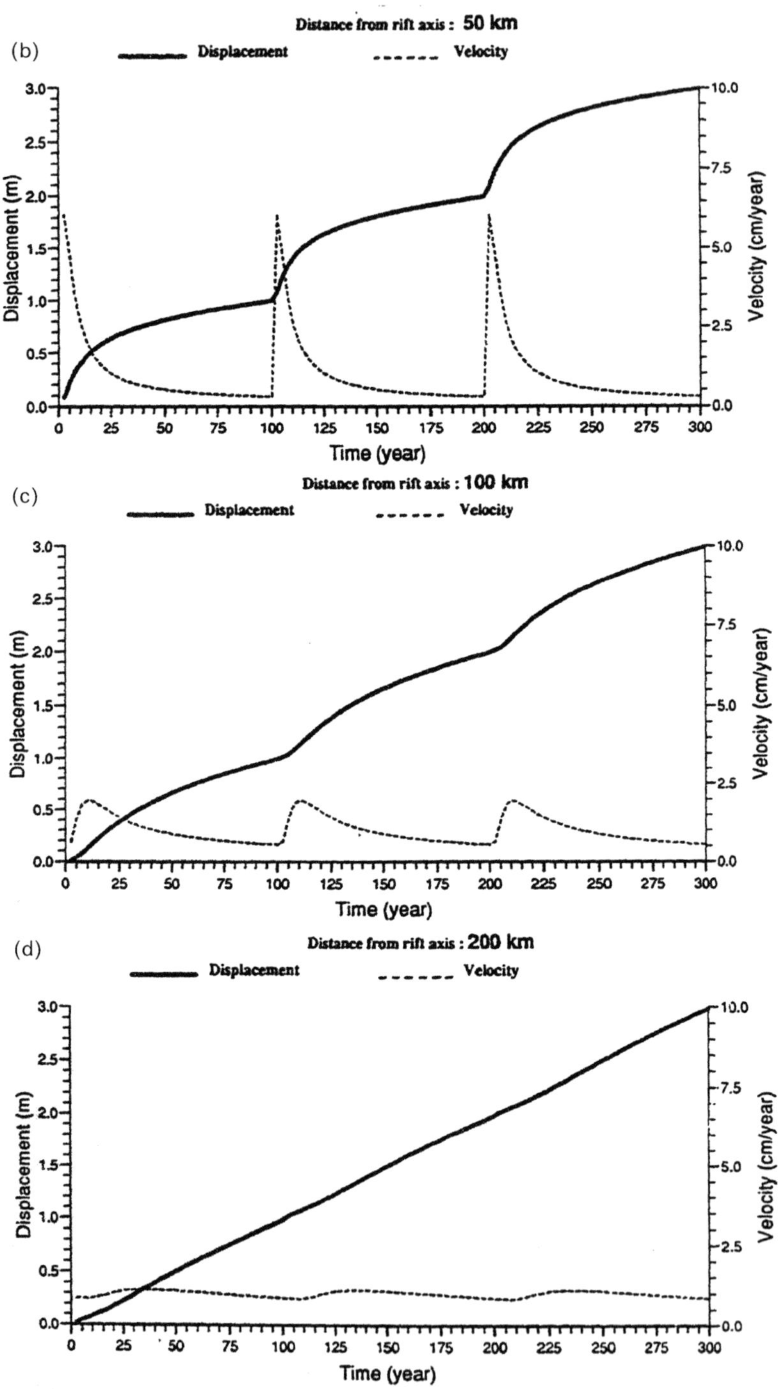

Figure 6.21 (*cont.*)

Considerable difference in plate-spreading patterns is to be expected between the different segments of the plate boundary in Iceland. Obliquity of spreading is an important factor as outlined in Section 6.5. The closer the trend of a rift zone is to the direction of far-field plate motion, then the lower the rate of extension across the rift and higher the shearing. The lower extensional strain may lead to longer times between major rifting episodes. This is in harmony with apparently longer non-eruptive periods at volcanoes on the Reykjanes Peninsula compared with the EVZ and the Northern Volcanic Zone in Iceland.

Dike and fissure swarms form under an angle with respect to the trend of rift zones; therefore they are arranged in an *en echelon* pattern along the plate boundary. Magma supply from depth is likely to determine which of these fissure swarms will be the site of the next rifting event. In the absence of magma, stretching across plate boundaries may need hundreds or thousands of years before tectonic stress reaches levels appropriate to trigger normal faulting (see Section 6.2.2). As rifts with magma at shallow depths fail at much lower stress levels, then inflow of magma from depth is likely to modulate the timing and location of rifting events.

7

Breaking the crust: Seismicity and faulting

Earthquakes are unevenly distributed along the plate boundary in Iceland. Seismicity in Iceland is focused on two main seismic zones, the Tjörnes Fracture Zone (TFZ) and the South Iceland Seismic Zone (SISZ). Both of these are transform zones associated with lateral offsets in the plate spreading. These zones have persistent microearthquake activity (e.g., Jakobsdóttir et al., 2002) and large earthquakes (which may exceed M_s7) occur there at intervals of $\sim$100 years (see Figure 3.20). Many of the volcanic systems are on the other hand characterized by little earthquake activity (see Figure 3.19). Some of the volcanic systems do, however, have high microearthquake activity as a result of magma movements and geothermal activity. The earthquakes in the volcanic systems are smaller than in the transform zones. This chapter focuses on earthquakes and tectonic deformation in the transform zones.

7.1 THE TJÖRNES FRACTURE ZONE

The TFZ is associated with a lateral shift of the spreading zone of 150 km, from the Northern Volcanic Zone (NVZ) in Iceland, to the submarine Kolbeinsey Ridge north of Iceland (Figure 7.1, see colour plates). The zone has existed for 4 million years since a rift jump initiated spreading in the NVZ (Sæmundsson, 1974). Earthquakes in the TFZ reach up to magnitude 7 or just above; the largest earthquakes in the last 200 years being an M7.1 earthquake in 1910 and an M7.0 earthquake in 1963. Tectonic structures in the zone are complex, reflecting its accommodation of both shearing and extension. Seismicity in the fracture zone occurs on three lineaments, the Grímsey Lineament, the Húsavík–Flatey Fault, and the Dalvík Lineament. Each of these lineaments has its separate style of seismicity, structures, and mode of deformation (e.g., Einarsson, 1991a; Rögnvaldsson et al., 1998).

Recently collected multibeam bathymetric data have revealed topographic and structural details of the offshore structures (Brandsdóttir et al., 2004; Brandsdóttir, pers. commun., 2005). An Edgetech SB0512 seismic sub-bottom profiling and sidescan-imaging system has been used as well. The system sweeps (chirps) across 1.0 to 6.0 kHz in 50 ms, yielding sub-bottom penetration on the order of 30–40 m with submetre resolution.

The Húsavík–Flatey Fault produces strike–slip earthquakes along its length. The fault is mostly offshore, but its easternmost part is onland. The fault has produced destructive earthquakes throughout the history of Iceland, the most recent one being an M6.3 event in 1872. The Grímsey Seismic Lineament lying to the north of the Húsavík–Flatey Fault is currently the most active of the three lineaments in the TFZ. It has a complex fault pattern, encompassing both north–south-oriented strike–slip faults and zones of crustal rifting. In 1976 a M6.4 strike–slip earthquake occurred at the eastern end of this lineament at its intersection with the Krafla Fissure Swarm. It occurred in association with a rifting event in the Krafla Fissure Swarm; the strike–slip earthquake occurred when rifting in the Krafla Fissure Swarm reached the intersection with the Grímsey Lineament (Einarsson, 1987). The earthquake in the fracture zone appears to have been triggered by the dike opening in the fissure swarm that amounted to about 1–2 m. The Dalvík Lineament is currently characterized by low diffuse seismic activity, but a damaging earthquake occurred in 1934 on this lineament near the town of Dalvík. The M7.0 earthquake in 1963 in the TFZ occurred furthermore on the offshore extension of the lineament (see Figure 3.20).

Although most of the TFZ is offshore, some constraints on deformation and strain accumulation are provided by campaign and continuous Global Positioning System (GPS) measurements at its onland exposures. Velocities estimated from observations at three continuous GPS stations in the area (Figure 7.2) show that majority of the plate spreading is accommodated in the area between the AKUR and RHOF stations, and that a station located between the Grímey and Húsavík–Flatey Lineament (ARHO) is moving at an intermediate velocity. About equal partition of strain between the Húsavík–Flatey Fault and the Grímsey Lineament is suggested (Geirsson et al., submitted). Campaign GPS measurements provide improved spatial resolution of the strain accumulation and suggest that the main part of the shearing occurs between the Húsavík–Flatey Fault and the Grímsey Lineament (Jouanne et al., submitted). Comparison of GPS results from 1997–1999 and 1999–2002 suggest a temporal evolution of the deformation field, that Jouanne et al. (submitted) relate to the decay of post-rifting deformation in North Iceland after the Krafla Rifting Episode and the return to "normal" background movements by 1999.

Structural studies of the onshore exposures of the TFZ reveal pervasive fracturing and alteration, rotation of spreading-related structures, reactivation of north–south-oriented faults in a bookshelf-faulting deformational style, cut by major west-northwest-trending dextral strike–slip fault zones (e.g., Gudmundsson, 1993; Karson et al., 2004). The linkage of the Húsavík–Flatey Fault with the Þeistareykir Fissure Swarm is particularly well exposed. Faulting at the junction occurs in an early Holocene lava flow (Figure 7.3). Here the Húsavík strike–slip

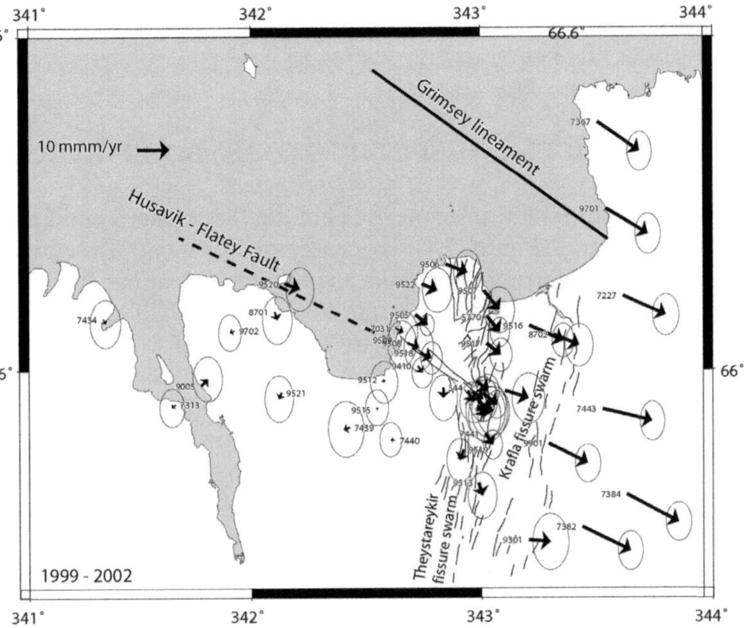

Figure 7.2. Velocities of GPS stations near the Tjörnes Fracture Zone 1999–2002 inferred from GPS data.

Courtesy of F. Jouanne and T. Villemin, Université de Savoie, France.

Figure 7.3. View over the Húsavík Fault.

Photo by Freysteinn Sigmundsson.

fault rotates and merges with a major normal fault of the Þeistareykir Fissure Swarm (Gudmundsson et al., 1993).

7.2 THE SOUTH ICELAND SEISMIC ZONE: "BOOKSHELF FAULTING"

The SISZ is an 80-km-long transform zone linking the Western and Eastern Volcanic Zones in South Iceland (Figure 7.4). It accommodates most of the plate movements in South Iceland, transferring them from the oblique spreading Reykjanes Peninsula Plate Boundary to the Eastern Volcanic Zone. Instead of one east–west-striking major transform fault, an array of north–south-striking faults accommodates left-lateral shear across the zone by bookshelf faulting (e.g., Sigmundsson et al., 1995). Persistent microearthquake activity occurs in the zone (e.g., Stefánsson et al., 1993; Jakobsdottir et al., 2002). Major earthquake sequences have occurred there in historical times at intervals ranging between 45 to 112 years, typically beginning with an ~M7 earthquake in the eastern part of the seismic zone, followed by similar or somewhat smaller events farther west (e.g., Einarsson et al., 1981). The most recent sequence occurred in 2000, with two M_s6.6 earthquakes on June 17 and 21. Large earthquakes also occur as isolated events, the most recent being an M7.0 earthquake in 1912 (e.g., Bjarnason et. al., 1993a).

7.2.1 Microearthquake activity and structure of the South Iceland Seismic Zone

The left-lateral shear accumulation across the SISZ and its east–west orientation suggest that a major east–west-oriented sinistral transform fault along the length of the zone would be the preferred way to accommodate movements across the zone. Faulting in the SISZ, however, occurs on an array of north–south-oriented dextral strike–slip faults (Figure 7.4). This style of faulting has been attributed to the transient nature of the zone, and the unstable tectonic environment caused by propagation of the Eastern Volcanic Zone towards the southwest.

The array of north–south faults has been mapped in considerable detail (e.g., Einarsson and Eiríksson, 1982; Einarsson, 1991a; Clifton and Einarsson, 2005). Spacing between the faults is 0.5–5 km, averaging 2 km. Their surface expression is gentle, consisting of an *en echelon* arrangement of open fissures and pushup structures, up to 20 km long. The width of the fissures is up to a few metres and the pushups are up to a few metres high. The open fissures typically take a northeast–southwest direction, opening up perpendicular to the direction of minimum stress. The fissures are left stepping with pushups located at the stepovers between fissures. Local crustal shortening causes the pushups to form, with their dimensions proportional to the amount of opening on adjacent fissures (e.g., Bjarnason et al., 1993a). The faults are segmented, each segment commonly deviating from the overall fault trend. The structures are consistent with right-lateral strike–slip fault movement on north–south-striking fault surfaces at depth.

The subsurface expressions of faults in the SISZ have been mapped by micro-earthquake studies. In particular, a large number of the faults experienced activity

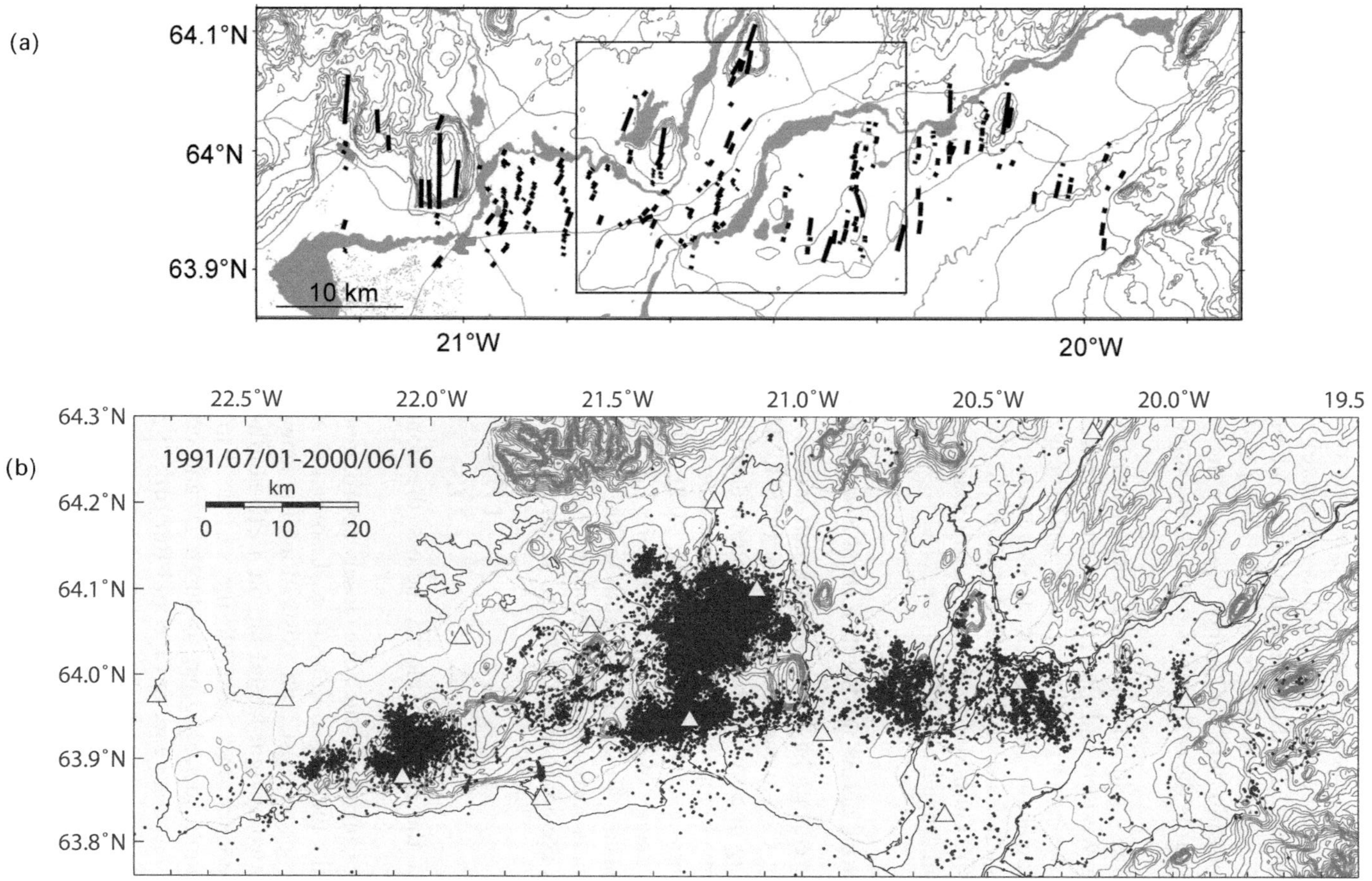

Figure 7.4. (a) Simplified map of the South Iceland Seismic Zone showing surface fractures associated with strike–slip faults. Background map shows 100-m elevation contours, with rivers and lakes shaded. The box shows the location of Figure 7.10. (b) Earthquakes in South Iceland from July 1, 1991 to June 16, 2000 recorded by the South Iceland Lowland seismic network of the Icelandic Meteorological Office. Large cluster near the centre of the image is at the Hengill Central Volcano; the Reykjanes Peninsula is to its west and the SISZ to its east. The displayed time period represents the final 9 years of an interseismic period in the SISZ with good seismic coverage by the SIL network. Triangles mark seismic stations.

(a) Reproduced from Clifton and Einarsson (2005) with permission from Elsevier. (b) Modified from Stefánsson and Guðmundsson (2005), courtesy of Gunnar Guðmundsson.

triggered by the June 17 and 21, 2000 earthquakes, outlining these faults. The fault planes appear to be mostly straight surfaces, corresponding in many cases to the mapped overall strike of surface expressions of faults. The length of the faults outlined by microearthquake activity compares well with the length of mapped surface breakage. Earthquake depth varies along the length of the SISZ, from about 8 km in the west to about 12 km in its eastern part (e.g., Stefánsson et al., 1993). The crust in the SISZ formed mostly in the Western Volcanic Zone and its age increases towards the east, consistent with thickening of the seismogenic crust in that direction.

The stress field in the SISZ depends highly on activity in the adjacent overlapping Western and Eastern Volcanic Zones. Activity within these zones has varied and the Eastern Volcanic Zone has propagated towards south through time, imposing significant variation in stresses on the SISZ. Evolution of stresses in the area has been revealed by structural studies and stress modelling – e.g., by Hackman et al. (1990), Gudmundsson and Brynjolfsson (1993), Luxey et al. (1997), and Bergerat and Angelier (2003).

7.2.2 Shearing across the South Iceland Seismic Zone

GPS measurements have revealed the left-lateral shearing taking place across the SISZ. Early GPS measurements from 1986 to 1992 suggested that $85 \pm 15\%$ of the spreading in South Iceland is accommodated by the zone (Sigmundsson et al., 1995). A much-improved view of the deformation field from subsequent GPS measurements (e.g., Perlt and Heinert, 2000; Árnadóttir et al., 2005, submitted) reveals the details of the shearing (Figure 7.5). The strain is focused on a central zone that has a ~40-km north–south width, over which about 10 mm/yr of relative plate motion is accommodated. Considering a wider area, a larger part of the plate velocities is accommodated. Average strain rate $\dot{\varepsilon}$ in the central area of the SISZ is given as:

$$\dot{\varepsilon} \approx \frac{1}{2}\frac{V}{L} \tag{7.1}$$

where V is the relative velocity across the zone and L its width. $\dot{\varepsilon}$ is about 0.1 μstrain/yr.

The shearing across the zone closely approximates that being observed at strike–slip faults like the San Andreas (e.g., Lisowski et al., 1991), despite the fact that no throughgoing east–west fault exists along the length of the SISZ and the motion is accommodated by right-lateral slip on north–south-oriented faults. In particular, the available geodetic data resemble the surface displacement field associated with a transform fault, locked down to depth D, and continuously sliding below with a velocity V, given as (e.g., Savage and Burford, 1973):

$$v(x) = \frac{V}{\pi}\arctan\frac{x}{D} \tag{7.2}$$

where $v(x)$ is the velocity parallel to the shear zone as a function of perpendicular distance, x, from the center of the zone (Figure 7.6). The GPS data from the SISZ

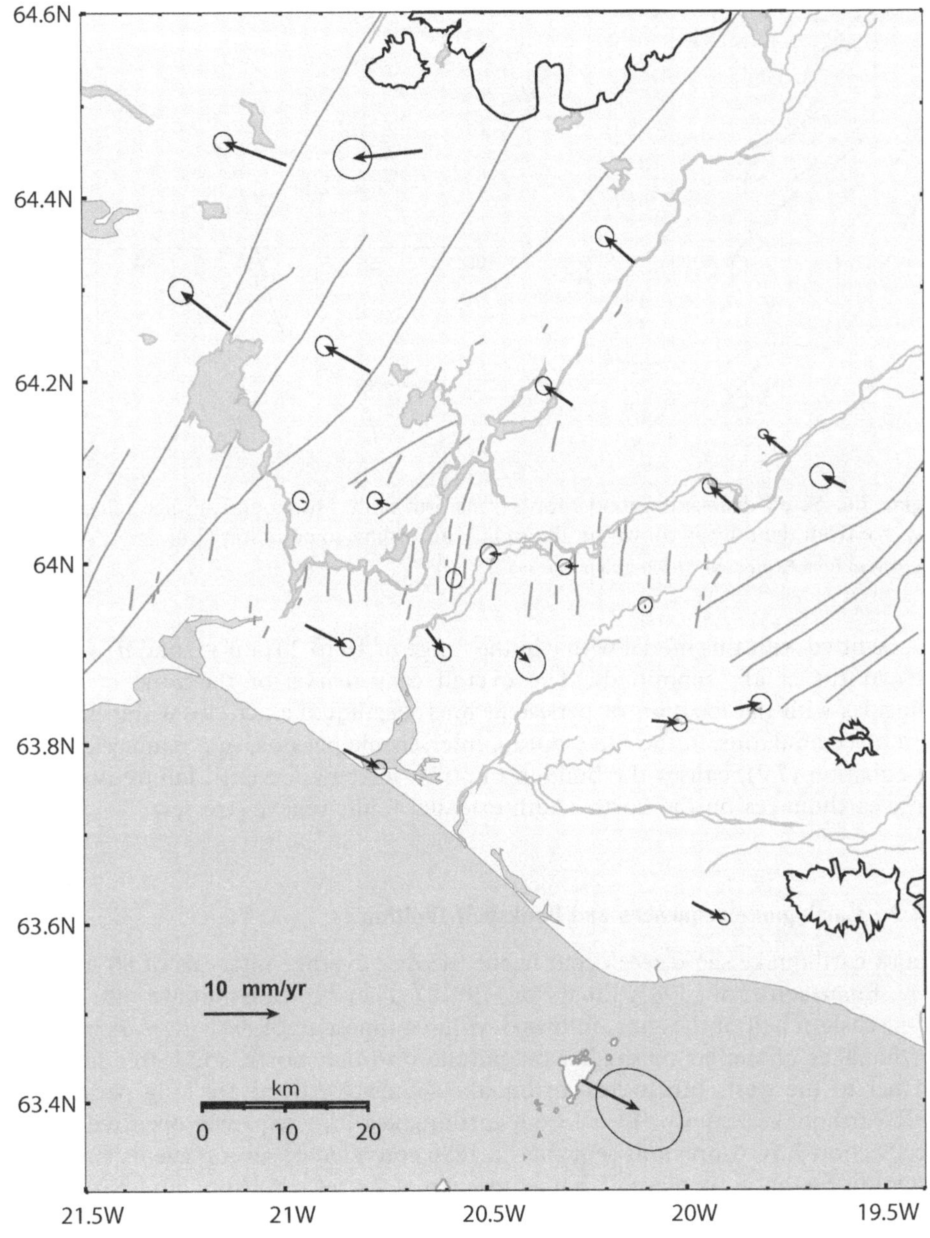

Figure 7.5. Horizontal GPS station velocities 1992–2000 across the South Iceland Seismic Zone with 95% confidence ellipses. The displacements are referred to the REYK reference station in Reykjavík on the North American Plate (outside the map). That station is assumed to move with a velocity of 10.5 mm/yr west and 1.8 mm/yr north, which is half of the plate separation rate across Iceland according to the REVEL model (Sella et al., 2002).

Courtesy of Þóra Árnadóttir, Nordic Volcanological Centre, Institute of Earth Sciences, University of Iceland.

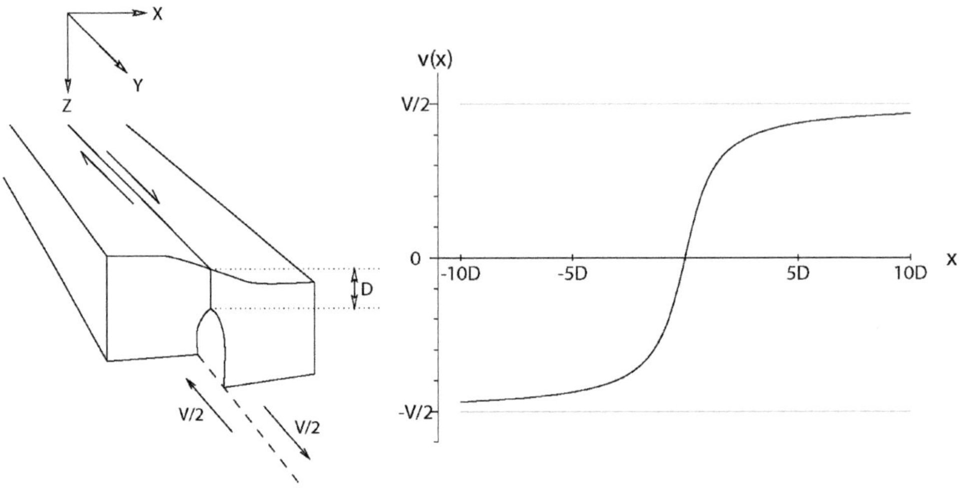

Figure 7.6. Screw dislocation model for a transform fault. Horizontal surface velocity versus distance from the fault is shown on the right – according to equation (7.2).
Reproduced from Hreinsdóttir (1999) with permission.

can be fitted with this model with v in the range of 16 to 20 mm/yr, and $d = 8$–10 km (Árnadóttir et al., submitted). The overall central axis of the zone of shearing coincides with the location of persistent microearthquake activity within the SISZ. Strain accumulation in the SISZ during interseismic periods, in a manner described by equation (7.2), causes the buildup of stress in the zone until failure occurs and large earthquakes on the north–south-oriented faults relieve stresses.

7.2.3 Earthquake sequences and bookshelf faulting

Major earthquake sequences occur in the SISZ at average intervals of 80–150 years (e.g., Einarsson et al., 1981; Einarsson, 1991a). They typically initiate with an event in the eastern half of the zone, followed within a timespan of days or years by several earthquakes of similar or smaller magnitude on other north–south-trending faults further to the west. Inbetween earthquake sequences, there are long periods with little earthquake activity. Two M_s6.6 earthquakes in a sequence occurred in 2000 (see Section 7.3). A previous sequence in 1896 consisted of several events larger than magnitude 6 that took place within a timespan of 2 weeks. A list of large earthquakes in Iceland is found on the homepage of the Icelandic Meteorological Office (*www.vedur.is*).

The style of observed faulting and seismicity patterns in the SISZ, together with measured left-lateral strain accumulation across the zone, show that it can be described kinematically as a zone of bookshelf faulting, where shear deformation is taken up by an array of faults trending perpendicular to the shear direction (e.g., Sigmundsson et al., 1995). In the bookshelf-faulting model (Figure 7.7) slip

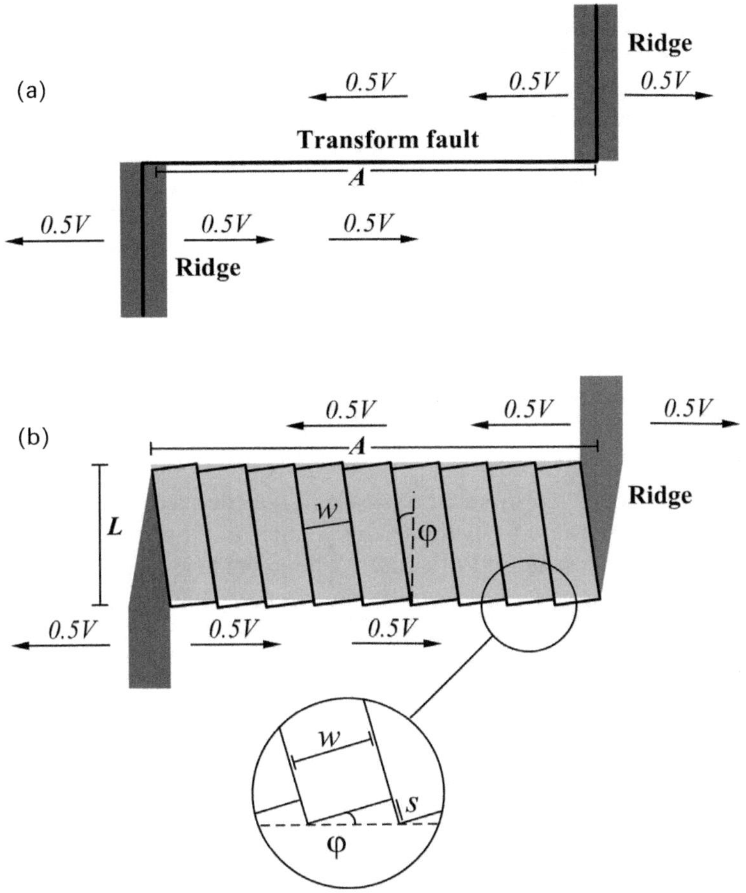

Figure 7.7. (a) A simple transform fault and (b) a bookshelf transform zone. See text for discussion.

on each fault in the array will depend on the overall plate velocity accommodated by the zone, V, the width of the zone, L, and the distance between the faults, ω. The crustal blocks between the faults will gradually rotate (like books in a bookshelf that are pushed from the side) at a rate of:

$$\dot{\varphi} = \tan^{-1}\frac{V}{L} \approx \frac{V}{L} \tag{7.3}$$

In South Iceland strain accumulation is focused on a zone about 40 km wide, with V about 10 mm/yr across this width. The rotation rate is therefore about 0.25 μrad/yr. If the plate motion is fully accommodated by bookshelf faulting on an array of north–south faults within the transform zone, the resulting average slip rate, $\dot{s}$, on each of the faults is:

$$\dot{s} = \omega \tan \dot{\varphi} \approx \omega\dot{\varphi} = V\frac{\omega}{L} \tag{7.4}$$

The distance between the north–south faults in South Iceland varies from 0.5 to 5 km, averaging about 2 km (Clifton and Einarsson, 2005). The average slip rate on each north–south fault in the SISZ is therefore on the order of 0.5 mm/yr. When the faults fail, M6–7 earthquakes with average slip on the order of 1 m occur. This yields an expected recurrence time for earthquakes on each of the faults of ~2,000 years. Within Postglacial times in Iceland (10,000 years), only few earthquakes would be expected on each of the faults.

The expected earthquake activity in a zone of bookshelf faulting can be compared with that of a simple transform fault by evaluating seismic moment release in each of the two cases. The seismic moment, M_0, is:

$$M_0 = \mu M'_0 \tag{7.5}$$

where M'_0 is the geometric moment release and μ is the shear modulus. For kinematic analysis, it is appropriate to evaluate the geometric moment release, assuming the shear modulus is the same in both cases. The geometric moment of an earthquake equals:

$$M'_0 = \bar{u}X \tag{7.6}$$

where $\bar{u}$ is the mean slip and X is the fault area. For a simple transform fault $X = AD$ where A is the length of the fault and D is the thickness of the brittle crust (the seismogenic layer). The long-term mean slip equals the applied plate velocity, $\bar{u} = V$. In this case the average rate of geometric moment release is:

$$\text{Transform fault} \quad \dot{M}_0 = V \cdot A \cdot D \tag{7.7}$$

In the case of bookshelf faulting the average rate of geometric moment release is found by summing up contributions from all the faults in the array accommodating the motion. We have:

$$\text{Bookshelf faulting} \quad \dot{M}_0 = LD\sum_{i=1}^{N} \dot{s}_i \tag{7.8}$$

where $\dot{s}_i$ is the slip rate on each of the faults and N is their number. The number of faults is $\approx A/\omega$, the length of the zone of the bookshelf faulting divided by the distance between faults. Inserting the slip rate from equation (7.4), it is indeed found that a similar moment release is to be expected in a zone of bookshelf faulting, as in a shear zone where stress is released by a simple transform fault:

$$\text{Bookshelf faulting} \quad \dot{M}_0 \approx LD\frac{A}{\omega}V\frac{\omega}{L} \tag{7.9}$$

and equation (7.7) is reproduced. This assumes that moment release associated with activity at the northern and southern ends of the blocks between the faults in the array is small compared with moment released on the main faults.

For South Iceland, $D = 10$–15 km and $A = 75$–85 km are appropriate values. $V = 16$–20 mm/yr is suggested from geodetic data, consistent with accommodation

80–100% of full plate velocities across the SISZ. The expected rate of geometric moment release is then $\sim 1.0\text{–}2.5 \times 10^7\ \mathrm{m^3/yr}$. This can be compared with moment release inferred to have happened in earthquake sequences during historical times in South Iceland. Based on the historical earthquake activity, Hackman et al. (1990) estimated $\dot{M}'_0 = 2.3 \times 10^7\ \mathrm{m^3/yr}$ for the period 1620–1912. A slightly different approach led Stefánsson and Halldórsson (1988) to conclude that moment release in a 140-yr period in South Iceland was 9.8×10^{26} dyn cm, corresponding to $2 \times 10^7\ \mathrm{m^3/yr}$. These two estimates agree, and correspond well to expected geometric moment release if the SISZ has accommodated the majority of the relative plate motion between the Eurasian and American Plates in historical times.

7.3 THE 2000 EARTHQUAKE SEQUENCE

In the year 2000, the celebration of Iceland's national day, June 17, was interrupted in the afternoon by a devastating widely felt M_s6.6 earthquake in the SISZ. It was the largest earthquake to occur in Iceland since 1912 and was the beginning of a sequence of events that followed a pattern similar to previous historic activity in the SISZ. The earthquake broke a north–south-oriented fault in the east–central part of the SISZ, and more earthquakes of similar magnitude were expected further to the west. On June 21 a second north–south fault failed, producing another M_s6.6 earthquake (Figure 7.8) that was also widely felt. Both events caused considerable damage (Figure 7.9) but no casualties. The events had been expected. An earthquake forecast given in 1985, based on the historical earthquake record, stated that there was more than 80% probability of a major earthquake sequence in the SISZ in the next 25 years (Einarsson, 1985). The actual location of the initial earthquake in a new earthquake sequence in South Iceland had also been forecast on the assumption that it would fill in a seismic gap. Historical strain release plotted as a function of longitude for events prior to 2000 show a marked minimum where the initial event in June 2000 occurred. Stefánsson and Halldórsson (1988) concluded that "there are strong indications that the next large earthquake of a size approaching 7 in this zone will take place near longitude 20.3–20.4°W."

The main shock on June 17, 2000, occurred at 15:40:41 GMT with a hypocentre at 63.975°N, 20.370°W, and an estimated depth of 6.3 km (Stefánsson et al., 2000). It was a right-lateral strike–slip earthquake with a minor component of dip slip, well detected by the global network of seismic stations. Extensive microearthquake activity followed the main shock on a north–south-oriented fault plane. The focal mechanism estimated by the United States Geological Survey has a fault plane corresponding to the aftershock activity with a strike of N5°E, dip of 83° towards east, and a rake of 175°. Aftershocks occurred at and adjacent to the fault plane, and triggered seismicity over a large area in South Iceland, mostly to the west of the main shock, up to a distance of 100 km. The triggered activity included three $M \geq 5$ events on the Reykjanes Peninsula (e.g., Árnadóttir et al., 2003). The aftershocks appear to have been triggered both by static stress increase and by dynamic stress associated with waves from the main shock. Aftershocks on the fault plane of the main event

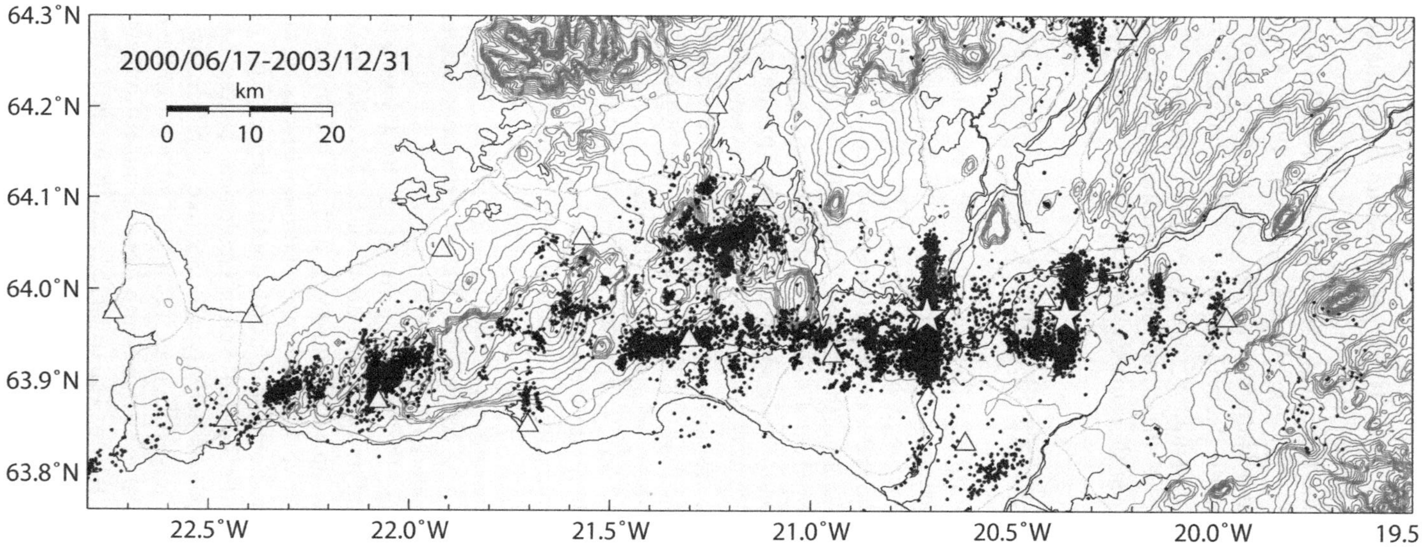

Figure 7.8. Earthquakes in South Iceland from June 17 to December 31, 2000, recorded by the South Iceland Lowland seismic network of the Icelandic Meteorological Office. Stars mark the epicentres of the June 17 (right) and June 21 (left) main shocks. Triangles show seismic stations. Aftershocks occur on the two fault planes, and smaller earthquakes line up as well along a number of other north–south faults. Modified from Stefánsson and Guðmundsson (2005). Courtesy of Gunnar Guðmundsson.

Figure 7.9. Surface rupture and damage from the June 21, 2000, earthquake. The open fracture is on an east–west segment near the centre of the surface rupture (see Figure 7.9). The maximum width of the fracture is about 2.3 m.

occurred down to a depth of 10 km along a 16-km-long plane (Stefánsson et al., 2000). The triggered activity to the west of the June 17 main shock focused on a number of north–south-striking faults. A second large earthquake occurred on June 21 at 00 : 51 : 47 on a north–south fault 17 km west of the main shock, at a depth of 5.1 km. It was of a very similar character to the June 17 main shock, with a strike–slip focal mechanism. Aftershocks outline an 18-km-long north–south-oriented fault plane extending to an 8-km depth. Triggered activity was not as pronounced as on June 17.

Both of the main shocks on June 17 and 21 caused considerable ground rupture, mapped by Clifton and Einarsson (2005). Surface faulting was observed along a 15–20-km length of each of the faults, consisting mostly of open fractures at different scales, arranged *en echelon*, with some pushups inbetween. Surface rupture on June 17 consisted of several segments, occurring in a north-northwest-directed zone up to 3 km wide. Surface rupture associated with the June 21 earthquake is more complex and includes a 2.5-km-long east–west segment near the middle of the fault (Figure 7.10). Some of the fine details of groundbreaks have been studied, including fracture pattern in an asphalted car park along the east–west segment near the middle of the June 21 fault, showing left-lateral strike–slip displacement along that segment (Angelier and Bergerat, 2002). Other segments of the June 17 and 21 faults were associated with right-lateral strike–slip movement. The June 2000 earthquakes were associated with extensive rockfalls and slope failures over wide areas in South Iceland, with the largest landslide occurring at the southern termination of the June 21 earthquake. Rockfalls occurred, for example, south of the seismic zone at the Westman Islands off the south coast,

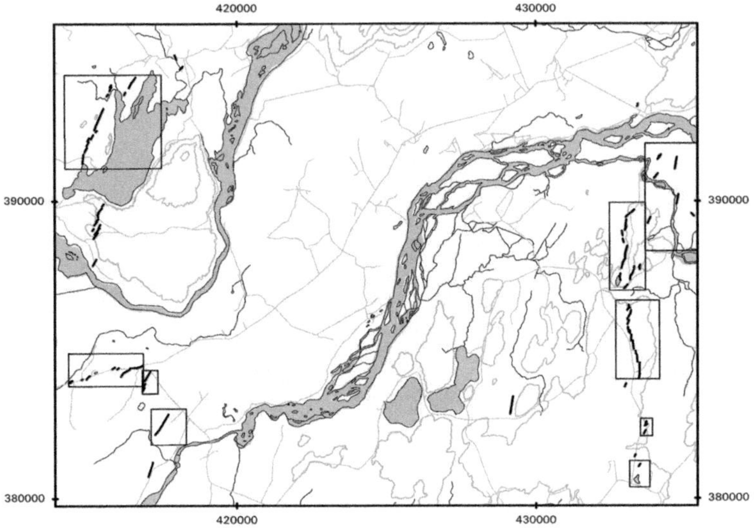

Figure 7.10. Map of the June 17 and June 21 earthquake areas. Bold lines mark surface rupture.

Modified from Clifton and Einarsson (2005) with permission from Elsevier.

and in many locations along the Reykjanes Peninsula, partly due to triggered earthquakes.

The co-seismic deformation field associated with the main shocks on June 17 and 21 was captured by SAR interferometry (InSAR) and GPS-geodetic measurements (Figure 7.11, see colour plates). Fortunately the ERS-2 satellite acquired an image of South Iceland on June 19, between the earthquakes, that could be processed interferometrically with an image acquired after the June 21 event. A series of interferograms spanning the earthquakes allowed resolution of the deformation produced by each of the main shocks.. The earthquakes were not well recorded by continuous GPS, as no stations were operating in the epicentral areas at the time of the earthquakes. The nearest station (VOGS) was at a distance of ~65 km from the June 17 event. It recorded co-seismic horizontal displacement of 19 mm towards the east and 10 mm towards the south, but the signal is also affected by activity along the Reykjanes Peninsula (Árnadóttir et al., 2004a). Campaign GPS data covering the earthquakes comes from a network of stations in South Iceland measured repeatedly in the years preceding the earthquakes. The few days between the earthquakes only allowed reoccupation of a few sites, but the complete network was remeasured after the June 21 event. The InSAR (Pedersen et al., 2001) and GPS data (Árnadóttir et al., 2001) spanning the earthquakes were initially evaluated independently to infer the co-seismic deformation. Agreement between the two datasets is good. A joint inversion of these data was also carried out by Pedersen et al. (2003), solving for distributed slip on the faults (Figure 7.12, see colour plates). Both faults are characterized by maximum slip in the uppermost 6 km of the crust, with model patches having up to 2.6 m of slip on the June 17 plane, and 2.9 m on the June 21 plane. The inferred co-seismic slip on the June 17 fault occurs along a 15-km-long plane, down to a 10-km depth. It has a total geodetic moment of 4.5×10^{18} Nm, corresponding to an earthquake of size $M_w6.4$. For the June 21 event, the total geodetic moment estimate is 4.5×10^{18} Nm, giving $M_w6.5$. Moment estimates from the geodetic data for the main shocks on June 17 and 21 are therefore in good agreement with the seismological estimates, and are also consistent with the distribution of aftershocks.

7.3.1 Hydrological signatures of earthquake strain

An immediate consequence of both the June 17 and 21 main shocks were major changes in groundwater level and pressure (Björnsson et al., 2001). Groundwater level rose as a result of increased water reservoir pressure in some areas, and dropped in other areas where water pressure was lowered. The changes occurred in a systematic pattern. For each of the main shocks, increase was observed in wells in two quadrants, and decrease in two quadrants relative to the fault plane. In both cases, the meeting point of these quadrants was at the earthquake epicentre (Figure 7.13). The wells of co-seismic increase in pressure correspond to areas compressed by the fault slip, and co-seismic lowering of pressure to areas dilated by the fault slip. The changes demonstrate directly the focal mechanism of the earthquakes. The changes were so large that many wells in compressive quadrants became artesian with water flowing out of the ground, and some productive wells in the areas of dilation dried out. Pressure changes occurred at least out to 75 km from the epicentres. The pressure changes were typically in the range of 0.1–1 bar (0.01–0.1 MPa), but may have exceeded 10 bar in a few cases (Björnsson et al., 2001). The co-seismic pressure offset appeared to be followed by 2–4 hours of additional pressure change, but then recovery to pre-seismic conditions began. In some cases, there were permanent changes to production wells, in most cases leading to an increase in their performance. The recovery of groundwater pre-seismic equilibrium conditions was achieved by flow of groundwater from areas of elevated pore pressures to areas of lower pressure. Exponential recovery of the earthquake-induced water level changes was typically observed, with new equilibrium being approached within 1–2 months (Björnsson et al., 2001; Jónsson et al., 2003).

Observations of hydrological signatures of earthquake strain have been evaluated by, for example, Muir-Wood and King (1993), who show that rise and decay time of water-level perturbations are critically dependent on the width of water-filled cracks in the crust.

7.3.2 Triggering of earthquakes

The initial earthquake on June 17 triggered widespread seismic activity in South Iceland (Figure 7.8). Coulomb failure calculations by Árnadóttir et al. (2003, 2004a) reveal that triggered earthquake activity occurred in areas of increased static Coulomb failure stress, ΔCFS, evaluated as:

$$\Delta CFS = \Delta\tau_s + \mu\left(\Delta\sigma_n - \frac{B}{3}\Delta\sigma_{kk}\right) \tag{7.10}$$

where $\Delta\tau_s$ is the change in shear stress resolved in the slip direction of a fault, $\Delta\sigma_n$ is the change in normal stress due to the first earthquake, $\Delta\sigma_{kk}$ is the volumetric stress,

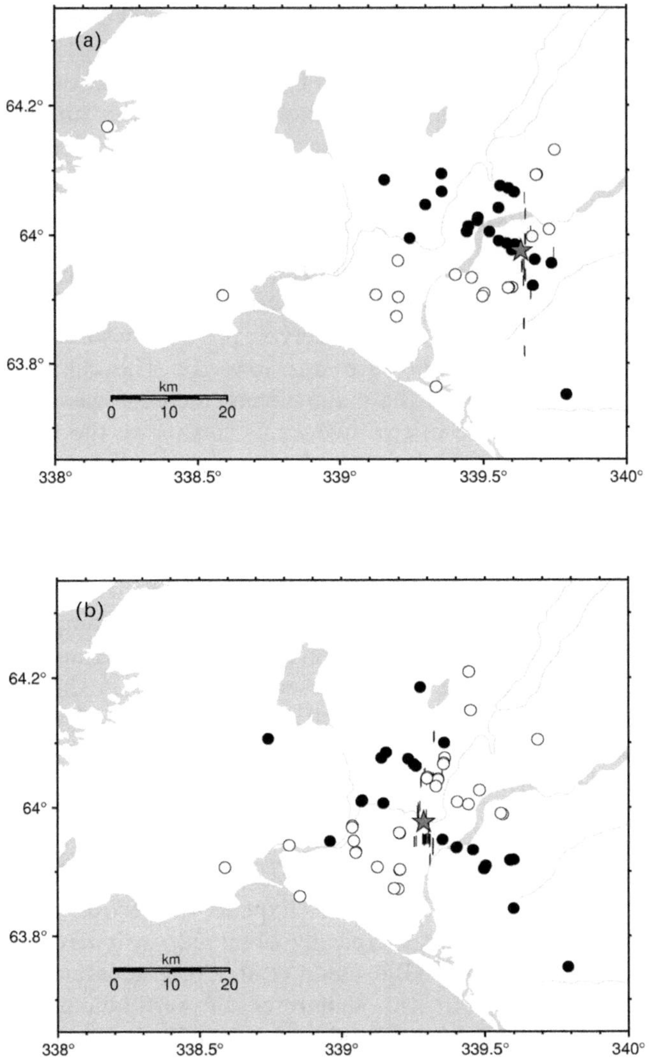

Figure 7.13. Water level change associated with June 17 and June 21 earthquakes, observed by the Iceland National Energy Authority. Wells of increased (bullets) and decreased pressure (circles) following the June 17 earthquake (a) and after the June 21 earthquake (b). Earthquake main shocks are indicated by stars.

Reproduced from Björnsson et al. (2001).

and B is the Skempton coefficient of the rock–fluid mixture. Calculation shows that the initial event on June 17 increased stress up to 0.2 MPa in the area of the June 21, 2000, earthquake (Figure 7.14, see colour plates).

The widespread seismic activity on June 17 extended much further to the west, to the Reykjanes Peninsula, than the static changes in ΔCFS. Activity there included three $M \geq 5$ events within 5 minutes of the main shock. Dynamic triggering of earthquakes by the passing of seismic waves from the main shock is suggested.

7.4 ASEISMIC SLIP: SLOW EARTHQUAKE AT KLEIFARVATN?

For the largest events of the June 2000 earthquake sequence in South Iceland there is a good correspondence between seismic and geodetic moment estimates, indicating insignificant aseismic slip for most of the events. There is one eventual exception, an earthquake at Kleifarvatn on the Reykjanes Peninsula that occurred on June 17, 30 seconds after the main shock (e.g., Clifton et al., 2003). In fact, this event was not recognized as a separate earthquake until its deformation was revealed by InSAR observations (Figure 7.15, see colour plates). The estimated geodetic moment of this event based on InSAR is 6.2×10^{17} N m, corresponding to an M_w5.8 event (Pagli et al., 2003a). Despite its size, the event does not appear in worldwide seismicity catalogues. Inclusion of GPS data in the inversion, as well as InSAR, results in geodetic moment estimate for the Kleifarvatn Event of $6.8–7.1 \times 10^{17}$ N m (Árnadóttir et al., 2004a). Seismic analyses of the Kleifarvatn Event are complicated by the fact that its waveforms are hidden in the wavetrain of the main shock on June 17, but initial values for its seismically determined moment were lower than from the geodetic inversion. Árnadóttir et al. (2004a) conclude that the Kleifarvatn Event had a significantly larger geodetic moment than seismic estimates, indicating a component of aseismic slip in this event.

Surface fractures were observed on part of the Kleifarvatn Fault, as well as surface disruption due to shaking. Shattered ground surface and moved boulders are indicative of acceleration, in agreement with part of the energy in the Kleifarvatn Event being released seismically (Clifton et al., 2003). The event at Kleifarvatn was associated with a dramatic 4-m drop in the water level of Lake Kleifarvatn which began on June 17, 2000, and continued gradually for a period of at least 4 months (Clifton et al., 2003). Water was observed flowing into a set of fractures which opened up on the lake bottom.

7.5 POST-SEISMIC DEFORMATION

The earthquake sequence in June 2000 provided the first opportunity to study post-seismic deformation in Iceland. Post-seismic deformation is modest but clearly detected in both InSAR and GPS data. Analysis of the deformation data suggests that more than one process was responsible for it, working on two different spatio-temporal scales (Árnadóttir et al., 2004b; in press). One of the processes is poroelastic (elastic material with pores) rebound associated with groundwater flow, due to pore–fluid flow in response to main-shock-induced pore pressure changes (Jónsson et al., 2003). The other relates to ductile flow in the lower crust and/or afterslip below the mainshock rupture.

InSAR data show post-seismic changes on the timescale of weeks, with amplitude up to ~5 cm, well observed around the fault that broke on June 17. Observed changes reveal a four-lobed deformation field, where changes in range from ground to satellite are positive in two lobes and negative in two lobes (Figure 7.16, see colour plates). The changes are opposite to those that occurred during the co-seismic right-lateral slip, but their amplitude is only a few percent of the co-seismic changes. The spatial extent, as well as the temporal decay of these changes, can be well explained as a result of deformation induced by flow of groundwater in the crust after the earthquakes (Jónsson et al., 2003). The evaluation of deformation associated with such flow requires consideration of the equation of motion for poroelastic material. Such material with water-filled pores is a good approximation to the water-saturated crust of the Earth. The initial and final response of such a material to an earthquake is different, as in the initial co-seismic response the water has no time to flow. Pore pressures are perturbed, increasing in volumes of crust compressed by the earthquake and decreasing in volumes that are dilated. The pore pressure gradients will induce groundwater flow and additional time-dependent strain (Jónsson et al., 2003). Consideration of the equations of motion show that the full poroelastic response can be estimated by calculating elastic deformation models with different effective Poisson's ratio, one being the initial "undrained" value, and the other the final "drained" value. The difference between the deformation fields calculated in this way is the expected poroelastic deformation signal.

Post-seismic deformation on longer timescales is evaluated by Árnadóttir et al. (in press), who invoke either afterslip or viscoelastic relaxation of the lower crust to explain small transient deformation seen in repeated GPS measurements after the 2000 earthquakes. Their optimal viscoelastic models have lower crustal viscosity of about 0.5–1×10^{19} Pa s and an upper mantle viscosity of about 3×10^{18} Pa s.

7.6 EARTHQUAKE PREDICTION RESEARCH

A number of projects related to earthquake prediction research have been carried out in Iceland. Some of the initial work included studies of radon anomalies. The Nordic South Iceland Lowland (SIL) project then initiated in 1988 (Stefánsson et al., 1993), with a major focus of seismological research on the SISZ. Some European-funded collaborative research projects have then been led by the Icelandic Meteorological Office relating to earthquake prediction research. In addition to efforts based on seismological methods, a number of other approaches have been used, including studies of strain anomalies, continued studies of radon changes, changes in water geochemistry, and changes in pore pressure preceding earthquakes. The aim has been to estimate time, location, and magnitude of a forthcoming earthquake.

Hydrogeochemical changes in relation to earthquake activity have also been studied both in the TFZ (Claesson, 2004; Claesson et al., 2004). Prior to and after an M5.8 earthquake in the TFZ on September 16, 2002, hydrogeochemical changes have been documented, including significant Cu, Zn, Mn, and Cr anomalies. The changes are interpreted to be caused by switching from younger to older ice age meteoric aquifers, possibly due to fault sealing and unsealing mechanisms.

8

Glacial isostasy and sea-level change: Rapid vertical movements and changes in volcanic production rates

Glacial rebound in Iceland at the end of the last glaciation around 10000 ^{14}C yr BP was exceptionally fast, having been completed in most coastal areas in about 1,000 years after the final ice retreat. Uplift rates may have exceeded 10 cm/yr. During glacial conditions, crustal subsidence under central Iceland may have been up to 500 m, under an icecap with maximum thickness of around 2 km. The observed rapid Postglacial rebound suggests a viscosity under Iceland of 10^{19} Pa s or less. A unique feature associated with the deglaciation of Iceland is a pulse in volcanic productivity associated with the ice unloading, which lasted for 1,000–2,000 years after the unloading. Volcanic production was up to 30 times higher than today's rate. Low viscosity under Iceland allows a rapid response to changes in the mass of current icecaps, with uplift ongoing in response to their recent thinning. The glaciers have retreated significantly after reaching a historic maximum in 1890 at the end of the Little Ice Age in Iceland. Inferring the Earth's response to recent and ongoing fluctuations in the extent of current icecaps in Iceland provides an important experiment in rheology. Available data and interpretation of this process suggest viscosity under Iceland is close to 10^{18}–10^{19} Pa s. Similar Earth response to ice unloading after the Little Ice Age has been documented in southern Alaska by Larsen et al. (2004). They find uplift rates as high as 2.5 cm/yr with peak amplitude at Glacier Bay, where an icefield has collapsed since the Little Ice Age.

8.1 SEA-LEVEL CHANGE IN ICELAND

Relative sea level in Iceland has varied greatly, having been both much higher and much lower than today. During glaciations, the eustatic fall in sea level contributed to larger exposure of Iceland's landmass above sea level than exists today. This facilitated the buildup of an ice sheet of larger dimensions than the current size of the country. Exploration of the ocean floor around Iceland has revealed evidence for glacial cover at the last glacial maximum extending much farther than the current coastal area (e.g., Andrews et al., 2000). New multibeam bathymetric and chirp sonar data from the northern insular margin have revealed structures and landforms interpreted as due to extensive glacial erosion on the submarine Kolbeinsey Ridge north of Iceland extending beyond 67°N. Observed features include structures now at a 400–500-m depth within a U-shaped valley that have initially been interpreted as multiple marginal moraines (Brandsdóttir, pers. commun., 2005; Helgadóttir et al., 2003).

There is also ample evidence for much higher relative sea level in the past than today. Marine sediments from the end of the last glaciation are found in lowlands all around Iceland, up to an elevation of over 135 m in western Iceland (Ingólfsson and Norddahl, 2001).

The Relative Sea Level (RSL) curve is best known from the end of the last glaciation. Two competing factors govern the RSL curve at each location: the change in land elevation due to glacial isostasy, and the eustatic changes in sea level due to variable volume of oceanwater. An example of an RSL curve from North Iceland is provided by results from Thors and Boulton (1991), who infer a low stand in RSL around 9000 ^{14}C yr BP (Figure 8.1). In southwestern Iceland, in the Reykjavík area, a similar RSL curve (Figure 8.2) has been inferred (e.g., Ingólfsson et al. 1995). A low stand in RSL around 9000 ^{14}C yr BP suggests that glacial rebound was completed around that time. Since then, transgression has prevailed, at least partly due to increasing eustatic sea-level rise.

Factors other than glacial isostasy which contribute to changes in sea level in Iceland include tectonic processes (e.g., Einarsson, 1994). Between rifting events, the rift zones subside in response to the plate spreading. Long-term subsidence is also expected as plate movements carry the crust out of the rift zones and it cools down. The age–depth relationship of the oceanic lithosphere of Parsons and Sclater (1977)

$$d = 2{,}500 + 350\sqrt{t} \tag{8.1}$$

where d is depth in metres and t is age in millions of years, suggests that considerable subsidence can take place due to cooling and thermal contraction. Although this general relationship is offset in Iceland because of excessive heat (the island being up to 2 km above sea level), thermal contraction can be expected to lead to a similar rate of subsidence. The cumulative subsidence in oceanic lithosphere from time $t = 0$ until $t = 1$ Myr (corresponding to conditions near the rift axis in Iceland) is 350 m according to equation (8.1), averaging at a subsidence rate of 0.35 mm/yr.

The longest time series of instrumentally recorded RSL change in Iceland comes from a hydrograph in Reykjavík Harbour (Figure 8.3). Analysis of data from this hydrograph 1956–1989 suggest RSL rise in Reykjavík of 3.4 mm/yr, with considerable yearly fluctuations (Einarsson, 1994). This is similar to the inferred geocentric average rate of global mean sea-level rise of 2.8 ± 0.4 mm/yr in 1993–2003 (Cazenave and Nerem, 2004). However, continuous Global Positioning System (GPS) measurements analysed in a global reference frame suggest that most of the relative increase in sea level in Reykjavík may be caused by land subsidence. When analysed in a global reference frame, the REYK GPS station in

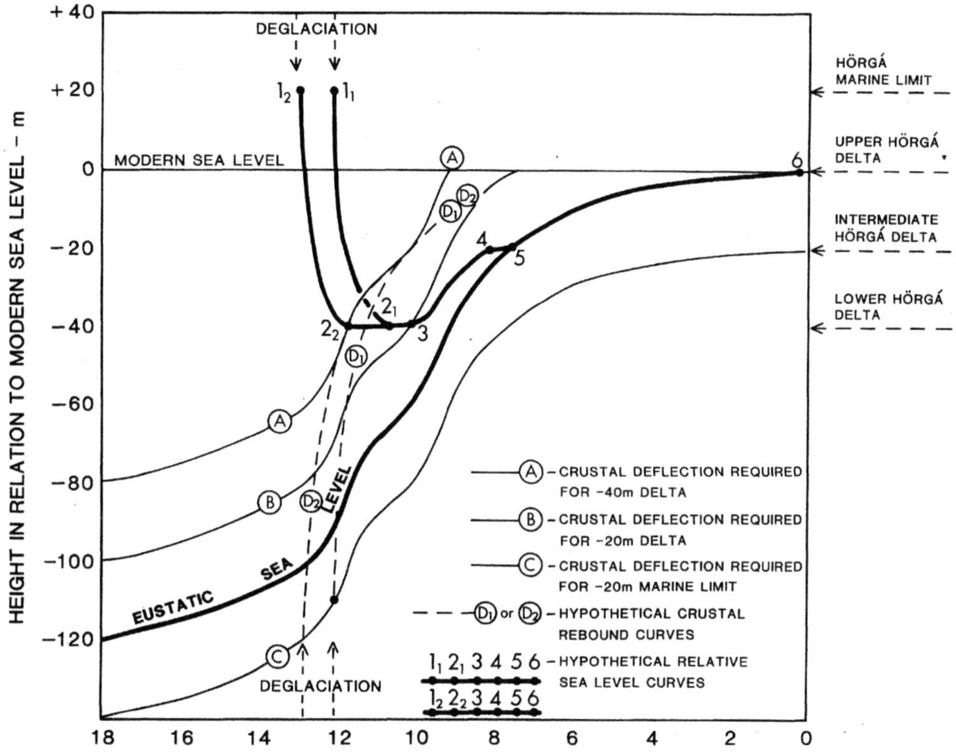

Figure 8.1. Relation between eustatic and isostatic component of sea level and relative sea-level change observed at Hörgá, Eyjafjörður, northern Iceland. Submerged deltas formed at lower relative sea level provide constraints on the relative sea-level curve, that is a combination of eustatic sea-level change (curve from Fairbanks, 1989) and crustal rebound.

Reproduced from Thors and Boulton (1991) with permission of Elsevier.

Reykjavík subsides 3 mm/yr (Sella et al., 2002). The analysis in the preceding section suggests that thermal contribution explains only a fraction of this observed subsidence, and other processes are needed to explain this relatively high rate of subsidence.

8.2 POSTGLACIAL REBOUND IN ICELAND

8.2.1 The glacial history

Studies of glacial deposits reveal that during the last glacial maximum, around 18,000 ^{14}C yr BP, Iceland was fully glaciated, with an ice front reaching well out on the current insular shelf (e.g., Andrews et al., 2000). Thereafter, the ice started to retreat and had mostly disappeared from coastal areas by around 9,000–10,000 ^{14}C yr BP. The retreat during this time period was not uniform, but rather was

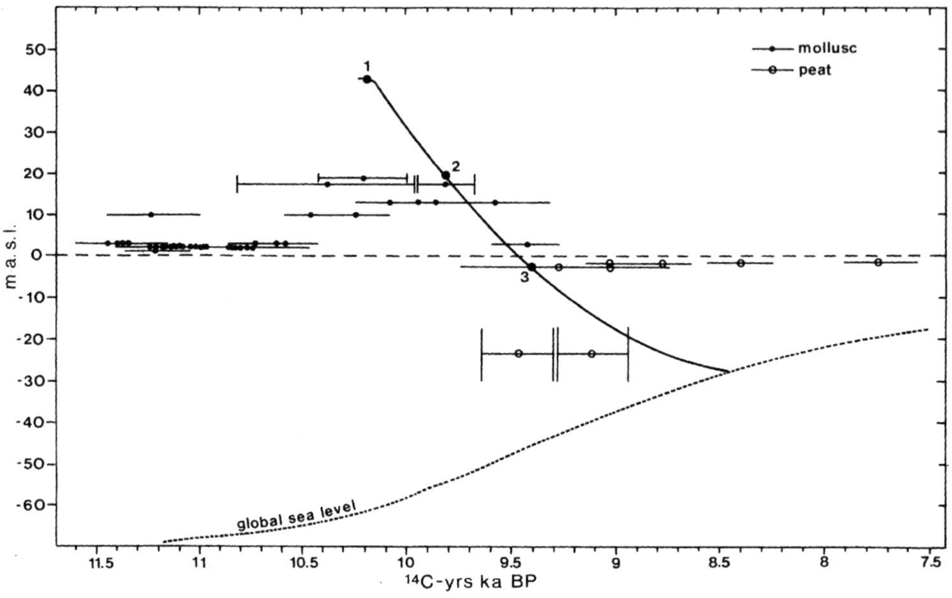

Figure 8.2. Tentative curve for relative sea-level curve in the Faxaflói area, southwestern Iceland.

Reproduced from Ingólfsson et al. (1995) with permission of Taylor & Francis.

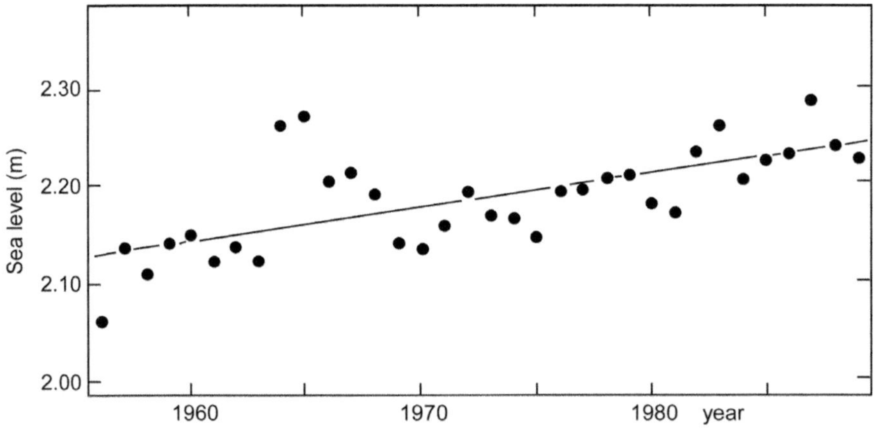

Figure 8.3. Sea level in Reykjavík 1956–1989 from tide gauge. Yearly averages are plotted as a function of time.

Reproduced from Einarsson (1994).

interrupted by ice advances during cold periods. The climatic fluctuations are well recorded in isotopic data from Greenland ice cores (Figure 8.4). The geographical proximity of Iceland and Greenland results in close correlation of their climate change. The last major advance during the final part of the last glaciation

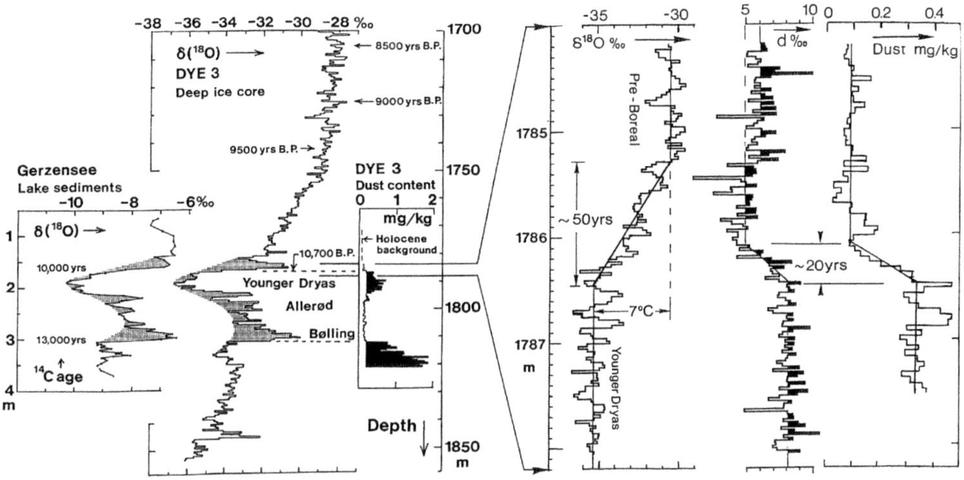

Figure 8.4. Climate constraints from ice cores and sediments. (a) Radiocarbon-dated $\delta^{18}O$ profile along a 4-m-long lake core from Switzerland. (b) $\delta^{18}O$ profile along 150-m-deep ice core from Dye 3 in southern Greenland. (c) Continental dust changes in antiphase with the $\delta^{18}O$. (d) Detailed $\delta^{18}O$ record through the Younger Dryas–Preboreal transition, during which the South Greenland temperature increased by 7°C in about 50 years. (e) Deuterium excess. (f) Dust concentration.

Reproduced from Dansgaard et al. (1989) with permission of *Nature*, London. See also Sveinbjörnsdóttir and Johnsen (1990).

occurred during the Younger Dryas period, from 10000 to 11000 ^{14}C yr BP, or about 11500–12650 cal yr BP (e.g., Alley, 2000). Prior to Younger Dryas, the most significant period of warm climate was the Bolling/Allerod period from 11000 to 13000 ^{14}C yr BP.

Geological evidence of isostatic rebound after the Younger Dryas suggests that it was completed in about 1,000 years. That length of time is equal to the duration of the Younger Dryas glaciation, and subsidence during this period is therefore likely to have reached equilibrium values. As a consequence, the isostatic rebound after the Younger Dryas therefore depends mainly on the glacial history during that 1,000-yr cold period, but is little influenced by the earlier glacial conditions. Therefore glacial conditions during the Younger Dryas are of prime importance for the modelling of rebound that took place after it.

The extent of glacial coverage in Iceland during the Younger Dryas is uncertain, and different ice models have been suggested. In particular, different interpretations have been given to an extensive terminal moraine complex in South Iceland, the Búði terminal moraine complex. Hjartarson and Ingólfsson (1988) argue that it is of Preboreal age (around 9700 ^{14}C yr BP), and that during the Younger Dryas the ice front was outside the current coastline, with most of Iceland covered by a glacier. On the other hand, Geirsdottir et al. (1997, 2000) suggest that the Búði moraine complex marks the advance of both the Preboreal as well as the Younger

Dryas Icecap. Sediment studies at Lake Hestvatn in the South Iceland Lowland are providing important constraints on the deglaciation history. The Vedde Tephra (11,890 cal yr) has been found within the marine unit of its southern basin, and it is suggested that an outlet glacier from the Younger Dryas Icecap occupied the northern basin of Hestvatn while the southern one was still submerged by seawater (Áslaug Geirsdóttir, pers. commun., 2005; Geirsdóttir et al., 2000; Hardardottir et al., 2001). In eastern Iceland, Norddahl and Einarsson (2001) argue that Younger Dryas glaciation extended to the current coastal areas. In the area near Mt. Akrafjall in southwestern Iceland, Magnúsdóttir and Norðdahl (2000) argue from radiocarbon age and stratigraphic position of whalebones and seashells that glaciers retreated inside the present coast more than 12600 ^{14}C yr BP and that the area has not been overridden by glaciers since then.

A simple model for the Younger Dryas glaciation has been used for modelling purposes, consisting of an axisymmetric icecap with a radius of 160 km (see Sigmundsson, 1991). It reconciles the suggestion that large parts of Iceland were covered by the Younger Dryas Icecap. This approximates the situation for South and West Iceland, where geological observations related to the rebound are most abundant. Eventual ice load on the mountainous Reykjanes Peninsula is ignored in this model, but is counterbalanced by an excessive model load on the South Iceland Lowlands, which may have been mostly ice-free. The axisymmetric ice model used for modelling has a centre in the current Hofsjökull Icecap (Figure 8.5). It should be regarded as a crude approximation to the real extent of the Younger Dryas Icecap.

The thickness of the Younger Dryas Icecap can be inferred from models of perfectly plastic icecaps with a basal shear stress of 1 bar, which appear to mimic well the shapes of icecaps (Paterson, 1983). A radial profile of ice thickness takes the form of a parabola. Ice thickness, h, is given by:

$$h = \sqrt{\frac{2\tau_0}{\rho_{ice} g}(R - r)} \tag{8.2}$$

where τ_0 is the shear stress at the bottom of the icecap, R is the radius of the icecap, r is the distance from the icecap centre, and ρ_{ice} is the density of ice. Taking shear stress at the bottom of the icecap as 0.1 MPa (Paterson, 1983), $\rho_{ice} = 920\,\mathrm{kg/m^3}$, and $g = 9.8\,\mathrm{m/s^2}$, equation (8.1) becomes:

$$h = 4.7\sqrt{(R - r)} \tag{8.3}$$

For $R = 160$ km, the maximum thickness of the icecap is 1,880 m. The shape of such an axisymmetric icecap with a parabolic radial profile has been modelled as a series of disks of constant thickness with different radius (Sigmundsson, 1991).

8.2.2 Observations of glacio-isostatic rebound

Marine sediments are common in Iceland's lowlands above the present sea level. Radiocarbon dates of marine shells from these sediments reveal that they are from the end of the Weichselian Glaciation, with most dates falling in the interval 9,400–

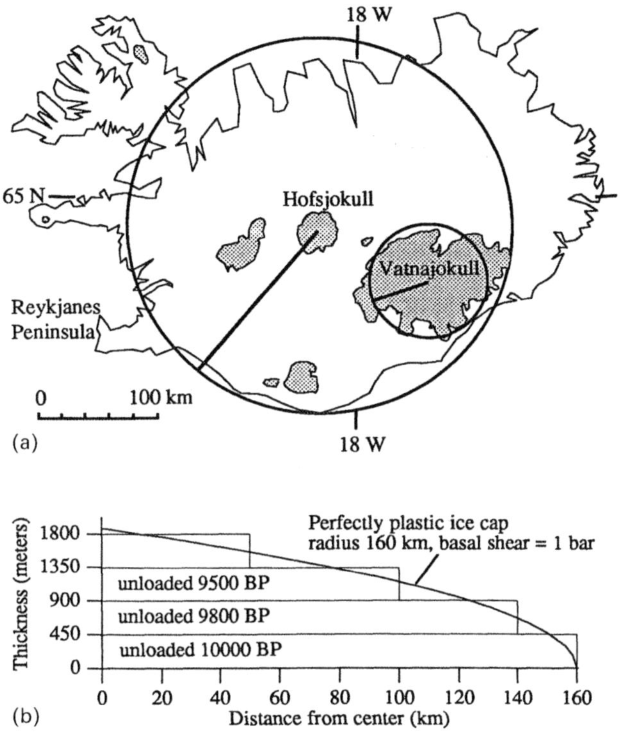

Figure 8.5. Ice model for Postglacial rebound studies. (a) The Younger Dryas icecap is modelled as a circular load with a center in Hofsjökull and radius 160 km. (b) Cross section of the ice model used. The Younger Dryas icecap in Iceland is approximated with four discs unloaded at different times.

12,700 ^{14}C yr BP (e.g., Norðdahl and Pétursson, 2005). This age distribution suggests that only after 12700 ^{14}C yr BP had the ice front retreated inside the current coastline of Iceland. In western Iceland, data from the Bolling interstadial period reveal marine deposits at elevations between 105 and 148 m a.s.l. formed around 12600 BP (Ingólfsson and Norddahl, 2001). In addition to radiocarbon dating of marine shells, a whalebone dated at 12575 ± 80 ^{14}C yr BP (sea-reservoir-corrected age) constrains the age of the associated marine limit shoreline (Magnúsdóttir and Norðdahl, 2000). At this time, eustatic sea level was about 100 m lower than it is today (Fairbanks, 1989). Crustal depression at this time in western Iceland is therefore inferred to have been 250 ± 20 m. More than one set of marine terraces and strandlines marking the marine limit can be identified in many areas of Iceland. In western Iceland, a second set of raised beach deposits from around 10300 ^{14}C yr BP is found 40–80 m below the highest limit (Ingólfsson and Norddahl, 2001). In eastern Iceland, Norddahl and Einarsson (2001) infer a minimum rate of uplift of 7.3 cm ^{14}C yr^{-1} in the period between 10300 and 9900 ^{14}C yr BP, based on inferred ages of two sets of shorelines. One of the best constrained RSL curves in Iceland has been inferred at the Skagi Peninsula, northern Iceland, using opportunities provided by near-coastal series of lakes at different elevations (Figure 8.6). Marine, brackish, and freshwater phases can be identified and dated in sediment cores, allowing reconstruction of the RSL change (Rundgren et al., 1997). The change in the area was rapid between two transgressions (10,000–9,850), with inferred mean absolute crustal uplift rate in this period of about 15 cm ^{14}C yr^{-1}.

The coastal terraces and strandlines associated with higher sea levels, as well as strandlines associated with ice-marginal lakes of the Weichselian Icecap, are often tilted. They reveal increasing uplift towards the centre of Iceland, resulting from an increasing ice load towards central Iceland, as well as flexure of the crust in areas close to the edge of an ice load. Inferred strandline gradients are 2.3 m/km for the uppermost strandline in the Borgarfjörður area (Ingólfsson and Norddahl, 2001). Strandlines formed at an ice-dammed marginal lake formed during the end of the Weicheselian glaciation in Fnjóskadalur, North Iceland (Figure 8.7) are tilted up to 2.6 m/km (Norddahl, 1983). Strandline tilt as high as 8–9 m/km has been measured in the tectonically active South Iceland Seismic Zone (Hjartarson, 1985; Árni Hjartarson, pers. commun., 2005).

The rebound after the disappearance of the Younger Dryas Icecap is a key constraint when inferring the viscosity under Iceland. Marine deposits formed at the end of or after the Younger Dryas period in South Iceland are now up to 100 m a.s.l., and at a distance of over 50 km from the coast (e.g., Norðdahl and Pétursson, 2005). Considering that global sea level (eustatic sea level) at this time was about 60 m lower than it is today (Fairbanks, 1989), the rebound in the South Iceland Lowland may have amounted to ~160 m. Absence of confirmed radiocarbon-dated shells younger than 9000 ^{14}C yr BP in Iceland suggest that all current coastal areas had risen from the sea by that time. Furthermore, RSL may have been considerably lower around that time.

In the Reykjavík area, Ingólfsson et al. (1995) infer a change in RSL from +43 m a.s.l. to at least −2 m over a period of 900 ^{14}C years from 10300–9400 ^{14}C yr BP. Their RSL curve is based on radiocarbon-dated shells in raised marine deposits as well as tephrostratigraphically controlled and radiocarbon-dated submerged peat deposits (Figure 8.2). However, the curve is uncertain and the +43 m a.s.l. raised beach deposits may be older. Shells near these deposits in the Fossvogur marine sediments are dated around 11000 ^{14}C yr BP (Sveinbjörnsdóttir et al., 1993; Geirsdóttir and Eiríksson, 1994). RSL could also have been significantly lower than −2 m at 9400 ^{14}C yr BP. Constraints are provided by studies of Thors and Helgadóttir (1991) of submerged landforms and dating of submerged peat from a nearby submarine locality in Faxaflói Bay. Their interpretation of seismic reflection profiles indicates the presence of flooded coastal features and an erosional unconformity associated with a lower RSL. At this locality, submerged peat dredged from a 17–30-m depth has been dated at an average ^{14}C age of 9300 BP. Ingólfsson et al. (1995) suggest this to be an absolute maximum age for the lowermost position of sea level in the area, arguing the peat could have been rafted from emerged land.

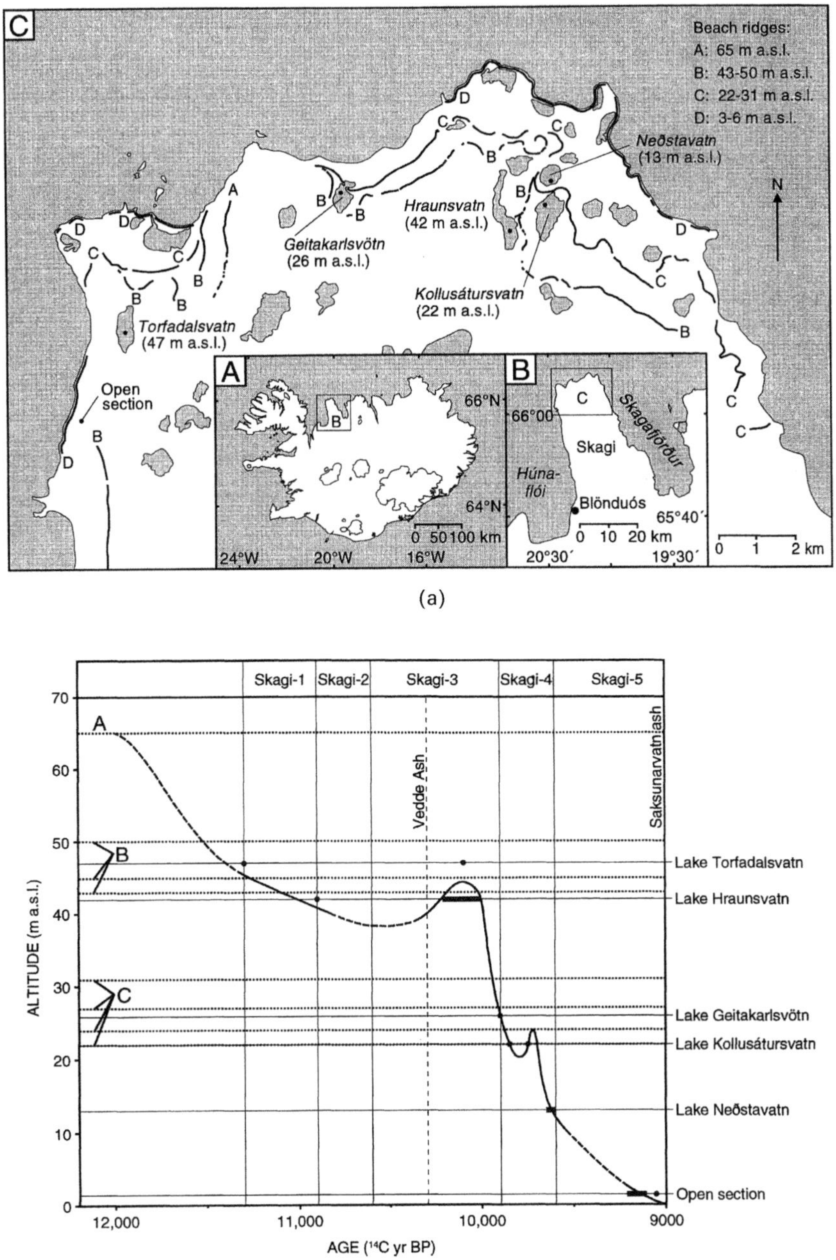

Figure 8.6. Study of relative sea-level change at Skagi, northern Iceland. (a) Location map – lakes and beach ridges marked. (b) Tentative RSL curve fixed by analysis of lake sediments and raised beaches.

Figure 8.7. Fnjóskadalur, northern Iceland. Tilted strandlines on each side of the valley originate from an ice-dammed lake that occupied the valley at the end of the Weichselian glaciation.

Photo courtesy of Oddur Sigurðsson.

8.2.3 Modelling

Postglacial rebound in Iceland was initially modelled by Einarsson (1966). It was further modelled by Sigmundsson (1990, 1991) using an Earth model consisting of a Newtonian viscous half-space overlain by an elastic plate (Figure 8.8). Analytic solutions to the response of loads on the surface of such a model exist only for periodic loads. The isostatic response to an ice load can be calculated if the load

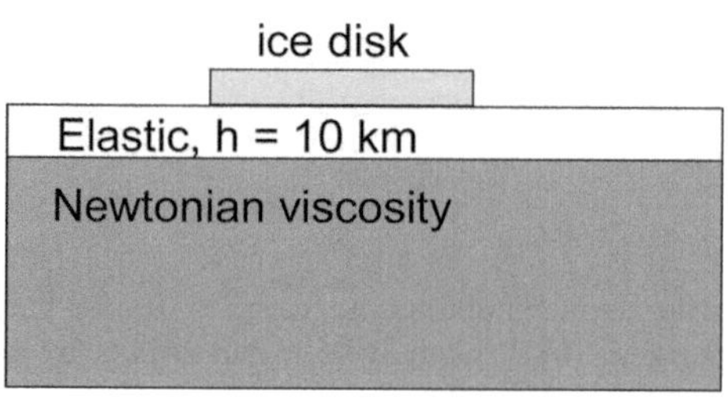

Figure 8.8. Simple Earth model used for modelling of postglacial rebound.

is decomposed into its spectral components (Fourier analysis). The timescale of the adjustment to loading is determined by the viscosity of the Newtonian half-space, while the flexural rigidity of the elastic plate determines the spatial decay of the load signal away from the edge of the load.

The removal of a disk load from the surface of a Newtonian half-space will result in uplift at the centre of the disk load, $u(t)$, that can be written as (Cathles, 1975):

$$u(t) = \frac{\rho_{ice} a}{\rho_{earth}} [1 - \exp(-t/\tau_r)] \tag{8.4}$$

where t is the time since the ice unloading, a is the thickness of the disk load removed, ρ_{ice} is the density of ice, and ρ_{earth} the density of the Earth. The effective relaxation time, τ_r, for the disk load is:

$$\tau_r = \frac{2\eta \bar{k}}{\rho_{earth} g} \tag{8.5}$$

where $\bar{k}$ is the effective wavelength of a disk load, equal to $1.2/R$ where R is the radius of the load (Cathles, 1975). These equations can be used to derive the order of magnitude of the viscosity under Iceland. A single icecap covering the current size of Iceland has a mean thickness of about 1.5 km, and a radius, R, around 160 km. This can be considered a crude approximation to the glacial load during the Younger Dryas period. The densities can be taken as $\rho_{earth} = 3{,}200\,\text{kg/m}^3$ and $\rho_{ice} = 920\,\text{kg/m}^3$. Observations of Postglacial rebound in Iceland suggest that after the Younger Dryas period the rebound was completed in about 1,000 years or even less. Assuming 1,000 years equal three times the relaxation time (95% of the isostatic response completed), the viscosity can be estimated from equations (8.4) and (8.5). Inserting the numbers above, one finds a viscosity of 1.5×10^{19} Pa s. Because of uncertainties regarding the glacial retreat history, this number should be considered a maximum viscosity value. Stepwise or gradual ice retreat over a period of centuries, rather than instantaneous unloading, may have limited the uplift rates.

The uppermost part of the Earth does not respond in a fully ductile manner to load changes; rather the response of the Earth approximates that of a ductile half-space overlain by an uppermost elastic layer. This layer is the elastic lithosphere, the uppermost part of the Earth which responds in an elastic manner on long timescales to loading. Vertical response outside the geographical extent of icecaps manifests this behaviour. The elastic uppermost crust behaves as a lowpass filter on the load effects and causes surface flexure to extend far outside the extent of icecaps. Tilted strandlines witness the flexure of the elastic lithosphere. When isostatic rebound is completed, the amount of flexure will depend on the flexural rigidity of the elastic crust.

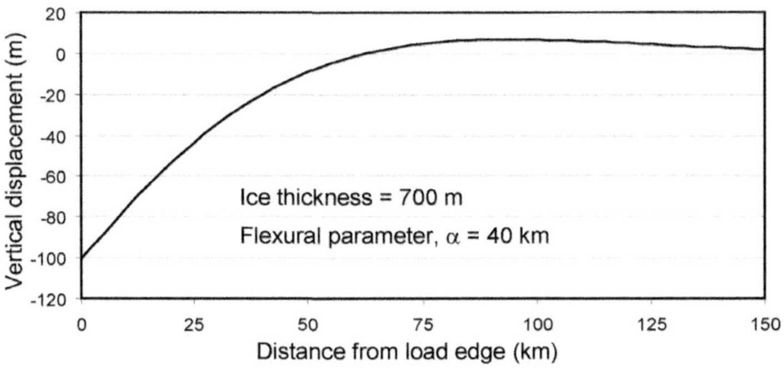

Figure 8.9. Subsidence near the edge of a load of uniform thickness on a thin elastic plate, according to equation (8.8). The calculations are based on $\alpha = 40$ km and $h = 700$ m.

For a thin-plate approximation, the flexural rigidity, D, of an elastic plate is given by (Watts, 2001):

$$D = \frac{ET_e^3}{12(1 - \nu^2)} \tag{8.6}$$

where T_e is the elastic thickness of the plate, E is the Young's modulus, and ν is the Poisson's ratio. The flexural rigidity influences how abruptly vertical movement decay away from the edge of a load. Another related parameter for elastic plates is the flexural parameter, α, given by:

$$\alpha = \left[\frac{4D}{\rho_{earth}g}\right]^{1/4} \tag{8.7}$$

Here it is assumed that there will be no infill of material in the surface deflection formed by the load. The flexural parameter has the units of distance. In two dimensions, the front of a large icecap of uniform thickness will lead to equilibrium subsidence, u_z, which can be written as:

$$u_z(x) = \frac{\rho_{ice}}{\rho_{earth}} \frac{h}{2} e^{-x/\alpha} \cos(x/\alpha) \tag{8.8}$$

where h is the ice thickness and x is distance from the edge of the ice load (Figure 8.9). This formula is valid outside the load edge, and assumes that the load is much wider than α (Watts, 2001, p. 102). Differentiation of this formula with respect to x gives the expected strandline tilt after isostatic rebound is complete. At the former ice edge ($x = 0$) we have:

$$\text{Strandline tilt} \quad \frac{-\partial u_z(x)}{\partial x} = \frac{1}{\alpha} \frac{\rho_{ice}}{\rho_{earth}} \frac{h}{2} \tag{8.9}$$

For a typical strandline tilt in Iceland of 2.5 m/km, and assuming the ice field thickness close to the edge of an icecap of 700 m (representing Younger Dryas load), one finds that the value of the flexural parameter is 40 km. Using equations (8.7) and (8.6), the flexural rigidity is found to be $D = 20 \times 10^9$ Nm. For a Young's

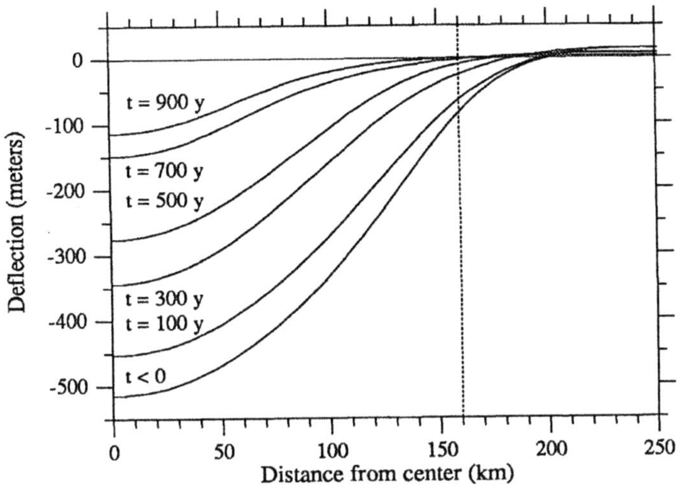

Figure 8.10. Model prediction for rebound following the disappearance of the Younger Dryas icecap in Iceland, the ice load being modelled as shown in Figure 8.5. A 10-km-thick elastic layer is assumed, and Newtonian viscous fluid with viscosity of 1×10^{19} Pa s below it.

modulus of 100 GPa and Poisson's ratio of 0.25, the derived elastic thickness, T_e, is 13 km. The above simplified analyses give the order of magnitude for the viscosity under Iceland and the elastic thickness of the lithosphere inferred from studies of glacial isostasy.

A more detailed modelling of the Postglacial rebound in Iceland (Sigmundsson 1990, 1991) draws similar conclusions (Figure 8.10). The viscosity under Iceland has to be 10^{19} Pa s or lower to match the observed short duration of Postglacial rebound. The thickness of the elastic lithosphere is not well constrained, but a thickness of 10 km is consistent with observations. An alternative model of isostatic rebound on a viscoelastic half-space leads essentially to the same conclusion regarding viscosity (Jull and McKenzie, 1996). The inferred viscosity under Iceland is lower than the inferred global average which is to be expected, as Iceland is a hotspot located on the Mid-Atlantic Ridge.

Coverage of Iceland with one main icecap having a radius near 160 km requires an ice thickness in central Iceland of up to about 1,800 m (Sigmundsson, 1991). Isostatic balance requires subsidence of around 500 m due to this ice load. Rates of uplift after sudden removal of the icecap could therefore have resulted in uplift rates up to 50 cm/yr in central Iceland, but around 10 cm/yr in the coastal areas. This large response was not the only effect of the ice unloading on the geodynamic environment in Iceland. The ice unloading also caused a major pulse of volcanic activity! It appears that the unloading did actually trigger increased mantle melting beneath Iceland as discussed in the following section, with this exceptional phenomenon being due to the setting of Iceland above an extensive melting regime in the mantle.

8.3 VARIABLE VOLCANIC PRODUCTION RATES AT THE END OF THE LAST GLACIATION

The neovolcanic zone in Iceland is covered by lavas erupted after the end of the last glaciation. These lavas have not erupted evenly in time and major changes in volcanic production rates occurred in the Holocene. The majority of the lavas were erupted during the few thousand years immediately after the end of the last glaciation. Compared with current volcanic production rates, an order of magnitude increase is observed in that period. In the Askja region in the northern rift zone, Sigvaldason et al. (1992) documented a 30-fold increase in lava production at the onset of the Holocene compared with present times. A similar pattern is found in the Western Volcanic Zone (Sinton et al., 2005). On the Reykjanes Peninsula, an estimated $2.3\,km^3$ of magma have erupted in historical times (last 1,000 years), whereas the total Postglacial production is about $40\,km^3$ (Jakobsson et al., 1978; Gudmundsson, 1986). These patterns (Figure 8.11) suggest a direct link between deglaciation and increased volcanic activity. The present landscape of Iceland contains much more extensive Holocene lava fields than present rates of volcanic activity would produce if they had prevailed throughout the Holocene. Not only has the production rate varied in the volcanic zones, it appears that the ice melting has also influenced the chemical composition of lavas. This has been documented, for example, by Hardarson and Fitton (1991) for the Snæfellsjökull Volcano and in the Northern Volcanic Zone by Slater et al. (1998). An overview is given by Maclennan et al. (2002).

Several suggestions have been proposed to explain the link between deglaciation and the pulse of volcanic activity, invoking either changing crustal conditions or, alternatively, changes in mantle-melting conditions. Ice load on volcanoes may inhibit magma from erupting, because of higher overburden pressure and induced stresses in the crust that inhibit magma propagation through the crust. When ice melts, the opposite situation arises. When analysing eruption frequency on the Reykjanes Peninsula, Gudmundsson (1986) suggested that stresses induced in roofs above shallow magma chambers during ice unloading were the cause of more frequent eruptions in the early Postglacial period. Sigvaldason et al. (1992) argue that the increase in melt output following the deglaciation resulted from either the release of accumulated magma from decreased overburden pressure, or by the opening of crustal pathways by differential tectonic movements during glacial rebound. In this model, the ice load inhibits eruption of magma but generation of magma in the mantle is unaffected.

An alternative view is that the ice load actually influenced the melting regime in the mantle. An ice load of similar dimensions to Iceland is capable of influencing melting conditions in the mantle (Jull and McKenzie, 1996). Melting under mid-ocean ridges occurs because of decompression as material moves closer to the surface of the Earth. If the melting column is situated under a retreating icecap, decompression will occur as well because of ice thinning. The ice thinning will have the same influence as upwelling of material towards the surface.

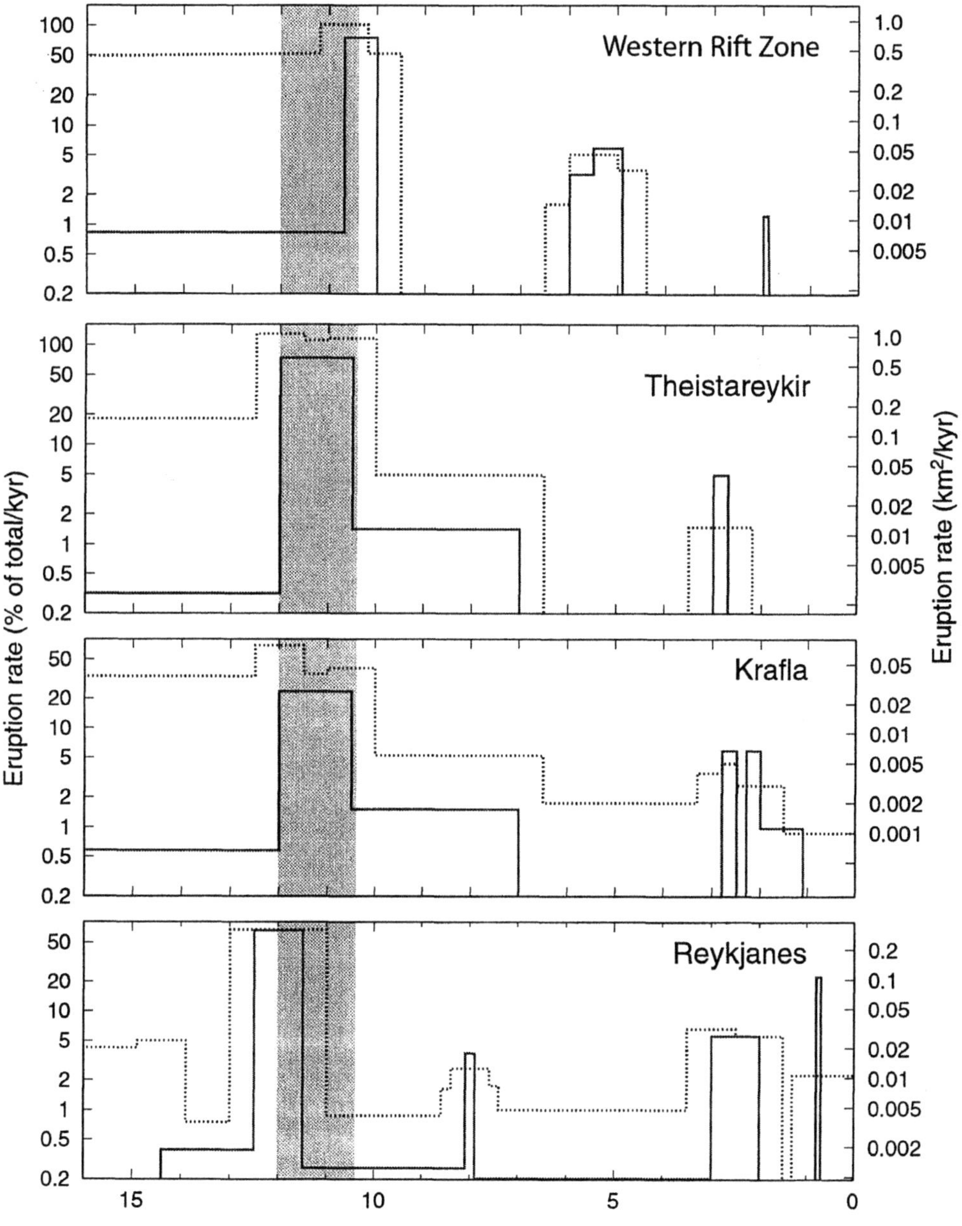

Figure 8.11. Eruption rate in different parts of Iceland. Note the vertical axis is a log scale. Shaded area shows the period of Postglacial rebound. Solid line shows average eruption rate and dotted line the maximum bound on the eruption rate.

Reproduced from Maclennan et al. (2002). Copyright by the American Geophysical Union.

In order to model the influence of ice unloading on melting rate, X, it has to be considered in a reference frame fixed to an initial position at the Earth's surface. Then the substantive derivative of the melting rate, DX/Dt, may be written in Eulerian form as:

$$\frac{DX}{Dt} = \frac{\partial X}{\partial t} + \bar{V} \cdot \nabla X \tag{8.10}$$

where $\bar{V}$ is the velocity field. Considering that decompression melting under a spreading ridge is isentropic, $X = X(P)$, equation (8.10) can be written as (Jull and McKenzie, 1996):

$$\frac{DX}{Dt} = \left(\frac{\partial X}{\partial P}\right)_S \left(\frac{\partial P}{\partial t} + \bar{V} \cdot \nabla P\right) \tag{8.11}$$

where P is pressure and S is entropy. The partial derivative of the melting rate with respect to pressure, for a constant entropy, $(\partial X/\partial P)_S$, is given by McKenzie (1984). The pressure field under an axisymmetric retreating icecap is evaluated by Jull and McKenzie (1996), to derive the melting rate under Iceland during ice unloading. If no unloading is occurring, then $\partial P/\partial t$ is zero, and melting occurs because material moves closer to the surface. If the upwelling occurs at about the same rate as the full spreading rate, 2 cm/yr, then the decompression rate is about 600 Pa/yr. If unloading is occurring, then $\partial P/\partial t$ is nonzero. In the case of Iceland, the Younger Dryas icecap was on the order of 1–2 km thick, and disappeared in about 1,000 years or less. This corresponds to an average thinning of 1–2 m of ice per year or even more. For an ice-thinning rate of 2 m/yr, the corresponding decompression rate is about 18,000 Pa/yr, equivalent to removal of 60 cm of rock each year. During the period of ice unloading, the decompression rate due to ice unloading is then about 30 times larger than the decompression rate due to mantle upwelling. Melt production rates will accordingly increase by a factor of 30 (Maclennan et al., 2002). This assumes the pressure change at the surface will lead to similar pressure change in the melting regime.

A complete model of the above process considers a wedge-shaped melting regime under Iceland and a gradual decay of the ice load (Jull and McKenzie, 1996). A circular icecap with radius of 180 km was assumed. Melting in the model occurs in a triangular wedge-shaped region with an angle of 45° at a constant upwelling rate (Figure 8.12). In this model, the steady-state melt production rate is $\sim 0.1\ \text{km}^3/\text{yr}$. During the unloading period, melt generation increases to $\sim 3\ \text{km}^3/\text{yr}$. The observed length of the excessive magma production period after ice unloading constrains the transfer time of melt from mantle to surface. The pulse of high production after the ice unloading lasted roughly about 2,000 years. Because melting occurs down to 100 km, a melt extraction velocity in the mantle of >50 m/yr is suggested (Figure 8.13).

8.4 HISTORICAL ICE VOLUME CHANGES AND RECENT FLUCTUATIONS IN LAND ELEVATION

Low viscosity under Iceland makes land elevation around Iceland's current icecaps sensitive to changes in ice volume. The rheological conditions can therefore be

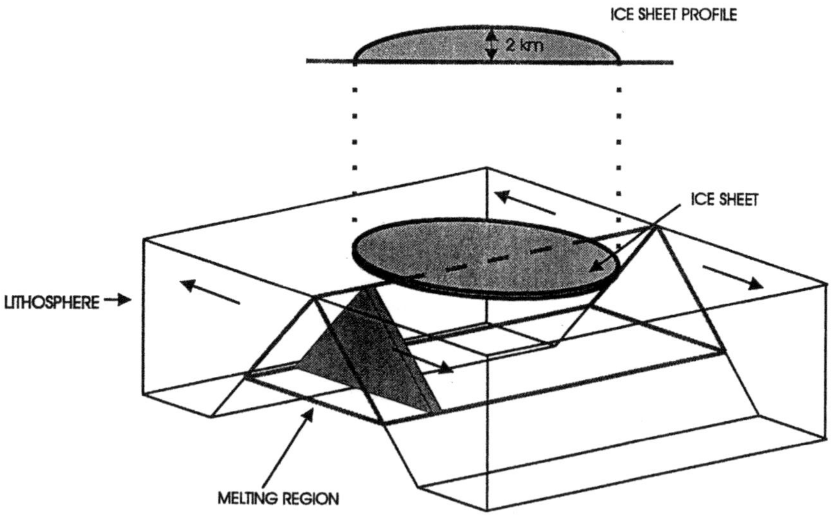

Figure 8.12. Schematic illustration of the deglaciation and melting model for Iceland. The ice sheet is shown as a circular disk above a wedge-shaped melting region.

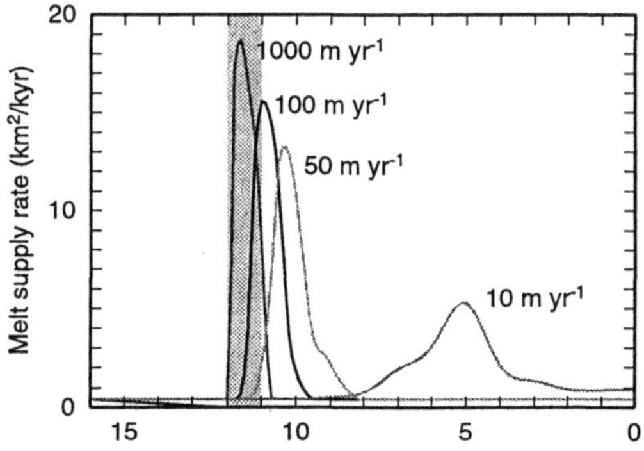

Figure 8.13. Predicted melt supply rates from the mantle to the crust when finite melt extraction velocities are incorporated in a melt generation model. The light-grey box shows the period of ice unloading.

further constrained by studying the ongoing elevation change around the icecaps and its correlation with ice volume change (Sigmundsson, 1992a; Einarsson et al., 1996).

The ice volume change has been extensive during historical time in Iceland as a result of mean temperature variations of 1–2°C (Figure 8.14). At the time of settlement, the temperature was similar to today's, but climate conditions deteriorated during the Middle Ages (the Little Ice Age), improving again only last century. At the Vatnajökull Icecap, the estimated ice volume loss from 1890 to 1978 was 182 km^3 (Figure 8.15).

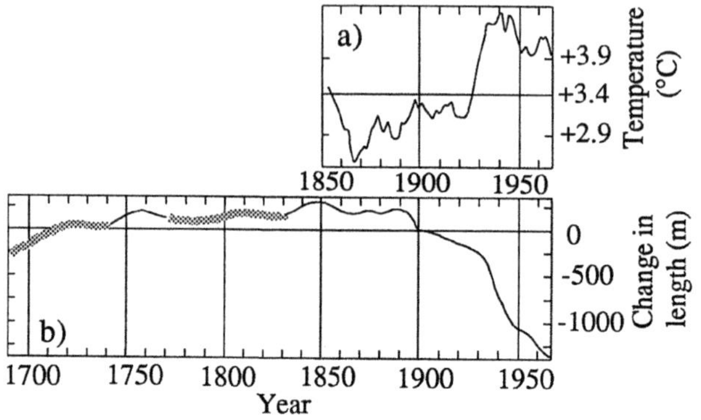

Figure 8.14. Atmospheric temperature and change in length of outlet glaciers at Vatnajökull, Iceland.

Reproduced from Sigmundsson and Einarsson (1992). Copyright by the American Geophysical Union..

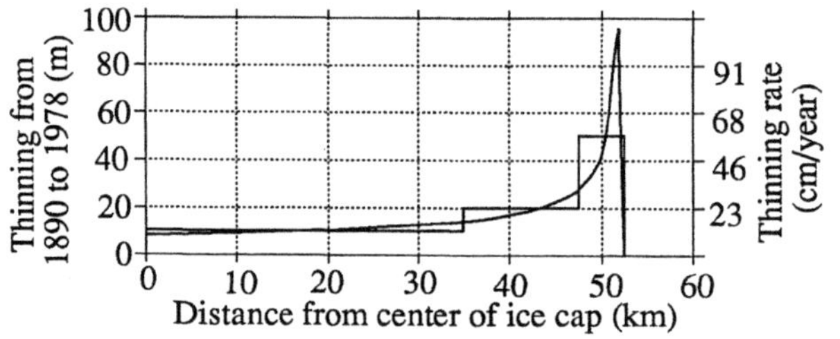

Figure 8.15. Model for thinning of the Vatnajökull Icecap 1890–1978 used to model glacio-isostatic response to reduction in the load of Vatnajökull.

Reproduced from Sigmundsson and Einarsson (1992). Copyright by the American Geophysical Union.

Several types of data reveal the Earth's response to this loading and unloading. Geologic records of changing conditions in southeastern Iceland reflect sea-level subsidence of a few meters that can be linked to increased ice load of Vatnajökull during the Middle Ages. Tephra from the 1362 eruption of Öræfajökull found within submarine freshwater peat in southeastern Iceland has been used to argue for at least a 3-m rise in RSL from 1362 to 1951 (Jonsson, 1957). Historical records suggest a fall in RSL (crustal uplift) in southeastern Iceland last century at a rate of 1–2 cm/yr (Imsland, 1992). Various types of geodetic observations suggest ongoing uplift around the edges of Vatnajökull.

LAKE LEVELLING Initial geodetic observations demonstrating rebound around the Vatnajökull Icecap consist of repeated lake-level measurements at Lake Langisjór, a 20-km-long lake perpendicular to the southwestern edge of the Vatnajökull Icecap. Benchmarks at each end of the lake have been used as reference, with their elevation above the lake level measured initially in 1959. The measurements have been repeated several times since and indicate a varying

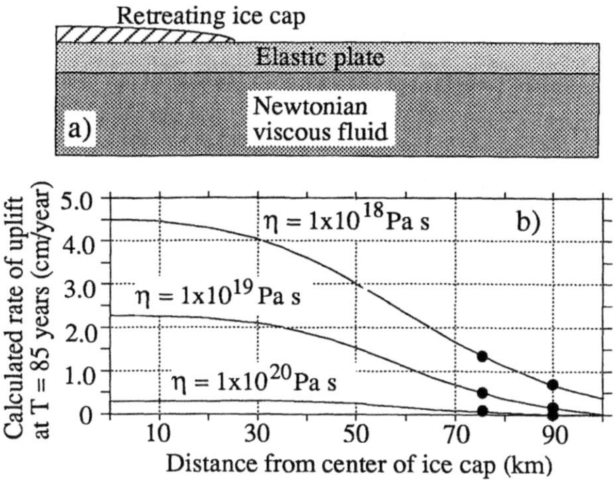

Figure 8.16. Model uplift rates versus distance from the centre of the Vatnajökull Icecap assuming thinning rates shown in Figure 8.15. Black dots mark the ends of Lake Langisjór. See text for discussion.

Reproduced from Sigmundsson and Einarsson (1992). Copyright by the American Geophysical Union.

amount of water in the lake, but also differential movement of one end of the lake relative to the other (the change in benchmark elevation relative to the lake is not the same at both ends of the lake). The measurements show that the end of the lake near the icecap is rising relative to its other end by 3.9 ± 0.9 mm/yr. Modelling of these data (Figure 8.16) suggests a sublithospheric viscosity in the range of 1×10^{18}–5×10^{19} Pa s (Sigmundsson and Einarsson, 1992).

CONTINUOUS GPS Ongoing uplift is evident from continuous GPS observations at the HOFN station which has been in operation since 1997 (Figure 8.17). The HOFN station is in the global network of continuous GPS sites, and a number of GPS data analysis centres process data from it. The REVEL model for recent plate velocities from space geodesy data (Sella et al., 2002) gives a yearly uplift rate of 4.0 ± 2.3 mm/yr for the HOFN station. For comparison, the fixed station at Reykjavík (REYK) is subsiding 3.4 ± 2.3 mm/yr according to the same model. Independent analysis of data from the HOFN and REYK stations, with a different approach, at the Icelandic Meteorological Office (Geirsson et al., submitted) gives uplift of HOFN relative to REYK as 9.7 ± 1.4 mm/yr in the 1999–2004 period, consistent with the REVEL model.

CAMPAIGN GPS A network of GPS stations was established in 1991 at the southeastern edge of the Vatnajökull Icecap to study ongoing glacial rebound (Sigmundsson, 1991; Einarsson et al., 1996). Repeated measurements in the area by Sjöberg et al. (2000, 2004) and Pagli et al. (2005) are consistent with uplift of 5–19 mm/yr for stations close to the ice edge, and uplift decreasing with distance

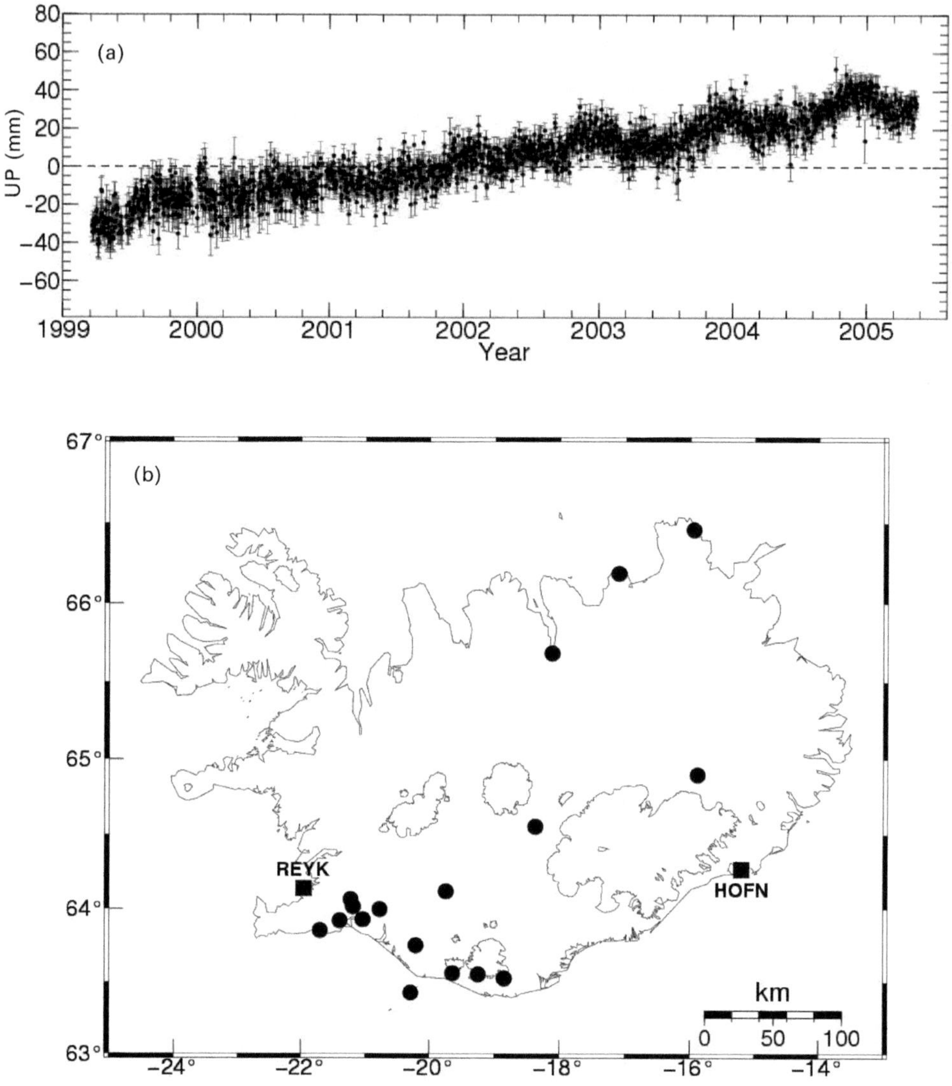

Figure 8.17. (a) Vertical displacement of the HOFN continuous GPS station in southeastern Iceland relative to the REYK station in Reykjavík, southwestern Iceland. (b) Location of continuous GPS sites.

Courtesy of Halldór Geirsson, Icelandic Metorological Office.

from the icecap (Figure 8.18). Part of this dataset has been modelled by Thoma and Wolf (2001) who argue it is consistent with an elastic lithosphere thickness of 10–20 km and underlying viscosity of 7×10^{16}–3×10^{18} Pa s. Geodetic data from other areas surrounding the icecap exist as well and can be used to constrain the rheological parameters.

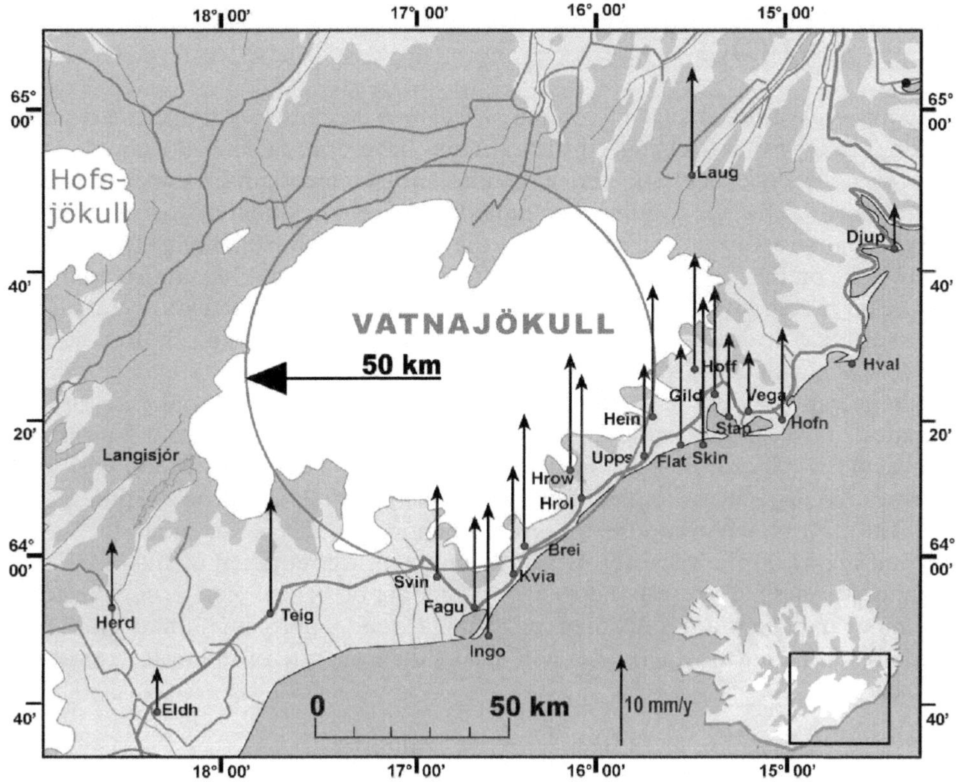

Figure 8.18. Rates of uplift in the 1992–1999 period near the southeastern edge of the Vatnajökull Icecap.

Reproduced from Sjöberg et al. (2004).

GRAVITY OBSERVATIONS A network of gravity stations in southeastern Iceland has been repeatedly measured by Jacoby et al. (2001). The results suggest gravity decrease and uplift. The change in gravity at Höfn in southeastern Iceland relative to Reykjavík is inferred to be $-3 \pm 2\,\mu\text{Gal/yr}$ in 1968–1996, corresponding to a relative uplift rate of 1.5 ± 1 cm/yr. Within uncertainties, this is the same rate as inferred from continuous GPS measurements in the 1999–2004 period.

In summary, various types of evidence suggest uplift near the edge of the Vatnajökull Icecap of about 5–20 mm/yr, decaying over a distance of 25 km from the icecap edge to about 5 mm/yr. Modelling suggests viscosity lower than 10^{19} Pa s. Further measurements should be able to resolve in more detail the viscosity structure, in particular measurements of eventual uplift associated with enhanced retreat of Iceland's icecaps in response to global warming.

8.5 MELTING OF ICECAPS BY GLOBAL WARMING: AN EXPERIMENT IN RHEOLOGY

Ongoing and future change in volume of Iceland's icecaps may induce deformation in the same way past changes in ice volume have done. Establishment of a still-improved geodetic network and a regular measurement programme, as well as monitoring of the ice volume fluctuations, holds the potential to constrain the rheological structure much better than the currently available data. The available knowledge on the rheological structure allows an estimation of the anticipated uplift at icecaps in response to global warming. Mass balance measurements of Vatnajökull are indicative of decrease in the ice load in the last decades (e.g., Björnsson et al., 2002; Magnússon et al., 2005). For the time period 1996–2004 the average thinning of the icecap per year is on the order of 1 m. If this thinning would continue for decades, it will be an addition to the ongoing natural experiment in rheology provided by glacial isostasy in Iceland. The resulting uplift would also have practical consequences for conditions along the southeast coast of Iceland.

The anticipated response to future thinning of the Vatnajökull Icecap will depend highly on the viscosity. Let us assume that ice retreat at Vatnajökull in the coming decades will correspond to a 1-m thinning of ice per year. A circular icecap thinning at a constant rate then provides a good approximation. If the Earth responds as a Newtonian viscous half-space, then the rate of uplift at its centre, $\dot{u}$, will be:

$$\dot{u}(t) = \frac{\rho_{ice}\dot{a}}{\rho_{earth}}[1 - \exp(-t/\tau_r)] \tag{8.12}$$

where t is the time since thinning of the icecap began, $\dot{a}$ is rate of thinning of the icecap, ρ_{ice} is the density of ice, and ρ_{earth} the density of the Earth (Sigmundsson and Einarsson, 1992). The effective relaxation time, τ_r, is given by equation (8.5). For Vatnajökull, $R = 50$ km is a good approximation. If $\eta = 5 \times 10^{18}$ Pa s, and ρ_{earth} is 3,200 kg/m^3, then τ_r is 240 years. Predicted rates of uplift for different viscosity values are shown in Figure 8.19.

A more refined model must take into account viscoelasticity and the elastic lithosphere. Uplift rate of edge of the icecap will however be of similar functional form as predicted by equation (8.12). Addition of an ~10-km-thick elastic surface layer to the above model will cause uplift at the edge of the icecap to be about 3/4 of the uplift at the centre, and at a distance of 20 km from the ice edge about 1/3 of the maximum uplift value (Sigmundsson, 1990). According to the previous analysis, the rate of uplift at station HOFN should increase markedly within a timespan of 10 years if the viscosity is any lower than 10^{18} Pa s, if Vatnajökull continues to thin. Vertical velocities predicted from thinning since 1996 would add to uplift due to thinning last century. Monitoring of contemporary changes in uplift rates around the Vatnajökull Icecap can therefore better resolve the viscosity structure, and also allow more precise prediction of future vertical changes around Iceland's icecaps. Future melting of Vatnajökull can influence magmatic systems and eventually trigger eruptions, but not necessarily cause increased mantle melting. Main stress change

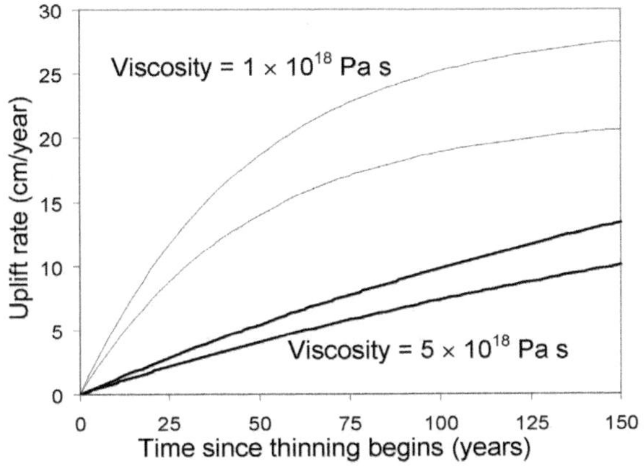

Figure 8.19. Predicted uplift rates at Vatnajökull using equation (8.12), assuming uniform thinning of the icecap of 1 m per year. Curves show predicted rate of uplift at the centre of the icecap and near its edge. The uplift rate increases with time as the thinning begins.

associated with the reduction of the icecap is likely to take place in the crust (~30 km thick in southeastern Iceland, see Figure 4.3) rather than in the mantle, because of the much smaller spatial dimension of Vatnajökull than the icecap covering Iceland during the Weichselian glaciation.

9

Iceland geodynamics: Outlook

The study of crustal deformation is a "slow" business, as observations normally require long periods of time in order to reveal significant deformation signals. Iceland is, however, unique with its many deformation processes and high rate of occurrence of earthquakes, eruptions, and magmatic movements. The high rate of geologic hazards has been key to the success of recent deformation studies that have added to understanding of Iceland geodynamics in the last decades. We are now in a position to ask ever-more-detailed questions. With appropriate instrumentation and preparedness for next events to happen in the crust in the Iceland geo-laboratory, we should still be able to significantly advance our understanding of how the Earth moves and deforms at divergent plate boundaries.

Remaining questions are many, and ongoing discussion, debate, and research are taking place in most of the fields touched upon in this book. Even the fundamental cause of Iceland is being debated with an alternative idea to mantle plume being proposed as an explanation for the existence of Iceland. More extensive seismic surveys including seismic stations on the ocean floor may be needed to resolve these issues. Regarding the geological history of Iceland, more radiometric datings and information on rift jumps will be important to understand plume–ridge interaction. For crustal structure, there is still one set of observations that has not been fully understood in context of crustal models developed in the 1990s. Magnetotelluric measurements reveal a low resistivity layer that was interpreted as partial melt, but on the contrary seismic studies are in favour of a cold crust. Full integration of magnetotelluric data into existing models is needed. For volcano dynamics, detailed joint interpretation of deformation and seismic patterns have the potential to reveal in greater detail how magma movements take place inside the crust. Gravimetric observations, when combined with seismic and other geodetic data, are likely to provide more constraints on deformation processes on volcanoes. Further understanding of the plate-spreading process can be achieved through more modelling of available data on the style of inter-rifting deformation. A new rifting episode will then provide unprecedented possibilities to monitor how the ocean crust is formed. In the transform zones, earthquake prediction research may advance with concerted efforts in the search for earthquake precursors. Future studies of the rheological structure of the crust and mantle in Iceland may be able to resolve lateral variations in crust and mantle properties, and provide a unified rheological model.

The future is before us, full of new geological events likely to happen in coming years. By making full use of present capabilities and extended, interdisciplinary research programmes we are likely to see new exciting research results in coming years from the study of Iceland geodynamics.

Appendix A

The Icelandic language

The Icelandic language is a Germanic language within the Indo-European family of languages, related to Faroese, Norwegian, Danish, and Swedish. Information on the language and its pronunciation can, for example, be found on the Internet homepage of the Icelandic Language Institute at *www.islenskan.is* and in Kristinsson (1988). Material on the homepage of the Icelandic Language Institute includes the brochure *Icelandic: At Once Ancient and Modern*, published by the Ministry of Education, Science, and Culture in Iceland, and serves as an excellent introduction to the language.

The Icelandic alphabet has ten characters which are different from those used in English. In the international geoscience literature, these letters are sometimes used for writing original Icelandic words and names, but they are often substituted by English letters. The same word may therefore be written differently in the international literature. The lists below give the special Icelandic characters, a list of some words appearing in the book in Icelandic writing and their transliterated form using English letters.

People's names in Icelandic are such that the last name of a person has the ending *-son* or *-dóttir*, meaning "son" and "daughter". The last name is normally formed from the Christian name of the father (its possessive form), with the attached ending *-son* or *-dóttir*. Last names within families therefore vary from one generation to the next. Women's names remain unchanged upon marriage. The different usage of Icelandic characters and their equivalent Latin transliteration for names of Icelandic authors in the international literature causes some confusion. When searching publications of these authors, often both versions of their names need to be considered when locating their work. Within Iceland, Christian names are used in communication (individuals are, e.g., arranged alphabetically according to their Christian names in the Icelandic phonebook).

Non-English letters used in Icelandic

Á	á
Ð	ð
É	é
Í	í
Ó	ó
Ú	ú
Ý	ý
Þ	þ
Æ	æ
Ö	ö

The most unusual letters in Icelandic are þ and ð, with þ being pronounced as "th" in think and ð as "th" in they.

Icelandic words (examples)	**Transliterated form in English letters**
Surnames	
Árnadóttir	Arnadottir
Friðleifsson	Fridleifsson
Guðmundsson	Gudmundsson
Pálmason	Palmason
Sæmundsson	Saemundsson
Steingrímsson	Steingrimsson
Þórarinsson	Thorarinsson
Christian names	
Guðmundur	Gudmundur
Kristján	Kristjan
Jón	Jon
Sigurður	Sigurdur
Þóra	Thora
Icelandic place names	
Axarfjörður	Axarfjordur
Almannagjá	Almannagja
Búði	Budi

Icelandic words (examples)	Transliterated form in English letters (*cont.*)
Icelandic place names (*cont.*)	
Eldgjá	Eldgja
Esjufjöll	Esjufjoll
Bárðarbunga	Bardarbunga
Borgarfjörður	Borgarfjordur
Breiðabunga	Breidabunga
Brennisteinsfjöll	Brennisteinsfjoll
Faxaflói	Faxafloi
Fnjóskadalur	Fnjoskadalur
Fremri-Námar	Fremri-Namar
Fögrufjöll	Fogrufjoll
Gjálp	Gjalp
Grímsvötn	Grimsvotn
Grímsey	Grimsey
Gæsafjöll	Gaesafjoll
Herðubreið	Herdubreid
Hofsjökull	Hofsjokull
Hrómundartindur	Hromundartindur
Hrúthálsar	Hruthalsar
Húsavík	Husavik
Hvalhnúkar	Hvalhnukar
Kálfstindar	Kalfstindar
Kerlingarfjöll	Kerlingarfjoll
Krísuvík	Krisuvik
Kverkfjöll	Kverkfjoll
Langjökull	Langjokull
Ljósufjöll	Ljosufjoll
Lýsuskarð	Lysuskard
Mýrdalssandur	Myrdalssandur
Núpshlíðarháls	Nupshlidarhals
Prestahnjúkur	Prestahnjukur
Reykjavík	Reykjavik
Skeiðarársandur	Skeidararsandur

Icelandic words (examples)	**Transliterated form in English letters** (*cont.*)	
Icelandic place names (*cont.*)		
Snæfell	Snaefell	
Snæfellsnes	Snaefellsnes	
Snæfellsjökull	Snaefellsjokull	
Reyðarfjörður	Reydarfjordur	
Tindfjöll	Tindfjoll	
Torfajökull	Torfajokull	
Tungnafellsjökull	Tungnafellsjokull	
Tjörnes	Tjornes	
Vatnafjöll	Vatnafjoll	
Vatnajökull	Vatnajokull	
Vatnaöldur	Vatnaoldur	
Veiðivötn	Veidivotn	
Þeistareykir	Theistareykir	
Þeistareykjabunga	Theistareykjabunga	
Þingvellir	Thingvellir	
Þingvallavatn	Thingvallavatn	
Þórðarhyrna	Thordarhyrna	
Öskjuvatn	Oskjuvatn	
Öræfajökull	Oraefajokull	
Other words		**Meaning**
Alþingi	Althingi	Parliament
Fjall		Mountain
Fjöll	Fjoll	Mountains
Jökull	Jokull	Glacier, icecap
Jökulhlaup	Jokulhlaup	Glacial outburst flood
Móberg	Moberg	Hyaloclastite

Appendix B

Notation

A	Thermal conductivity
A	Amplitude of seismic waves
A	Fault length
a	Radius of a spherical deformation source (Mogi model)
a	Semi-major axis of an ellipsoidal deformation source
a	Radius of pressurized pipe deformation source
a	Radius of the Earth
a	Load thickness
$\dot{a}$	Rate of thinning of ice load
B	Skempton coefficient
b	Semi-minor axis of an ellipsoidal deformation source
b	Thickness of a viscous layer
b	Dike width
C	Strength parameter of a Mogi source
ΔCFS	Change in static Coulomb failure stress
c_p	Specific heat at constant pressure
D	Flexural rigidity
D	Locking depth
D	Thickness of brittle crust (the seismogenic layer)
d	Depth
E	Young's modulus
E_{hmax}	Direction of maximum horizontal stress
f	Wave frequency
g	Acceleration of gravity, 9.8 m/s^2
G	The gravitational constant
h	Isostatically compensated topography
h	Elevation above sea level
h	Thickness of an elastic layer

h	Ice thickness
h_0	Maximum vertical displacement directly above a Mogi source
k	Bulk modulus
k	Effective bulk modulus
k	Stress amplification factor
$\bar{k}$	Effective wavelength
L	Width of a bookshelf-faulting zone, length of faults
L	Dike length, sill length
M	Elastic modulus for plane stress conditions
M_0	Seismic moment
M'_0	Geometric moment
$\dot{M}_0$	Rate of geometric moment release
$\dot{M}_p$	Mantle plume buoyancy flux
N	Number
ΔN	Geoid anomaly
P	Pressure
ΔP	Change in pressure
ΔP_{sheet}	Pressure in planar sheet-like intrusion
ΔP_c	Critical overpressure for failure
dp/dx	Pressure gradient along a magmatic conduit
Q	Volumetric magma flow rate
Q_v	Mantle plume volume flux
Q	Heat flow
Q	Seismic quality factor
q	Lava deposition rate per unit length of rift in the Pálmason model
R	Radius of a magma conduit
R	Radius of a circular icecap
R	Rift trend
r	Horizontally radial distance in cylindrical polar coordinates
S	Entropy
S	Half spreading rate at a mid-ocean ridge
$\dot{s}$	Slip rate
T	Temperature
T	Recurrence interval of diking events
ΔT	Temperature difference
T	Travel time
T_e	Elastic plate thickness
T_s	Tensile strength
t	Time
t	Age
$U(x, t)$	Horizontal displacement field
U_0	Dike half-width
u	Thickness of a planar intrusion
u	Relative plate velocity
$u(t)$	Uplift as a function of time at a centre of a load

u_r	Horizontal radial displacement (Mogi model)
u_ϕ	Horizontal tangential displacement (Mogi model)
u_z	Vertical displacement
$\dot{u}$	Uplift rate
$\bar{u}$	Mean fault slip
$u(x, t)$	Horizontal surface velocity
X	Melting rate
X	Fault area
x	Horizontal distance from rift axis
z	Depth
z	Depth to Moho
$z(x)$	Trajectory of mass elements in the Pálmason model
V	Plate velocity (half spreading rate) in the Pálmason model
V	Relative velocity (full plate movement rate)
V	Volume
ΔV_{magma}	Magma volume
$\Delta V_{edifice}$	Integrated ground surface volume change
ΔV_{Mogi}	Volume change of a Mogi source
$\bar{V}$	Velocity field
v_a	Velocity of asthenospheric flow along a mid-ocean ridge
v_p	P-wave velocity
v_s	S-wave velocity
v_x	Horizontal velocity
v_z	Vertical velocity
$v(x)$	Velocity as a function of distance
W	Depth of compensation of topography
α	Flexural parameter
α	Coefficient of thermal expansion
α	Acute angle between a rift trend and plate movements in oblique spreading
Δ	Aerial strain
Δ	Angle subtended at centre of the Earth between pole of rotation and a site
δ	Radial tilt (Mogi model)
τ_0	Shear stress at bottom of an icecap
τ_r	Relaxation time
$\Delta\tau_s$	Change in shear stress resolved in slip direction of a fault
κ	Stress diffusivity
ε_r	Horizontal radial strain
ε_t	Horizontal tangential strain
ε_{xx}	Normal component of strain in x-direction
$\dot{\varepsilon}$	Average strain rate
$\dot{\varphi}$	Rotation rate
ϕ	Angle in cylindrical polar coordinates
θ_R	Angle between a limb of a V-shaped ridge and a spreading axis

λ	Lamé modulus
μ	Modulus of rigidity (shear modulus), Lamé modulus
η	Dynamic viscosity
ρ	Density
ρ_0	Reference density
ρ_0	Average upper crustal density
ρ_m	Mantle density
$\Delta\rho$	Density difference
ν	Poisson's ratio
σ_{xx}	Normal stress in an elastic layer
$\Delta\sigma_{xx}$	Horizontal tectonic stress
$\Delta\sigma_{kk}$	Volumetric stress
σ_1	Standard deviation of long-term average horizontal strain rate
σ_2	Standard deviation of lava deposition rate in the Pálmason model
$\Delta\sigma_3$	Deviatoric minimum compressive stress
$\Delta\sigma_n$	Change in normal stress
ω	Angular velocity of rotation in plate motion models
ω	Distance between faults

References

Agustsson, K., Stefansson, R., Linde, A. T., Einarsson, P., Sacks, I. S., Gudmundsson, G. B., and Thorbjarnardottir, B. (2000) Successful prediction and warning of the 2000 eruption of Hekla based on seismicity and strain changes, *EOS Trans. AGU*, **81**, Fall Meet. Suppl., Abstract V11B-30.

Ágústsson, K., and Flóvenz, Ó. G. (2005) The thickness of the seismogenic crust in Iceland and its implications for geothermal systems, paper 0743, *Proceedings of the World Geothermal Congress 2005, Antalya, Turkey, April 24–29*, International Geothermal Association.

Aki, K., and Richards, P. G. (1980) *Quantitative Seismology*, Vol. I, Freeman, New York.

Allen, R. M., Nolet, G., Morgan, W. J., Vogfjörd, K., Nettles, M., Ekström, G., Bergsson, B. H., Erlendsson, P., Foulger, G. R., Jakobsdottir, S. et al. (2002a) Plume-driven plumbing and crustal formation in Iceland, *J. Geophys. Res.*, **107**, doi:10.1029/2001JB000584.

Allen, R. M., Nolet, G., Morgan, W. J., Vogfjörd, K., Bergsson, B. H., Erlendsson, P., Foulger, G. R., Jakobsdottir, S., Julian, B., Pritchard, M. et al. (2002b) Imaging the mantle beneath Iceland using integrated seismological techniques, *J. Geophys. Res.*, **107**, doi:10.1029/2001JB000595.

Alley, R. B. (2000) The Younger Dryas cold interval as viewed from central Greenland, *Quaternary Science Reviews*, **19**, 213–226.

Anderson, E. M. (1936) The dynamics of the formation of cone-sheets, ring-dykes and cauldron subsidence, *Proc. R. Soc. Edinburgh*, **56**, 128–157.

Andrews, J. T., Hardardóttir, J., Helgadóttir, G., Jennings, A. E., Geirsdóttir, A., Sveinbjornsdóttir, A. E., Schoolfield, S., Kristjánsdóttir, G. B., Smith, L. M., Thors, K. et al. (2000) The N and W Iceland Shelf: Insights into Last Glacial Maximum ice extent and deglaciation based on acoustic stratigraphy and basal radiocarbon AMS dates, *Quaternary Science Reviews*, **19**, 619–631.

Angelier, J., and Bergerat, F. (2002) Behaviour of a rupture of the 21 June 2000 earthquake in South Iceland as revealed in an asphalted car park, *J. Struct. Geol.*, **24**, 1925–1936.

Árnadóttir, Th., Sigmundsson, F., and Delaney, P. T. (1998) Sources of crustal deformation associated with the Krafla, Iceland, eruption of September 1984, *Geophys. Res. Lett.*, **25**, 1043–1046.

Árnadóttir, Th., Hreinsdóttir, S., Gudmundsson, G., Einarsson, P., Heinert, M., and Völksen, C. (2001) Crustal deformation measured by GPS in the South Iceland Seismic Zone due to two large earthquakes in June 2000, *Geophys. Res. Lett.*, **28**, 4031–4033.

Árnadóttir, Th., Jónsson, S., Pedersen, R., and Gudmundsson, G. (2003) Coulomb stress changes in the South Iceland Seismic Zone due to two large earthquakes in June 2000, *Geophys. Res. Lett.*, **30**, doi:10.1029/2002GL016495.

Árnadóttir, Th., Geirsson, H., and Einarsson, P. (2004) Coseismic stress changes and crustal deformation on the Reykjanes Peninsula due to triggered earthquakes on 17 June 2000, *J. Geophys. Res.*, **109**, doi:10.1029/2004JB003130.

Arnadottir, Th., Jónsson, S, Pollitz, F. Jiang, W., Feigl, K. L., Sturkell, E., and Geirsson, H. (2004b) Post-seismic deformation following the June 2000 earthquake sequence in Southwest Iceland, *EOS Trans. AGU*, **85**, Fall. Meet. Suppl., Abstract G12A-06.

Árnadóttir, Th., Jónsson, S, Pollitz, F. Jiang, W., and Feigl, K. L. (in press) Post-seismic deformation following the June 2000 earthquake sequence in the south Icelandic seismic zone, *J. Geophys. Res.*

Árnadóttir, Th., Weiping, J., Feigl, K. L., Geirsson, H., and Sturkell, E. (submitted), Kinematic models of plate boundary deformation in the southwest Iceland derived from GPS observations, *J. Geophys. Res.*

Båth, M. (1960) Crustal structure of Iceland, *J. Geophys. Res.*, **65**, 1793–1807.

Bergþórsson, P. (1969) An estimate of drift ice and temperature in Iceland in 1000 years, *Jökull*, **19**, 94–101.

Bergerat, F., and Angelier, J. (2003) Mechanical behaviour of the Árnes ND Hestfjall faults of the June 2000 earthquakes in Southern Iceland: Inferences from surface traces and tectonic model, *J. Struct. Geol.*, **25**, 1507–1523.

Beblo, M., and Bjornsson, A. (1980) A model of electrical resistivity beneath NE-Iceland, correlation with temperature, *J. Geophys.*, **45**, 184–190

Bijwaard, H., and Spakman, W. (1999) Tomographic evidence for a narrow whole mantle plume below Iceland, *Earth Plan. Sci. Lett.*, **166**, 121–126.

Bjarnason, I., Cowie, P. A., Anders, M. H., Seeber, L., and Scholz, C. H. (1993a) The 1912 Iceland earthquake rupture: Growth and development of a nascent transform system, *Bull. Seis. Soc. Am.*, **83**, 416–435.

Bjarnason, I. Th., Menke, W., Flóvenz, Ó., and Caress, D. (1993b) Tomographic image of the Mid-Atlantic Plate Boundary in Southwestern Iceland, *J. Geophys. Res.*, **98**, 6607–6622.

Bjarnason, I. Th., Silver, P. G., Rümpker, and Solomon, S. C. (2002) Shear wave splitting across the Iceland hot spot: Results from the ICEMELT experiment, *J. Geophys. Res.*, **107**, doi:10.1029/2001JB000916.

Björnsson, A. (1985) Dynamics of crustal rifting in NE Iceland, *J. Geophys. Res.*, **90**, 10151–10162.

Björnsson, A., and Eysteinsson, H. (1998) *Changes in Land Elevation at Krafla 1974–1995: Summary of Measurements* (Orkustofnun Report OS-98002), National Energy Authority, Reykjavik, Iceland [in Icelandic].

Björnsson, A., Saemundsson, K., Einarsson, P., Tryggvason, E., and Grönvold, K. (1977), Current rifting episode in North Iceland, *Nature*, **266**, 318–323.

Björnsson, A., Johnsen, G., Sigurdsson, S., Thorbergsson, G., and Tryggvason, E. (1979) Rifting of the plate boundary in North Iceland 1975–1978, *J. Geophys. Res.*, **84**, 3029–3038.

Björnsson, G., Flóvenz, Ó. G., Sæmundsson, K., and Haraldsson, E. H. (2001) Pressure changes in Icelandic geothermal reservoirs associated with two large earthquakes in June 2000, *Proc. 26th Workshop on Geothermal Reservoir Engineering* (SPG-TR-168, pp. 327–334), Stanford University, Stanford.

Björnsson, H. (1992) Jökulhlaups in Iceland: Prediction, characteristics and simulation, *Ann. Glaciology*, **16**, 95–106.

Björnsson, H., and Einarsson, P. (1990) Volcanoes beneath Vatnajökull, Iceland: Evidence from radio echo-sounding, earthquakes and jökulhlaups, *Jökull*, **40**, 147–168.

Björnsson, H., and Guðmundsson, M. T. G. (1993) Variations in the thermal output of the subglacial Grímsvötn Caldera, Iceland, *Geophys. Res. Lett.*, **19**, 2127–2130.

Björnsson, H., Pálsson, F., and Guðmundsson, M. T. (2000) Surface and bedrock topography of the Mýrdalsjökull ice cap, *Jökull*, **49**, 29–46.

Björnsson, H., Pálsson, F., and Haraldsson, H. H. (2002) Mass balance of Vatnajökull (1991–2001) and Langjökull (1996–2001), Iceland, *Jökull*, **51**, 75–78.

Björnsson, S., and Einarsson, P. (1981) Jarðskjálftar (Earthquakes), in: H. Þórarinsdóttir and and S. Steinþórsson (eds), *Nattura Islands* (pp. 121–155), Almenna bókafélagið, Reykjavík [in Icelandic].

Bonaccorso, A., and Davis, P. M. (1999), Models of ground deformation from vertical volcanic conduits with application to eruptions of Mount St. Helens and Mount Etna, *J. Geophys. Res.*, **104**, 10531–10542.

Brander, J. L., Mason, R. G., and Calvert, R. W. (1976) Precise distance measurements in Icleand, *Tectonophysics*, **31**, 193–206.

Brandsdóttir, B. (1992) Historical accounts of earthquakes associated with eruptive activity in the Askja volcanic system, *Jökull*, **42**, 1–12.

Brandsdóttir, B., Menke, W., Einarsson, P., White, R. S., and Staples, R. K. (1997) Faroe–Iceland ridge experiment. 2: Crustal structure of the Krafla central volcano, *J. Geophys. Res.*, **102**, 7867–7886.

Brandsdottir, B., Richter, B., Riedel, C., Dahm, T., Helgadóttir, G., Kjartanson, E., Detrick, R., Magnusson, A., Asgrimsson, A. L., Palsson, B. H. et al. (2004) Tectonic details of the Tjornes Fracture Zone, and onshore–offshore ridge-transform in N-Iceland, *EOS Trans. AGU*, **85**, Fall Meet. Suppl., Abstract T41A-1172.

Braun, M. G., Hirth, G., and Parmentier, E. M. (2000) The effects of deep damp melting on mantle flow and melt generation beneath mid-ocean ridges, *Earth Planet. Sci. Lett.*, **176**, 339–356.

Camitz, J., Sigmundsson, F., Foulger, G., Jahn, C.-H., Völksen, C., and Einarsson, P. (1995), Plate boundary deformation and continuing deflation of the Askja volcano, North Iceland, determined with GPS, 1987–1993, *Bull. Volcanol.*, **57**, 136–145.

Cathles, L. M. (1975) *The Viscosity of the Earth's Mantle*, Princeton University Press, Princeton, 386 pp.

Cazenave, A., and Nerem, R. S. (2004) Present-day sea level change: Observations and causes, *Rev. Geophys.*, **42**, RG3001, doi:10.1029/2003RG000139

Claesson, L. (2004) Water–rock interaction in active fault zones: A potential earthquake precursor, Fil. Lic. Thesis, Univ. of Stockholm.

Claesson, L., Skelton, A., Graham, C., Dietl, C., Mörth, M., Torssander, P., and I Kockum (2004) Hydrogeochemical changes before and after a major earthquake, *Geology*, **32**, 641–644.

Clifton, A. E., and Schlische, R. W. (2003) Fracture populations on the Reykjanes Peninsula, Iceland: Comparison with experimental clay models of oblique rifting, *J. Geophys. Res.*, **108**, doi:10.1029/2001JB000635.

Clifton, A., and Einarsson, P. (2005) Styles of surface rupture accompanying the June 17 and 21, 2000 earthquakes in the South Iceland Seismic Zone, *Tectonophysics*, **396**, 141–159.

Clifton, A., Sigmundsson, F., Feigl, K., Gunnarsson, G., and Árnadóttir, Th. (2002) Surface effects of faulting and deformation resulting from magma accumulation at the Hengill triple junction, SW Iceland, 1994–1998, *J. Volcanol. Geotherm. Res.*, **115**, 233–255.

Clifton A. E., Pagli, C., Jónsdóttir, J. F., Eythorsdóttir, K., and Vogfjörd, K. (2003) Surface effects of triggered fault slip on Reykjanes Peninsula, SW-Iceland, *Tectonophysics*, **369**, 145–154.

Conrad, C. P., Litgow-Bertelloni, C., and Louden, K. (2004) Iceland, the Farallon slab, and dynamic topography of the North Atlantic, *Geology*, **32**, 177–180, doi:10.1130/G20137.1.

Dansgaard, W., White, J. W. C., and Johnsen, S. J. (1989) The abrupt termination of the Younger Dryas climate event, *Nature*, **339**, 532–534.

Darbyshire, F. A., Bjarnason, I. T., White, R. S., and Flóvenz, Ó. G. (1998) Crustal structure above the Iceland mantle plume imaged by the ICEMELT refraction profile, *Geophys. J. Int.* **135**, 1131–1149.

Darbyshire, F. A., White, R. S., and Priestley, K. F. (2000) Structure of the crust and uppermost mantle of Iceland from a combined seismic and gravity study, *Earth Planet. Sci. Lett.*, **181**, 409–428.

Decker, R., W., Einarsson, P., and Mohr, P. A. (1971) Rifting in Iceland: New geodetic data, *Science*, **173**, 530–533.

Decker, R., W., Einarsson, P., and Plumb, R. (1976) Rifting in Iceland: Measuring horizontal movements, *Greinar*, **5**, 61–71 [Societas Scientiarium Islandica, Reykjavík].

Delaney, P. T., and McTigue, D. F. (1994) Volume of magma accumulation and withdrawal estimated from surface uplift or subsidence, with application to the 1960 collapse of Kilauea volcano, *Bull. Volcanol.*, **56**, 417–424.

DeMets, C., Gordon, R. G., Argus, D. F., and Stein, S. (1994) Effect of recent revisions to the geomagnetic reversal time scale on estimates of current plate motions, *Geophys. Res. Lett.*, **21**, 2191–2194.

DeMets, C., Gordon, R. G., Argus, D. F., and Stein, S. (1990) Current plate motions, *Geophys. J. Int.*, **101**, 425–478.

de Zeeuw-van Dalfsen, E., Pedersen, R., Sigmundsson, F., and Pagli, C. (2004) Satellite radar interferometry 1993–1999 suggests deep accumulation of magma near the crust–mantle boundary at the Krafla volcanic system, Iceland, *Geophys. Res. Lett.*, **31**, L13611, doi:10.1029/2004GL020059.

de Zeeuw-van Dalfsen, E., Rymer, H., Sigmundsson, F., and Sturkell, E. (2005) Net gravity decrease at Askja volcano, Iceland: Constraints on processes responsible for continuous caldera deflation, 1988–2003, *J. Volcanol. Geotherm. Res.*, **139**, 227–239.

Einarsson, Þ. (1994) *Myndun og mótun lands, jarðfræði* (301 pp.), Mál og Menning, Reykjavík, Iceland [in Icelandic].

Einarsson, P. (1978) S-wave shadows in the Krafla caldera in NE-Iceland, evidence for a magma chamber in the crust, *Bull. Volcanol.*, **41**, 1–9.

Einarsson, P. (1985) Jarðskjálftaspár [Earthquake predictions], *Natturufraedingurinn*, **55**, 9–28 [in Icelandic with English abstract].

Einarsson, P. (1986) Seismicity along the eastern margin of the North American Plate, in: P. R. Vogt and B. E. Tucholke {eds), *The Geology of North America*, Volume M: *The Western North Atlantic Region* (pp. 99–116), Geological Society of America, Boulder, Colorado.

Einarsson, P. (1987) Compilation of earthquake fault plane solutions in the North Atlantic, in: K. Kasahara (ed.), *Recent Plate Movements and Deformation* (Geodyn. Ser. 20, pp. 47–62), American Geophysical Union, Washington, DC.

Einarsson, P. (1991a) Earthquakes and present-day tectonism in Iceland, *Tectonophysics*, **189**, 261–279.

Einarsson, P. (1994) Crustal movements and relative sea level changes in Iceland, in: G. Viggósson (ed.), *Proceedings of the Hornafjörður International Coastal Symposium* (pp. 24–34), The Icelandic Harbour Authority.

Einarsson, P. (1991b) *Umbrotin vid Kroflu 1975–89 [The 1975–89 activity at Krafla]*, in: A. Garðarson and Á. Einarsson (eds), *Náttúra Mývatns* (pp. 97–139), Hið åslenska Nattúrufræðifélag, Reykjavik [in Icelandic].

Einarsson, P. and Brandsdottir, B. (1980) Seismological evidence for lateral magma intrusion during the July 1978 deflation of the Krafla volcano in NE-Iceland, *J. Geophys.*, **47**, 160–165.

Einarsson, P. and Eiríksson, J. (1982) Earthquake fractures in the districts Land and Rangarvellir in the South Iceland Seismic Zone, *Jökull*, **32**, 113–120.

Einarsson, P. and Sæmundsson, K. (1987) Earthquake epicenters 1982–1985 and volcanic systems in Iceland, in: Th. Sigfusson (ed.), *I hlutarins edli: Festschrift for Thorbjorn Sigurgeirsson*, Menningarsjodur, Reykjavik [map].

Einarsson, P., Björnsson, S., Foulger, G., Stefánsson, R., and Skaftadóttir, T. (1981) Seismicity pattern in the South Iceland Seismic Zone, in: D. W. Simpson and P. G. Richards (eds), *Earthquake Prediction: An International Review* (Maurice Ewing Ser. Vol. 4, pp. 141–151), American Geophysical Union, Washington, DC.

Einarsson, P., Sigmundsson, F., Hofton, M. A., Foulger, G. R., and Jacoby, W. (1996) An experiment in glacio-isostasy near Vatnajökull, Iceland, 1991, *Jökull*, **44**, 29–39.

Einarsson, T. (1954) *Survey of Gravity in Iceland*, Societas Scientarium Islandica, Reykjavík.

Einarsson, T. (1966) Late and post-glacial rise in Iceland and sub-crustal viscosity, *Jökull*, **16**, 157–166.

Einarsson, Þ. (1991) *Myndun og mótun lands*, Mál og Menning, Reykjavik.

Eliasson, J., G. Larsen, M. T. Gudmundsson, and F. Sigmundsson (submitted) Probabilistic model for eruptions and associated flood events in the Katla caldera, Iceland, *Computational Geosciences*.

Eysteinsson, H. and Hermance, J. F. (1985) Magnetotelluric measurements across the eastern volcanic zone in South Iceland, *J. Geophys. Res.*, **87**, 10093–10103.

Eysteinsson, H. and Gunnarsson, K. (1995) *Maps of Gravity, Bathymetry and Magnetics for Iceland and Surroundings* (Rep. OS-95055/JHD-07), Orkustofnun, Reykjavik.

Ewart, J. A., Voight, B., and Björnsson, A. (1991) Elastic deformation models of Krafla volcano, Iceland, for the decade 1975 to 1985, *Bull. Volcanol.*, **53**, 436–459.

Fairbanks, R. G. (1989) A 17000-year glacio-eustatic sea level record: Influence of glacial melting rates on the Younger Dryas event and deep-ocean circulation, *Nature*, **342**, 637–642.

Feigl, K., Gasperi, J., Sigmundsson, F., and Rigo, A (2000) Crustal deformation near Hengill volcano, Iceland 1993–1998: Coupling between volcanism and faulting inferred from elastic modelling of satellite radar interferograms, *J. Geophys. Res.*, **105**, 25655–25670.

Flóvenz, Ó, G. and Gunnarsson, K. (1991) Seismic crustal structure in Iceland and surrounding area, *Tectonophysics*, **189**, 1–17.

Flóvenz, Ó., G. and Sæmundsson, K. (1993) Heat flow and geothermal processes in Iceland, *Tectonophysics*, **225**, 123–138.

Foulger, G. R. and Anderson, D. L. (2005) A cool model for the Iceland hotspot, *J. Volcanol. Geotherm. Res.*, **141**, 1–22.

Foulger, G. R., Jahn, C.-H., Seeber, G., Einarsson, P., Julian, B. R., and Heki, K. (1992) Post-rifting stress relaxation at the divergent plate boundary in Northeast Iceland, *Nature*, **358**, 488–490.

Foulger, G. R., Hofton, M. A., Julian, B. R., Jahn, C.-H., and Heki, K. (1994) Regional post-diking deformation in Northeast Iceland: A third epoch of GPS measurements in 1992, *Proceedings of the CRCM 1993, Kobe, December 6–11, 1993* (pp. 99–105).

Foulger, G. R., Pritchard, M. J., Julian, B. R., Evans, J. R., Allen, R. M., Nolet, G., Morgan, W. J., Bergsson, B. H., Erlendsson, P., Jakobsdottir, S. et al. (2000) The seismic anomaly beneath Iceland extends down to the mantle transition zone and no deeper, *Geophys. J. Int.*, **142**, F1–F5.

Foulger, G. R., Pritchard, M. J., Julian, B. R., Evans, J. R., Allen, R. M., Nolet, G., Morgan, W. J., Bergsson, B. H., Erlendsson, P., Jakobsdottir, S. et al. (2001) Seismic tomography shows that upwelling beneath Iceland is confined to the upper mantle, *Geophys. J. Int.*, **146**, 504–530.

Foulger, G. R., Natland, J. H., and Anderson, D. L. (2005) A source for Icelandic magmas in remelted Iapetus crust, *J. Volcanol. Geotherm. Res.*, **141**, 23–44.

Fowler, C. M. R. (2005) *The Solid Earth, an Introduction to Global Geophysics* (2nd edn), Cambridge University Press.

Friðleifsson, G. Ó. (1995) Miocene glaciation in eastern Iceland, in: B. Hróarson, D. Jónsson, and S. S. Jónsson (eds), *Eyjar í eldhafi* (pp. 77–85), Gott mál, Reykjavík [in Icelandic].

Gebrande, H., Miller, H., and Einarsson, P. (1980) Seismic profile of Iceland along the RRISP profile, *J. Geophys.*, **47**, 239–249.

Geirsdóttir, Á., Hardardóttir, J., and Eiríksson, J. (1997) The depositional history of the Younger Dryas–Preboreal Búdi moraines in south-central Iceland, *Arctic and Alpine Research*, **29**, 13–23.

Geirsdóttir, Á., Hardardóttir, J., and Sveinbjörnsdóttir, A. E. (2000) Glacial extent and catastrophic meltwater events during the deglaciation of Southern Iceland, *Quaternary Science Reviews*, **19**, 1749–1761.

Geirsson, H. (2003) Continuous GPS measurements in Iceland 1999–2002, M.S. thesis, University of Iceland [Rep. 03014, Icelandic Meteorological Office, Reykjavik].

Geirsson, H., Árnadóttir, Th., Völksen, C., Jiang, W., Sturkell, E., Villemin, T., Einarsson, P., Sigmundsson, F., and Stefánsson, R. (in press) Current plate movements across the Mid-Atlantic ridge determined from 5 years of continuous GPS measurements in Iceland, *J. Geophys. Res.*

Gudmundsson, A. Trausti (2001) *åslenskar eldstöðvar [Icelandic Volcanoes]* (320 pp.), Vaka-Helgafell, Reykjavik [in Icelandic].

Gudmundsson, A. (1986) Mechanical aspects of postglacial volcanism and tectonics of the Reykjanes Peninsula, Southwest Iceland, *J. Geophys. Res.*, **91**, 12711–12721.

Gudmundsson, A. (1988) Effect of tensile stress concentration around magma chambers on intrusion and extrusion frequencies, *J. Volcanol. Geotherm. Res.*, **35**, 179–194.

Gudmundsson, A. (1993) On the structure and formation of fracture zones, *Terra Nova*, **5**, 215–224.

Gudmundsson, A. and Brynjolfsson, S. (1993) Overlapping rift-zone segments and the evolution of the south Iceland seismic zone, *Geophys. Res. Lett.*, **20**, 1903–1906.

Gudmundsson, A., Brynjolfsson, S., and Jonsson, M. Th. (1993) Structural analysis of a transform fault-rift zone junction in North Iceland, *Tectonophysics*, **220**, 205–221.

Gudmundsson, A., Oskarsson, N., Gronvold, K., Saemundsson, K. Sigurdsson, O., Stefansson, R., Gislason, S. R., Einarsson, P., Brandsdottir, B., Larsen, G. et al. (1992) The 1991 eruption of Hekla, Iceland, *Bull. Volcanol.*, **54**, 238–246.

Gudmundsson, G. and Saemundsson, K. (1980) Statistical analysis of damaging earthquakes and volcanic eruptions in Iceland from 1550–1978, *J. Geophys.*, **47**, 99–109.

Gudmundsson, M. T. (2005) Subglacial volcanic activity in Iceland, in: Caseldyne et al. (eds.), *Iceland: Modern Processes and Past Environments*, Elsevier [in press].

Gudmundsson, M. T. and Milsom, J. (1997) Gravity and magnetic studies of the subglacial Grímsvötn volcano, Iceland: Implications for crustal and thermal structure, *J. Geophys. Res.*, **102**, 7691–7704.

Gudmundsson, M. T., Sigmundsson, F., and Bjornsson, H. (1997) Ice–volcano interaction of the 1996 Gjálp subglacial eruption, Vatnajökull, Iceland, *Nature*, **389**, 954–957.

Gudmundsson, M. T., Sigmundsson, F., Bjornsson, H., and Hognadottir, Th. (2004) The 1996 eruption at Gjálp, Vatnajökull ice cap: Course of events, efficiency of heat transfer, ice deformation and subglacial water pressure, *Bull. Volc.*, **66**, 46–65.

Gudmundsson, Ó. (2003) The dense root of the Iceland crust, *Earth Planet. Sci. Lett.*, **206**, 427–440.

Gudmundsson, O., Brandsdottir, B., Menke, W., and Sigvaldason, G. E. (1994) The crustal magma chamber of the Katla volcano in south Iceland revealed by 2-D seismic undershooting, *Geophys. J. Int.*, **119**, 277–296.

Gudmundsson, S., Sigmundsson, F., and Carstensen, J. M. (2002) Three-dimensional surface motion maps estimated from combined InSAR and GPS data, *J. Geophys. Res.*, **107**, doi:10.1029/2001JB000283.

Gripp, A. E. and Gordon, R. G. (2002) Young tracks of hotspots and current plate velocities, *Geophys. J. Int.*, **150**, 321–361.

Gronvold, K., Larsen, G., Einarsson, P., Thorarinsson, S., and Saemundsson, K. (1983) The Hekla eruption 1980–1981, *Bull. Volcanol.*, **46**, 349–363.

Grönvold, K., Óskarsson, N., Johnsen, S. J., Clausen, H. B., Hammer, C. U, Bond, G., and Bard, E. (1995) Ash layers from Iceland in the Greenland GRIP ice core correlated with oceanic and land sediments, *Earth Planet. Sci. Lett.*, **135**, 149–155.

Gunnarsson, B., Marsh, B. D., and Taylor, Jr. H. P. (1998) Generation of Icelandic rhyolites: Silicic lavas from the Torfajökull central volcano, *J. Volcanol. Geotherm. Res.*, **83**, 1–45.

Hackman, M. C., King, G. C. P., and Bilham, R. (1990) The mechanics of the south Iceland seismic zone, *J. Geophys. Res.*, **95**, 17339–17351.

Hardarson, B. S. and Fitton, J. G. (1991) Increased mantle melting beneath Snaefellsjökull volcano during late Pleistocene deglaciation, *Nature*, **353**, 62–64.

Hardarson, B. S., Fitton, J. G., Ellam, R. M., and M. S. Pringle (1997) Rift relocation: A geochemical and geochronological investigation of a palaeo-rift in northwest Iceland, *Earth Planet. Sci. Lett.*, **153**, 181–196.

Hauksson, E. (1983) Episodic rifting and volcanism at Krafla in North Iceland, growth of large ground fissures along the plate boundary, *J. Geophys. Res.*, **88**, 625–636.

Hauksson, E. and Goddard, J. G. (1981) Radon earthquake precursors studies in Iceland, *J. Geophys. Res.*, **86**, 7037–7054.

Haimson B. C. and Rummel F.(1982) Hydrofracturing stress measurements in the Iceland Research Drilling Project drill hole at Reydarfjordur, Iceland, *J. Geophys. Res.*, **87**, 6631–6649.

Hardardottir, J., Geirsdottir, A., and Sveinbjörnsdóttir, A. E. (2001) Seismostratigraphy and sediment studies of Lake Hestvatn, south Iceland: Implications for the deglacial history of the region, *Journal of Quaternary Science*, **16**, 167–179.

Heki, K., Foulger, G. R., Julian, B. R., and Jahn, C.-H. (1993) Plate dynamics near divergent boundaries: Geophysical implications of postrifting crustal deformation in NE Iceland, *J. Geophys. Res.*, **98**, 14279–14297.

Helgadóttir, G., Brandsdóttir, B., Detrick, R. S., and Driscoll, N. (2003) Glacial features on the northern insular margin of Iceland, *EOS Trans. AGU*, 84, Fall Meet. Suppl., Abstract PP32B-0286.

Hirth, G. and Kohlstedt, D. L. (1996) Water in oceanic upper mantle: Implications for rheology, melt extraction, and the evolution of the lithosphere, *Earth Planet. Sci. Lett.*, **144**, 93–108.

Hjartarson, Á. (1985) *Report OS-85044/VOD-19B*, Orkustofnun, Reykjavik.

Hjartarson, Á., and Ingólfsson, Ó. (1988) Preboreal glaciation of Southern Iceland, *Jökull*, **38**, 1–15.

Hjartarson, A. and Hafstað, Þ. (1997) *Sviðinhornahraun, berggrunnsrannsóknir og kort* (Report OS-97016), Orkustofnun, Reykjavik [in Icelandic]

Hofton, M. A. and Foulger, G. R. (1996a) Postrifting anelastic deformation around the spreading plate boundary, north Iceland. 1: Modeling of the 1987–1992 deformation field using a viscoelastic Earth structure, *J. Geophys. Res.*, **101**, 25403–25421.

Hofton, M. A. and Foulger, G. R. (1996b) Postrifting anelastic deformation around the spreading plate boundary, north Iceland. 2: Implications of the model derived from the 1987–1992 deformation field, *J. Geophys. Res.*, **101**, 25423–25436.

Hreinsdóttir, S. (1999) GPS geodetic measurements on the Reykjanes Peninsula, SW Iceland: Crustal deformation from 1993 to 1998, M.Sc. Thesis, University of Iceland, Reykjavik.

Hreinsdóttir, S., Einarsson, P., and Sigmundsson, F. (2001) Crustal deformation at the oblique spreading Reykjanes Peninsula, SW-Iceland: GPS measurements from 1993 to 1998, *J. Geophys. Res.*, **106**, 13803–13816.

Hung, S.-H., Shen, Y., and Chiao, L.-Y. (2004) Imaging the seismic velocity structure beneath the Iceland hot spot: A finite frequency approach, *J. Geophys. Res.*, **109**, doi:10.1029/2003JB002889.

Imsland, P. (1992) Sögur af Hellnaskeri, *Skaftfellingur*, **8**, 130–139 [in Icelandic].

Ingólfsson, Ó. and Norddahl, H. (2001) High relative sea level during the Bolling interstadial in Western Iceland: A reflection of ice-sheet collapse and extremely rapid glacial unloading, *Artic, Antarctic, and Alpine Research*, **33**, 231–243.

Ingólfsson, Ó., Norddahl, H., and H. Haflidason (1995) Rapid isostatic rebound in southwestern Iceland at the end of the last glaciation, *Boreas*, **24**, 245–259.

Ito, G. (2001) Reykjanes V-shaped ridges originating from a pulsing and dehydrating mantle plume, *Nature*, **411**, 681–684.

Ito, G., Lin, J., and Gable, C. W. (1996) Dynamics of mantle flow and melting at a ridge-centered hotspot: Iceland and the Mid-Atlantic Ridge, *Earth Planet. Sci. Lett.*, **144**, 53–74.

Ito, G., Shen, Y., Hirth, G., and Wolfe, C. (1999) Mantle flow, melting and dehydration of the Iceland mantle plume, *Earth Planet. Sci. Lett.*, **165**, 81–96.

Ito, G., Lin, J., and Graham, D. (2003) Observational and theoretical studies of the dynamics of mantle plume–mid-ocean ridge interaction, *Rev. Geophys.*, **41**(4), doi:10.1029/2002RG000117.

Jahn, C.-H. (1992) Untersuchungen über den Einsatz des Global Positioning Systems (GPS) zum nachweis recenter Erdkrustenbewegungen im spaltengebiet Nordost-Islands, Ph. D. thesis, Wissenschaftliche Arbeiten der Fachrichtung Vermessungswesen der Universität Hannover, Nr. 182.

Jacoby, W. R., Bürger, S., Smilde, P., and Wallner, H. (2001) Temporal gravity variations observed in SE Iceland, *InterRidge News*, **10**(1), 52–55.

Jakobsdóttir, S. S., Gudmundsson, G. B., and Stefansson, R. (2002) Seismicity in Iceland 1991–2000 monitored by the SIL system, *Jökull*, **51**, 87–94.

Jakobsson, S. P. (1972) Chemistry and distribution pattern of recent basaltic rocks in Iceland, *Lithos*, **5**, 365–386.

Jakobsson, S. P. (1979a) Petrology of recent basalts of the Eastern Volcanic Zone, Iceland, *Acta Naturalia Islandica*, **26**, 1–103.

Jakobsson, S. P. (1979b) Outline of the petrology of Iceland, *Jökull*, **29**, 57–73.

Jakobsson, S. P., Jónsson, J., and Shido, F. (1978) Petrology of the western Reykjanes Peninsula, Iceland, *J. Petrol.*, **19**, 669–705.

Jeffrey, G. B. (1920) Plane stress and strain in bipolar co-ordinates, *Trans. R. Soc. London*, Ser. A, **221**, 265–293.

Jóhannesson, H. (1980) Jarðlagaskipan og þróun rekbelta á Vesturlandi [Evolution of rift zones in western Iceland], *Náttúrufræðingurinn*, **50**, 13–31 [in Icelandic with English summary].

Jóhannesson, H. and Sæmundsson, K. (1998) *Geological Map of Iceland, 1:500 000, Bedrock Geology* (2nd edn), Icelandic Institute of Natural History, Reykjavik.

Johansen, B., Vogt., P. R., and Eldholm, O. (1984) Reykjanes Ridge: further analysis of crustal subsidence and time-transgressive basement topography, *Earth Planet. Sci. Lett.*, **68**, 249–258.

Johnsen, G., Bjornsson, A., and Sigurdsson, S. (1980) Gravity and elevation changes caused by magma movement beneath the Krafla caldera, North-east Iceland, *J. Geophys.*, **47**, 132–140.

Johnson, D. J. (1987) *Elastic and Inelastic Magma Storage at Kilauea Volcano* (USGS Prof. Paper 1350, pp. 1297–1306). U.S. Geological Survey, Reston, VA.

Johnson, D. J., Sigmundsson, F., and Delaney, P. T. (2000) Comment on volume of magma accumulation or withdrawal estimated from surface uplift or subsidence, with application to the 1960 collapse of Kilauea volcano by P. T. Delaney and D. F. McTigue, *Bull. Volc.*, **61**, 491–493.

Jones, S. M., White, N., and Maclennan, J. (2002) V-shaped ridges around Iceland: Implications for spatial and temporal patterns of mantle convection, *Geochem. Geophys. Geosyst.*, **3**, 1059, doi:10.1029/2002GC000361.

Jonsson, G., Kristjansson, L., and Sverrisson, M. (1991) Magnetic surveys of Iceland, *Tectonophysics*, **189**, 229–247.

Jonsson, J. (1957) Notes on changes of sea-level in Iceland: The Hoffellssandur, Part III, *Geografiska Annaler*, **39**, 143–212.

Jónsson, S. (1996) Crustal deformation across a divergent plate boundary, the Eastern Volcanic Rift Zone, south Iceland, 1967–1994, using GPS and EDM, M.Sc. thesis, University of Iceland, Reykjavik.

Jónsson, S., Einarsson, P., and Sigmundsson, F. (1997) Extension across a divergent plate boundary, the Eastern Volcanic Rift Zone, south Iceland, 1967–1994, observed with GPS and electronic distance measurements, *J. Geophys. Res.*, **102**, 11913–11929.

Jónsson, S., Segall, P., Pedersen, R., and Björnsson, G. (2003) Post-earthquake ground movements correlated to pore-pressure transients, *Nature*, **424**, 179–183.

Jouanne, F., Villemin, T., Berger, A., and Henriot, O. (submitted) Rift–transform junction in North-Iceland: Rigid blocks and narrow accommodation zones revealed by GPS 1997–1999–2002, *Geophys. J. Int.*

Jull, M. and McKenzie, D. (1996) The effect of deglaciation on mantle melting beneath Iceland, *J. Geophys. Res.*, **101**, 21815–21828.

Kaban, M. K., Flóvenz, Ó. G., and Pálmason, G. (2002) Nature of the crust–mantle transition zone and the thermal state of the upper mantle beneath Iceland from gravity modeling, *Geophys. J. Int.*, **149**, 281–299.

Kampfmann, W. and Berckhemer, H. (1985) High temperature experiments on the elastic and anelastic behaviour of magmatic rocks, *Phys. Earth Planet. Interior*, **40**, 223–247.

Karson, J. A., Brandsdottir, B., and Saemundsson, K. (2004) Upper crustal deformation in onshore exposures of the Tjornes Fracture Zone, Northern Iceland, *EOS Trans. AGU*, **85**, Fall. Meet. Suppl., Abstract T43D-07.

Kjartansson, E. and Gronvold, K. (1983) Location of a magma reservoir beneath Hekla volcano, Iceland, *Nature*, **301**, 139–141.

Kristinsson, A. P. (1988) *The Pronunciation of Modern Icelandic: A Brief Course for Foreign Students* (3rd edn, 67 pp.), The Icelandic Language Institute, Reykjavík.

Kristjánsson, L. and Jónsson, G. (1998) Aeromagnetic results and the presence of an extinct rift zone in western Iceland, *J. Geodynamics*, **2**, 99–108.

Kristjansson, L., Fridleifsson, I. B., and Watkins, N. D. (1980) Stratigraphy and paleomagnetism of the Esja, Eyrarfjall and Akrafjall Mountains, *J. Geophys.*, **47**, 31–42.

Kristjánsson, L., Jónsson, G., and Sverrisson, M. (1989) *Magnetic Surveys at the Science Institute* (Report RH01.89, 44 pp. and colour map at a scale 1 : 1,000,000), Science Institute, University of Iceland.

Kristjansson, L., Gudmundsson, A., and Haraldsson, H. (1995) Stratigraphy and paleomagnetism of a 3-km-thick Miocene lava pile in the Mjoifjördur area, eastern Iceland, *Geol. Rundsch.*, **84**, 813–830.

Larsen, C. F., Motyka, R. J., Freymuller, J. T., Echelmeyer, K. A., and Ivins, E. R. (2004) Rapid uplift of southern Alaska caused by recent ice loss, *Geophys. J. Int.*, **158**, 1118–1133.

Larsen, G. (2000) Holocene eruptions within the Katla volcanic system, south Iceland: Characteristics and environmental impact, *Jökull*, **49**, 1–28.

Larsen, G., Gudmundsson, M. T., and H. Björnsson (1998) Eight centuries of periodic volcanism at the center of the Iceland hotspot revealed by glacier tephrostratigraphy, *Geology*, **26**, 943–946.

Larsen, G., Dugmore, A., and Newton, A. (1999) Geochemistry of historical-age silicic tephras in Icleand, *The Holocene*, **9**, 463–471.

La Femina, P. C., Dixon, T. H., Malservisi, R., Árnadóttir, Th., Sturkell, E., Sigmundsson, F., and Einarsson, P. (in press) Geodetic GPS measurements in South Iceland: Strain accumulation and partitioning in a propagating ridge system, *J. Geophys. Res.*

Laundau, L. D. and Lifshitz, E. M (1986) *Theory of Elasticity* (3rd edn), Pergamon Press.

Lawver, L. A. and Müller, R. D. (1994) Iceland hotspot track, *Geology*, **22**, 311–314.

Linde, A., T., Agustsson, K., Sacks, I. S., and Stefansson, R. (1993) Mechanism of the 1991 eruption of Hekla from continuous borehole strain monitoring, *Nature*, **365**, 737–740.

Lisowski, M., Savage, J. C., and Prescott, W. H. (1991) The velocity field along the San Andreas Fault in Central and Southern California, *J. Geophys. Res.*, **96**, 8369–8389.

Lister, J. R. and Kerr, R. C. (1991) Fluid-mechanical models of crack propagation and their application to magma transport in dykes, *J. Geophys. Res.*, **96**, 10049–10077.

Luxey, P., Blondel, P., and Parson, L. M. (1997) The tectonic significance of the Central and South Iceland Seismic Transform Zone, *J. Geophys. Res.*, **102**, 17967–17980.

Maclennan, J., Jull., M, McKenzie, D., Slater, L., and Grönvold, K. (2002) The link between volcanism and deglaciation in Iceland, *Geochem. Geophys. Geosyst.*, **3**, 1062, doi:10.1029/2001GC000282.

Magnúsdóttir, B. and Norðdahl, H. (2000) Age of a whalebone and highest marine limit in Mt. Akrafjall, *Náttúrufræðingurinn*, **69**, 177–188 [in Icelandic with English summary].

Magnússon, E., Björnsson, H., Dall, J., and Pálsson, F. (2005) Volume changes of the Vatnajökull ice cap, Iceland, due to surface mass balance, ice flow, and subglacial melting at geothermal areas, *Geophys. Res. Lett.*, **32**, doi:10.1029/2004GL021615.

Marquart, G. and Schmeling, H. (2004) A dynamic model for the Iceland Plume and the North Atlantic based on tomography and gravity data, *Geophys. J. Int.*, **159**, 40–52.

Massonnet, D. and Sigmundsson, F. (2000) Remote sensing of volcano deformation by radar interferometry from various satellites, *Remote Sensing of Volcanoes*, AGU Geophysical Monograph 116, pp. 207–221). American Geophysical Union, Washington, DC.

McKenzie, D. (1984) The generation and compaction of partially molten rock, *J. Petrol.*, **25**, 713–765.

McTigue, D. F. (1987) Elastic stress and deformation near a finite spherical magma body: Resolution of the point source paradox, *J. Geophys. Res.*, **92**, 12391–12940.

Menke, W. (1999) Crustal isostasy indicates anomalous densities beneath Iceland, *Geophys. Res. Lett.*, **26**, 1215–1218.

Menke, W. and Levin, V. (1994) Cold crust in a hot spot, *Geophys. Res. Letters*, **21**, 1967–1970.

Menke, W. and Sparks, D. (1995) Crustal accretion model for Iceland predicts 'cold' crust, *Geophys. Res. Lett.*, **22**, 1673–1676.

Menke, W., Levin, V., and Sethi, R. (1995) Seismic attenuation in the crust at the mid-Atlantic plate boundary in south-west Iceland, *Geophys. J. Int.*, **122**, 175–182.

Menke, W., Brandsdóttir, B., Einarsson, P., and Bjarnason, I. Th. (1996) Reinterpretation of the RRISP-77 Iceland shear-wave profiles, *Geophys. J. Int.*, **126**, 166–172.

Menke, W., West, M., Brandsdóttir, B., and Sparks, D. (1998) Compressional and shear velocity structure of the lithosphere in Northern Iceland, *Bull. Seis. Soc. Am.*, **88**, 1561–1571.

Mogi, K. (1958) Relations between eruptions of various volcanoes and the deformations of the ground surface around them, *Bull. Earthquake Res. Inst. Univ. Tokyo*, **36**, 99–134.

Montelli, R., Nolet, G., Dahlen, F. A., Masters, G., Engdahl, E. R., and Hung, Shu-Huei (2004a) Finite-frequency tomography reveals a variety of plumes in the mantle, *Science*, **303**, 338–343.

Montelli, R., Nolet, G., Masters, G., Dahlen, F. A., and Hung, S.-H. (2004b) Global P and PP traveltime tomography: Rays versus waves, *Geophys. J. Int.*, **158**, 637–654.

Montelli, R., Nolet, G., Dahlen, F. A., and Masters, G. (2004c) Plumes or not. Yes, and plenty!, *EOS. Trans. AGU*, **85**, Fall Meet. Suppl., Abstract V43G-05C.

Montelli R., Nolet G., Dahlen F. A., and Masters G. (in preparation) Deep mantle plumes imaged with finite-frequency S-wave tomography, to be submitted to *Geochem. Geophys. Geosyst.*

Morgan, W. J. (1971) Convection plumes in the lower mantle, *Nature*, **230**, 42–43.

Morgan, W. J. (1983) Hotspot tracks and the early rifting of the Atlantic, *Tectonophysics*, **94**, 123–139.

Muir-Wood, R. and King, G. C. P. (1993) Hydrological signatures of earthquake strain, *J. Geophys. Res.*, **98**, 22035–22068.

Norddahl, H. (1983) Late Quaternary stratigraphy of Fnjoskadalur central north Iceland, Lundqua thesis, 12, Lund University, Lund, Sweden.

Norddahl, H. and Einarsson, Th. (2001) Concurrent changes of relative sea-level and glacier extent at the Weichselian–Holocene boundary in Berufjordur, Eastern Iceland, *Quaternary Science Reviews*, **20**, 1607–1622.

Norðdahl, H. and Pétursson, H. G. (2005) Relative sea-level changes in Iceland, new aspects of the Weichselian deglaciation of Iceland, in: C. Caseldine, A. Russel, J. Hardardóttir, and O. Knudsen (Eds), *Iceland: Modern Processes and Past Environments*, Elsevier, Amsterdam [in press].

Niemczyk, O. (1943) *Spalten auf Island*, K. Wittwer, Stuttgart.

Okada, Y. (1985) Surface deformation due to shear and tensile faults in a half-space, *Bull. Seism. Soc. Am.*, **75**, 1135–1154.

Ólafsdóttir, R. Höskuldsson, A., and Grönvold, K. (2002) The evolution of the lava flow from Hekla eruption 2000, *Nordic Geological Winter Meeting, Reykjavik, 6–9 January* (Abstract Volume, p. 149).

Óskarsson, N. (1980) The interaction between volcanic gases and tephra: Fluorine adhering to tephra of the 1970 Hekla eruption, *J. Volcanol. Geotherm. Res.*, **8**, 251–266.

Óskarsson, N., Sigvaldason, G.E., and Steinthorsson, S. (1982) A dynamic model of rift zone petrogenesis and the regional petrology of Iceland, *J. Petrol.*, **23**, 28–74.

Oskarsson, N., Steinthorsson, S., and Sigvaldason, G. E. (1985) Iceland geochemical anomaly: Origin, volcanotectonics, chemical fractionation and isotope evolution of the crust, *J. Geophys. Res.*, **90**, 10011–10025.

Pagli, C., Pedersen, R., Sigmundsson, F., and Feigl, K. (2003) Triggered fault slip on June 17, 2000 on the Reykjanes Peninsula, SW-Iceland captured by radar interferometry, *Geophys. Res. Lett.*, **30**, doi:10.1029/2002GL015310.

Pagli, C., Sigmundsson, F., Árnadóttir, Th., and Einarsson, P. (2003) Subsidence of Askja volcano, North Iceland, *EOS Trans. AGU*, 84, AGU Fall Meet. Abstracts, p. F1520.

Pagli, C., Sigmundsson, F., Sturkell, E., Geirsson, H., Einarsson, P., Lund, B., and Árnadóttir, Th. (2005) Ongoing glacio-isostatic crustal deformation around Vatnajokull ice cap, Iceland due to ice retreat: Observations and finite element modeling, *EOS Trans. AGU*, **86**, AGU Fall Meet. Abstracts, No. G33A-0017.

Pagli, C., Sigmundsson, F., Árnadóttir, Th., Einarsson, P., and Sturkell, E. (in press) Deflation of the Askja volcanic system: Constraints on the deformation source from combined inversion of satellite radar interferograms and GPS measurements, *J. Volcanol. Geotherm. Res.*.

Pálmason, G. (1971) *Crustal Structure of Iceland from Explosion Seismology* (187 pp.), Societas Scientiarium, Iceland, XL, Reykjavík.

Pálmason, G. (1973) Kinematics and heat flow in a volcanic rift zone, with application to Icleand, *Geophys. J. R. Astron. Soc.*, **33**, 451–481.

Pálmason, G. (1980) A continuum model of crustal generation in Iceland: Kinematic aspects, *J. Geophys.*, **47**, 7–18.

Pálmason, G. (1981) Crustal rifting, and related thermo-mechanical processes in the lithosphere beneath Iceland, Geologische Rundschau, 70, 244-260.

Pálmason, G. (1986) Model of crustal formation in Iceland, and application to submarine mid-ocean ridge, in: P. R. Vogt and B. E.Tucholke (eds), *The Geology of North America*, Volume M: *The Western North Atlantic Region* (pp. 87–97), Geological Society of America, Boulder, CO.

Parsons, B. and Sclater, J. G. (1977) An analysis of the variation of ocean floor bathymetry and heat flow with age, *J. Geophys. Res.*, **82**, 803–827.

Paterson, W. S. B. (1983), *The Physics of Glaciers* (2nd edn), Pergamon Press, Oxford.

Pedersen, R. and Sigmundsson, F. (2004) InSAR based sill model links spatially offset areas of deformation and seismicity for the 1994 unrest episode at Eyjafjallajökull volcano, Iceland, *Geophys. Res. Lett.*, **31**, L14610, doi:10.1029/2004GL020368.

Pedersen, R. and Sigmundsson, F. (in press) Temporal development of the 1999 intrusive episode in the Eyjafjallajökull volcano, Iceland, derived from InSAR images, *Bull. Volc.*.

Pedersen, R., Sigmundsson, F., Feigl, K.L and Th. Árnadóttir (2001) Coseismic interferograms of two Ms = 6.6 earthquakes in the South Iceland Seismic Zone, June, 2000, *Geophys. Res. Lett.*, **28**, 3341–3344.

Pedersen, R., Jónsson, S., Árnadóttir, Th., Sigmundsson, F., and Feigl,K.L. (2003) Fault slip distribution of two June 2000 Mw6.4 earthquakes in South Iceland estimated from joint inversion of InSAR and GPS measurements, *Earth Planet. Sci. Lett.*, **213**, 487–502.

Perlt, J. and Heinert, M. (2000) Kinematic model of the South Iceland tectonic system, *Geophys. J. Int.*, **142**, 1–9.

Pilidou, S., Priestly, K., Gudmundsson, Ó., and Debayle, E. (2004) Upper mantle S-wave speed heterogeneity and anisotropy beneath the North Atlantic from regional surface wave tomography: The Iceland and Azores plumes, *Geophys. J. Int.*, **159**, 1057–1076.

Pinel, V. and Jaupart, C. (2004) Magma storage and horizontal dyke injection beneath a volcanic edifice, *Earth Planet. Sci. Lett.*, **221**, 245–262.

Pinel, V. and Jaupart, C. (2003) Magma chamber behaviour beneath a volcanic edifice, *J. Geophys. Res.*, **108**, 2072, doi:10.1029/2002JB001751.

Pollitz, F. F. and Sacks, I. S. (1996) Viscosity structure beneath northeast Iceland, *J. Geophys. Res.*, **101**, 17771–17793.

Ribe, N. M., Christensen, U. R., and Theissing, J. (1995) The dynamics of plume–ridge interaction, 1: Ridge-centered plumes, *Earth Planet. Sci. Lett.*, **134**, 155–168.

Ritter, B. (1982) *Untersuchungen geodätischer Netze in Island zur Analyse von Deformationen von 1965 bis 1977* (Reihe C, p. 271), Deutche Geodätische Kommision bei der Bayerischen Akademie der Wissenschaften, München.

Rögnvaldsson, S. T., Gudmundsson, A., and Slunga, R. (1998) Seismotectonic analysis of the Tjörnes Fracture Zone, an active transform fault in north Iceland, *J. Geophys. Res.*, **103**, 30117–30129.

Rundgren, M., Ingólfsson, Ó., Björck, S., Jiang, H., and Hafliðason, H. (1997) Dynamic sea-level change during the last glaciation of northern Iceland, *Boreas*, **26**, 202–215.

Rymer, H., J. Cassidy, C. A. Locke, and F. Sigmundsson (1998) Post-eruptive gravity changes from 1990 to 1996 at Krafla volcano, Iceland, *J. Volcanol. Geotherm. Res.*, **87**, 141–149.

Saemundsson, K. (1967) An outline of the structure of SW-Iceland, *Iceland and Mid-Ocean Ridges* (Publication 38, pp. 151–161), Societas Scientiarium Islandica, Reykjavik.

Sæmundsson, K. (1974) Evolution of the axial rifting zone in Northern Iceland and the Tjörnes Fracture Zone, *Bull. Geol. Soc. Am.*, **85**, 495–504.

Saemundsson, K. (1978) Fissure swarms and central volcanoes of the neovolcanic zones of Iceland, in: D. R. Bowes and B. E. Leake (eds), Crustal evolution in NW-Britain and adjacent regions, *Geol. J.*, Spec. Issue, **10**, 415–432.

Saemundsson, K. (1979) Outline of the geology of Iceland, *Jökull*, **29**, 7–28.

Saemundsson, K. (1986) Subaerial volcanism in the western North Atlantic, in: P. R. Vogt and B. E. Tucholke (eds), *The Geology of North America*, Volume M: *The Western North Atlantic Region* (pp. 69–86), Geological Society of America, Boulder, CO.

Sæmundsson, K. (1991) Jardfraedi Kroflukerfisins [The geology of the Krafla volcanic system], in: A. Garðarson and Á. Einarsson (eds), *Náttúra åslands* (pp. 25–95), Hið åslenska Nattúrufræðifélag, Reykjavik [in Icelandic].

Saemundsson, K. (1991) Geology of the Thingvallavatn area, *OIKOS*, **64**, 40–68.

Saemundsson, K. and Noll, H. (1974) K/Ar ages of rocks from Husafell, Western Iceland, and the development of the Husafell central volcano, *Jökull*, **24**, 40–59.

Sæmundsson, K. and Friðleifsson, G. Ó. (2001) *Torfajökull* (geological and geothermal map of the Torfajökull area, OS-2001/036), Orkustofnun, Reykjavik.

Sæmundsson, K., Jóhannesson, H., and Grönvold, K. (2005) Hrúthálsar: A central volcano in Ódáðahraun, *Spring Meeting* (abstract volume, p. 47), Geoscience Society of Iceland [in Icelandic].

Sandwell, D. T. and MacKenzie, K. R. (1989) Geoid height versus topography for oceanic plateaus and swells, *J. Geophys. Res.*, **94**, 7043–7418.

Sandwell, D. and Smith, W. H. F. (1997) Marine gravity from Geosat and ERS-1 altimetry, *J. Geophys. Res.*, **102**, 10039–10054.

Savage, J. C., and Burford, R. O. (1973) Geodetic determination of relative plate motion in central California, *J. Geophys. Res.*, **78**, 832–845.

Schilling, J.-G. (1973a) Iceland mantle plume: Geochemical study of Reykjanes Ridge, *Nature*, **242**, 565–571.

Schilling, J.-G. (1973b) Iceland mantle plume, *Nature*, **246**, 141–143.

Schilling, J.-G. (1986) Geochemical and isotopic variation along the Mid-Atlantic Ridge axis from 79°N to 0°N, in: P. R. Vogt and B. E. Tucholke (eds), *The Geology of North America*, Volume M: *The Western North Atlantic Region* (pp. 137–156), Geological Society of America, Boulder, CO.

Schubert, G., Turcotte, D. L., and Olson, P. (2001) *Mantle Convection in the Earth and Planets* (940 pp.), Cambridge University Press, Cambridge.

Sella, G. F., Dixon, T., Mao, A. (2002), REVEL: A model for recent plate velocities from space geodesy, *J. Geophys. Res.*, **107**, doi:10.1029/2000JB000033.

Shen, Y., Solomon, S. C., Bjarnason, I. Th., and Purdy, G. M. (1996) Hot mantle transition zone beneath Iceland and the adjacent Mid-Atlantic Ridge inferred from P-to-S conversions at the 410- and 660-km discontinuities, *Geophys. Res. Lett.*, **23**, 3527–3530.

Shen, Y., Salomon, S. C., Bjarnason, I. Th., and Wolfe, C. J. (1998) Seismic evidence for a lower-mantle origin of the Iceland plume, *Nature*, **395**, 62–65.

Shen, Y., Salomon, S. C., Bjarnason, I. Th., Nolet, G., Morgan, W. J., Allen, R. M., Vogfjörd, K., Jakobsdóttir, S., Stefánsson, R., Julian, B. R. et al. (2002) Seismic evidence for a tilted mantle plume and north–south mantle flow beneath Iceland, *Earth. Planet. Sci. Lett.*, **197**, 261–272.

Sigmarsson, O., Condomines, M., and Fourcade, S. (1992) A detailed Th, Sr and O isotope study of Hekla: Differentiation processes in an Icelandic volcano, *Contrib. Mineral. Petrol.*, **112**, 20–34.

Sigmundsson, F. (1990), Viscosity of the Earth beneath Iceland: Comparison of model calculations with geological data, M.S. thesis, University of Iceland, 121 pp. [in Icelandic with English abstract].

Sigmundsson, F. (1991) Post-glacial rebound and asthenosphere viscosity in Iceland, *Geophys. Res. Lett.*, **18**, 1131–1134.

Sigmundsson, F. (1992a) Crustal deformation studies in sub-aerial parts of the World Oceanic Rift System: Iceland and Afar, Ph. D. thesis, University of Colorado, Boulder, CO.

Sigmundsson, F. (1992b) Tectonic implications of the 1989 Afar earthquake sequence, *Geophys. Res. Lett.*, **19**, 877–880.

Sigmundson, F. (1996) Crustal deformation at volcanoes, in: F. Barberi and R. Casale (eds), *Proceedings of the Course: The Mitigation of Volcanic Hazards, European School of Climatology and Natural Hazards, held on Vulcano, Sicily, 12–18 June 1994* (ECSC-EC-EAEC, pp. 237–257), Office for Official Publications of the EC, Belgium.

Sigmundsson, F. and Einarsson, P. (1992) Glacio-isostatic crustal movements caused by historical volume change of the Vatnajökull ice cap, Iceland, *Geophys. Res. Lett.*, **19**, 2123–2126.

Sigmundsson, F., Einarsson P., and Bilham, R. (1992) Magma chamber deflation recorded by the Global Positioning System: The Hekla 1991 eruption, *Geophys. Res. Lett.*, **19**, 1483–1486.

Sigmundsson, F., Einarsson, P., Bilham, R., and Sturkell, E. (1995) Rift–transform kinematics in south Iceland: Deformation from Global Positioning System measurements, 1986–1992, *J. Geophys. Res.*, **100**, 6235–6248.

Sigmundsson, F., Einarsson, P., Rögnvaldsson, S. Th., Foulger, G. R., Hodgkinson, K. M., and Thorbergsson, G. (1997) The 1994–1995 seismicity and deformation at the Hengill triple junction, Iceland: Triggering of earthquakes by minor magma injection in a zone of horizontal shear stress, *J. Geophys. Res.*, **102**, 15151–15161.

Sigmundsson F., H. Vadon, and D. Massonnet (1997) Readjustment of the Krafla spreading segment to crustal rifting measured by satellite radar interferometry, *Geophys. Res. Lett.*, **24**, 1843–1846.

Sigmundsson, F., Pedersen, R., Feigl, K., Sturkell, E., Einarsson, P., Jonsson, S., Agustsson, K., Linde, A., and Bretar, F. (2001) Joint interpretation of geodetic data for volcano studies: Evalution of pre- and co-eruptive deformation for the Hekla 2000 eruption from InSAR, tilt, GPS and strain data, *EOS Trans. AGU*, **82**, Fall. Meet. Suppl., Abstract V21E-E07.

Sigurðsson, H. (1966) Geology of the Setberg Area, Snæfellsnes, Western Iceland, *Greinar*, **IV**, 2 [Societas Scientiarium Islandica, Reykjavik].

Sigurdsson, H. (1970) Structural origin and plate tectonics of the Snaefellsnes volcanic zone, Western Iceland, *Earth Planet. Sci. Lett.*, **10**, 129–135.

Sigurdsson, H. and Sparks, S. (1978) Rifting episode in North Iceland in 1874–1875 and the eruptions of Askja and Sveinagja, *Bull. Volcanol.*, **41**, 149–167.

Sigurgeirsson, Þ. (1975-1980) *Aeromagnetic Profile Maps of Iceland in Scale 1:250,000* (Sheets 1 to 9), Science Institute, University of Iceland.

Sigvaldason, G. E. (1974) The petrology of Hekla and origin of silicic rocks in Iceland, *Societas Scientiarium Islandica*, **5**, 1–44.

Sigvaldason, G. E. (1979) *Rifting, Magmatic Activity and Interaction between Acid and Basic Liquids: The 1875 Askja Eruption in Iceland* (Report 7903), Nordic Volcanological Institute, Reykjavík.

Sigvaldason, G. E., Steinthorsson, S., Oskarsson., N., and Imsland, P. (1974) Compositional variation in recent Icelandic tholeiites and the Kverkfjöll hot spot, *Nature*, **251**, 579–582.

Sigvaldason, G. E., Annertz, K., and Nielsson, M. (1992) Effect of glacier loading/deloading on volcanism: Postglacial volcanic production rate of the Dyngjufjoll area, central Iceland, *Bull. Volcanol.*, **54**, 385–392.

Sigvaldason, G. E. (2002) Volcanic and tectonic processes coinciding with glaciation and crustal rebound: An early Holocene rhyolitic eruption in the Dyngjufjöll volcanic centre and the formation of the Askja caldera, north Iceland, *Bull. Volcanol.*, **64**, 192–205.

Sinton, J., Grönvold, K., and Sæmundsson, K. (2005) Post-glacial eruptive history of the Western Volcanic Zone, Iceland, *Geochemistry, Geophysics, Geosystems*, doi:10.1029/2005GC0010121.

Sjöberg, L. E., Pan., M., Asenjo, E., and Erlingsson, S. (2000) Glacial rebound at Vatnajökull, Iceland, studied by GPS campaigns in 1992 and 1996, *J. Geodynamics*, **29**, 63–70.

Sjöberg, L. E., Pan, M., Erlingsson, S., Asenjo, E., and Arnason, K. (2004) Land uplift near Vatnajökull, Iceland, as observed by GPS in 1992, 1996 and 1999, Geophys. J. Int., 159, 943-948.

Slater, L., Jull, M., McKenzie, D., and Grönvold, K. (1998) Deglaciation effects on mantle melting under Iceland: Results from the northern volcanic zone, *Earth. Planet. Sci. Lett.*, **164**, 151–164.

Sleep, N. H. (1990) Hotspots and mantle plumes: Some phenomenology, *J. Geophys. Res.*, **95**, 6715–6736.

Soosalu, H. and Einarsson, P. (1997) Seismicity around the Hekla and Torfajökull volcanoes, Iceland, during a volcanically quit period, 1991–1995, *Bull. Volcanol.*, **59**, 36–48.

Soosalu, H. and Einarsson, P. (2002) Earthquake activity related to the 1991 eruption of the Hekla volcano, Iceland, *Bull. Volcanol.*, **63**, 536–544.

Soosalu, H. and Einarsson, P. (2004) Seismic constraints on magma chambers at Hekla and Torfajökull volcanoes, Iceland, *Bull. Volcanol.*, **66**, 276–286.

Soosalu, H., Einarsson, P., and Þorbjarnardóttir, B. (2005) Seismic activity related to the 2000 eruption of the Hekla volcano, Iceland, *Bull. Volcanol.* (in press).

Sparks, R. S. J., Sigurdsson, H., and L. Wilson (1977) Magma mixing: A mechanism for triggering acid explosive eruption, *Nature*, **267**, 315–318.

Sparks, R. S. J., Wilson, L., and Sigurdsson, H. (1981) The pyroclastic deposits of the 1875 eruption of Askja, Iceland, *Philos. Trans. R. Soc. London*, **299**, 241–273.

Staples, R. K., White, R. S., Brandsdottir, B., Menke, W., Maguire, P. K. H., and McBride, J. H. (1997) Faeroe–Iceland ridge experiment, 1: Crustal structure of northeastern Iceland, *J. Geophys. Res.*, **102**, 7849–7866.

Stefánsson, R. and Halldórsson, P. (1988) Strain release and strain build-up in the south Iceland seismic zone, *Tectonophysics*, **152**, 267–276.

Stefánsson, R. and Guðmundsson, G. B. (2005) *About the State-of-the-art in Providing Earthquake Warnings in Iceland* (Report Vå-ES-03), Icelandic Meteorological Office, Reykjavík.

Stefánsson, R., Bodvarsson, R., Slunga, R., Einarsson, P., Jakobsdottir, S., Bungum, H., Gregersen, S., Havskov, J., Hjelme, J., and Korhonen, H. (1993) Earthquake prediction research in the south Iceland seismic zone and the SIL project, *Bull. Seis. Soc. Am.*, **83**, 696–716.

Stefánsson, R., Guðmundsson, G., and Halldórsson, P. (2000) The two large earthquakes in the South Iceland Seismic Zone on June 17 and 21, 2000, Online report of the Icelandic Meteorological Office, Iceland, available at *http://hraun.vedur.is/ja/skyrslur/June17and21_2000/index.html*

Steingrímsson, J. (1998) *Fires of the Earth: The Laki Eruption 1783–1784* (93 pp.), English translation by K. Kunz of *Fullkomið rit um Síðueld*, University of Iceland Press and the Nordic Volcanological Institute, Reykjavik,

Stone, R. (2004) Iceland's doomsday scenario news focus, *Science*, **306**, 1278–1281.

Stothers, R. B. (1998) Far reach of the tenth century Eldgjá eruption, Iceland, *Climatic Change*, **39**, 715–726.

Sturkell, E. and Sigmundsson, F. (2000) Continuous deflation of the Askja caldera Iceland, during the 1983–1998 noneruptive period, *J. Geophys. Res.*, **105**, 25671–25684.

Sturkell, E., Sigmundsson, F., Einarsson, P., and Bilham, R. (1994) Strain accumulation 1986–1992 across the Reykjanes Peninsula plate boundary, Iceland, determined from GPS measurements, *Geophys. Res. Lett.*, **21**, 125–128.

Sturkell, E., Sigmundsson, F., and Slunga, R. (submitted) 1983–2003 decaying rate of deflation at Askja caldera: Pressure decrease in an extensive magma plumbing system at a spreading plate boundary, *Bull. Volcanol.*

Sturkell, E., Einarsson, P., Sigmundsson, F., Hreinsdottir, S., Geirsson, H. (2003a) Deformation of Grímsvötn volcano, Iceland: 1998 eruption and subsequent inflation, *Geophys. Res. Lett.*, **30**, 1182, doi:10.129/2002GL016460, 2003.

Sturkell, E., Sigmundsson, F., and Einarsson, P. (2003b) Recent unrest and magma movements at Eyjafjallajökull and Katla volcanoes, Iceland, *J. Geophys. Res.*, **108**, 2369, doi:10.1029/2001JB000917.

Sturkell, E., Einarsson, P., Sigmundsson, F., Geirsson, H., Olafsson, H., Pedersen, R., de Zeeuw-van Dalfsen, E., Linde, A. L., Sacks, I. S., and Stefánsson, R. (2005) Volcano geodesy and magma dynamics in Iceland, *J. Volcanol. Geotherm. Res.* (in press).

Sun, S. S., Tatsumoto, M., and Schilling, J.-G. (1975) Mantle plume mixing along the Reykjanes Ridge Axis: Lead isotopic evidence, *Science*, **190**, 143–147.

Sveinbjörnsdóttir, Á., E. and Johnsen, S. J. (1990) The late glacial history of Iceland: Comparison with isotopic data from Greenland and Europe, and deep sea sediments, *Jökull*, **40**, 83–96.

Sveinbjörnsdóttir, Á., E., Eiríksson, J., Geirsdóttir, Á., Heinemeir, J., and Rud, N. (1993) The Fossvogur marine sediments in SW-Iceland: Confined to the Allerød/Younger Dryas transition by AMS ^{14}C dating, *Boreas*, **22**, 147–157.

Thoma, M. and Wolf, D. (2001) Inverting land uplift near Vatnajökull, Iceland, in terms of lithosphere thickness and viscosity stratification, in: M. G. Sideris (ed.), *Gravity, Geoid, and Geodynamics 2000* (pp. 97–102), Springer-Verlag, Berlin.

Thorarinsson, S. (1944) *Tefrokronologiska studier pa Island*, Munksgaard, Copenhagen.

Thorarinsson, S. (1953) Some new aspects of the Grímsvötn problem, *J. Glaciol.*, **2**, 267–275.

Thorarinsson, S. (1958) The Oraefajökull eruption of 1362, *Acta Naturalia Islandica* (Vol. II, No. 2). Natturugripasafn Islands, Reykjavik

Thorarinsson, S. (1967) The eruptions of Hekla in historical times, *The Eruption of Hekla 1947–1948* (Vol. I, pp. 1–170), Societas Scientiarium Islandica, Reykjavik.

Thorarinsson, S. and Saemundsson, K. (1979) Volcanic activity in historical time, *Jökull*, **29**, 29–32.

Thorarinsson, S. and Sigvaldason, G. E. (1972) The Trollagigar eruption 1862–1864, *Jökull*, **22**, 12–26 [in Icelandic with English summary].

Thordarson, Th. and Self, S. (1993) The Laki (Skaftar Fires) and Grimsvotn eruptions in 1783–1785, *Bull. Volcanol.*, **55**, 233–263.

Thordarsson, T., Miller, D. J., Larsen, G., Self, S., and Sigurdsson, H. (2001) New estimates of sulfur degassing and atmospheric mass-loading by the 934 AD Eldgja eruption, Iceland, *J. Volcanol. Geotherm. Res.*, **108**, 33–54.

Thordarson, T. and Hoskuldsson, A. (2002) *Iceland: Classic Geology in Europe* (Vol. 3, 200 pp.). Terra, UK.

Thors, K. and Boulton, G. S. (1991) Deltas, spits and littoral terraces associated with rising sea level: Late Quaternary examples from northern Iceland, *Marine Geology*, **98**, 99–112.

Thors, K. and Helgadóttir, G. (1991) Evidence from south west Iceland of low sea level in early Flandrain time, in: J. K. Maizels and C. Caseldine (eds), *Environmental Change in Iceland: Past and Present* (pp. 93–104), Kluwer Academic Publishers, Dordrecht.

Tiampo, K. F., Rundle, J. B., Fernandez, J., and Lanbein, J. O (2000) Spherical and ellipsoidal volcanic sources at Long Valley caldera, California, using a genetic algorithm inversion technique, *J. Volcanol. Geotherm. Res.*, **102**, 189–206

Tryggvason, E. (1964) Arrival times of P waves and upper mantle structure, *Bull. Seism. Soc. Am.*, **54**, 727–736.

Tryggvason, E. (1973) Seismicity, earthquake swarms, and plate boundaries in the Iceland region, *Bull. Seism. Soc. Am.*, **63**, 1327–1348.

Tryggvason, E. (1974) Vertical crustal movement in Iceland, in Kristjansson (ed.), *Geodynamics of Iceland and the North Atlantic Area* (pp. 241–262), Reidel, Dordrecht.

Tryggvason, E. (1980) Subsidence events in the Krafla area, North Iceland, 1975-1979, *J. Geophys.*, **47**, 141–153.

Tryggvason, E. (1984) Widening of the Krafla fissure swarm during the 1975–1981 volcano-tectonic episode, *Bull. Volcanol.*, **47**, 47–69.

Tryggvason, E. (1986) Multiple magma reservoirs in a rift-zone volcano: Ground deformation and magma transport during the September 1984 eruption of Krafla, Iceland, *J. Volcanol. Geotherm. Res.*, **28**, 1–44.

Tryggvason, E. (1987) Myvatn lake level observations 1984–1986 and ground deformation during a Krafla eruption, *J. Volcanol. Geotherm. Res.*, **39**, 61–71.

Tryggvason, E. (1989) Ground deformation in Askja, Iceland: Its source and possible relation to flow of the mantle plume, *J. Volcanol. Geotherm. Res.*, **39**, 61–71.

Tryggvason, E. (1991) Magnitude of deformation of active volcanoes in Iceland, *Cahiers du Centre Européen de Géodynamique et de Séismologie*, **4**, 37–49.

Tryggvason, E. (1994) Observed ground deformation at Hekla, Iceland prior to and during the eruptions of 1970, 1980–81 and 1991, *J. Volcanol. Geotherm. Res.*, **61**, 281–291.

Tryggvason, E. and Båth, M. (1961) Upper crustal structure of Iceland, *J. Geophys. Res.*, **66**, 1913–1925.

Tryggvason, K., Husebye, E. S., and Stefánsson, R. (1983) Seismic image of the hypothesized Icelandic hot spot, *Tectonophysics*, **100**, 97–188.

Turcotte, D. L., and Schubert, G. (1982) *Geodynamics*, John Wiley & Sons.

Vadon, H. and Sigmundsson, F. (1997) Crustal deformation from 1992-1995 at the Mid-Atlantic Ridge, SW Iceland, mapped by satellite radar interferometry, *Science*, **275**, 193–197.

Vine, F. J. (1966) Spreading of the ocean floor: New evidence, *Science*, **154**, 1405–1415.

Vine, F. J. and Matthews, D. H. (1963) Magnetic anomalies over ocean ridges, *Nature*, **199**, 947–949.

Vink, G. E. (1984) A hotspot model for Iceland and the Voring Plateau, *J. Geophys. Res.*, **89**, 9949–9959.

Vogfjörd, K. S., Jakobsdóttir, S. S., Gudmundsson, G. B., Roberts, M. J., Ágústsson, K., Arason, T., Geirsson, H., Karlsdóttir, S., Hjaltadóttir, S., Ólafsdóttir, U. et al. (2005) Forecasting and monitoring a subglacial eruption in Iceland, *EOS Trans. AGU*, **86**, 245–252.

Vogt, P. R. (1971) Asthenosphere motion recorded by the ocean floor south of Iceland, *Earth Planet. Sci. Lett.*, **13**, 153–160.

Vogt, P. R. (1983) The Iceland mantle plume; Status of the hypothesis after a decade of new work, in: M. H. P. Bott, S. Saxov, M. Talwani, and J. Thiede (eds), *Structure and Development of the Greenland–Scotland Ridge* (pp. 191–216), Plenum Press, New York.

Vogt, P. R. (1986a) The present plate boundary configuration, in: P. R. Vogt and B. E. Tucholke (eds), *The Geology of North America*, Volume M: *The Western North Atlantic Region* (pp. 189–204), Geological Society of America, Boulder, CO.

Vogt, P. R. (1986b) Plate kinematics during the last 20 m.y., and the problem of present motions, in: P. R. Vogt and B. E. Tucholke (eds), *The Geology of North America*,

Volume M: *The Western North Atlantic Region* (pp. 405–425), Geological Society of America, Boulder, CO.
Vogt, P. R. (1986c) Geoid undulations mapped by spaceborne radar altimetry, in: P. R. Vogt and B. E. Tucholke (eds), *The Geology of North America*, Volume M: *The Western North Atlantic Region* (pp. 215–228), Geological Society of America, Boulder, CO.
Vogt, P. R., Johnson, G. L., and Kristjansson, L. (1980), Morphology and magnetic anomalies north of Iceland, *J. Geophys.*, **47**, 67–80.
Völksen, C. (2000) Die Nutzung von GPS für die Deformationsanalyse in regionalen Netzen am Beispiel Island, Ph.D. thesis, Wissenschaftliche Arbeiten der Fachrichtung Vermessungswesen der Universität Hannover, Nr. 237.
Völksen, C. and Seeber, G. (1998) Nachweis von rezenten Krustendeformationen in Nordisland mit GPS, *Zeitschrift für Vermessungswesen*, **2**/1998, 68–75.
Walker, G. P. L. (1963) The Breiddalur central volcano, *Q. J. Geol. Soc. London*, **119**, 25–63.
Walker, G. P. L. (1965) Some aspects of Quaternary volcanism, *Trans. Leicester Lit. Phil. Soc.*, **59**, 25–40.
Watts, A. B. (2001) *Isostasy and Flexure of the Lithosphere*, Cambridge University Press, Cambridge.
Wendt, K., Moeller, D., and Ritter, B. (1985) Geodetic measurements of surface deformations during the present rifting episode in NE-Iceland, *J. Geophys. Res.*, **90**, 10163–10173.
Werner, R., Schmincke, H.-U., and Sigvaldason, G. (1996) A new model for the evolution of table mountains: Volcanological and petrological evidence from Herdubreid and Herdubreidartögl volcanoes (Iceland). *Geol. Rundsch.*, **85**, 390–397.
Weir, N. R. W., White, R. S., Brandsdottir, B., Einarsson, P., Shimamura, H., and Shiobara, H. (2001) Crustal structure of the northern Reykjanes ridge and the Reykjanes Peninsula, southwest Iceland, *J. Geophys. Res.*, **106**, 6347–6368.
White, N. and Lovell, B. (1997) Measuring the pulse of a plume with the sedimentary record, *Nature*, **387**, 888–891.
White, R. and McKenzie, D. (1989) Magmatism at rift zones: The generation of volcanic continental margins and flood basalts, *J. Geophys. Res.*, **94**, 7685–7729.
White, R. S. and McKenzie, D. (1995) Mantle plumes and flood basalts, *J. Geophys. Res.*, **100**, 17543–17585.
White, R., S., Spence, G. D., Fowler, S. R., McKenzie, D. P., Westbrook, G. K., and Bowen, A. N. (1987) Magmatism at rifted continental margins, *Nature*, **330**, 439–444.
Withjack, M. O. and Jamison, W. R. (1986) Deformation produced by oblique rifting, *Tectonophysics*, **126**, 99–124.
Wolfe, C.-J., Bjarnason, I. T., VanDecar, J. C., and Salomon, S. C (1997) Seismic structure of the Iceland mantle plume, *Nature*, **385**, 245–247.
Yang, X. M., Davies, P. M., and Dieterich, J. H. (1988) Deformation from inflation of a dipping finite prolate sheroid in an elastic half-space as a model for volcanic stressing, *J. Geophys. Res.*, **93**, 4249–4257.

Index

Colour section

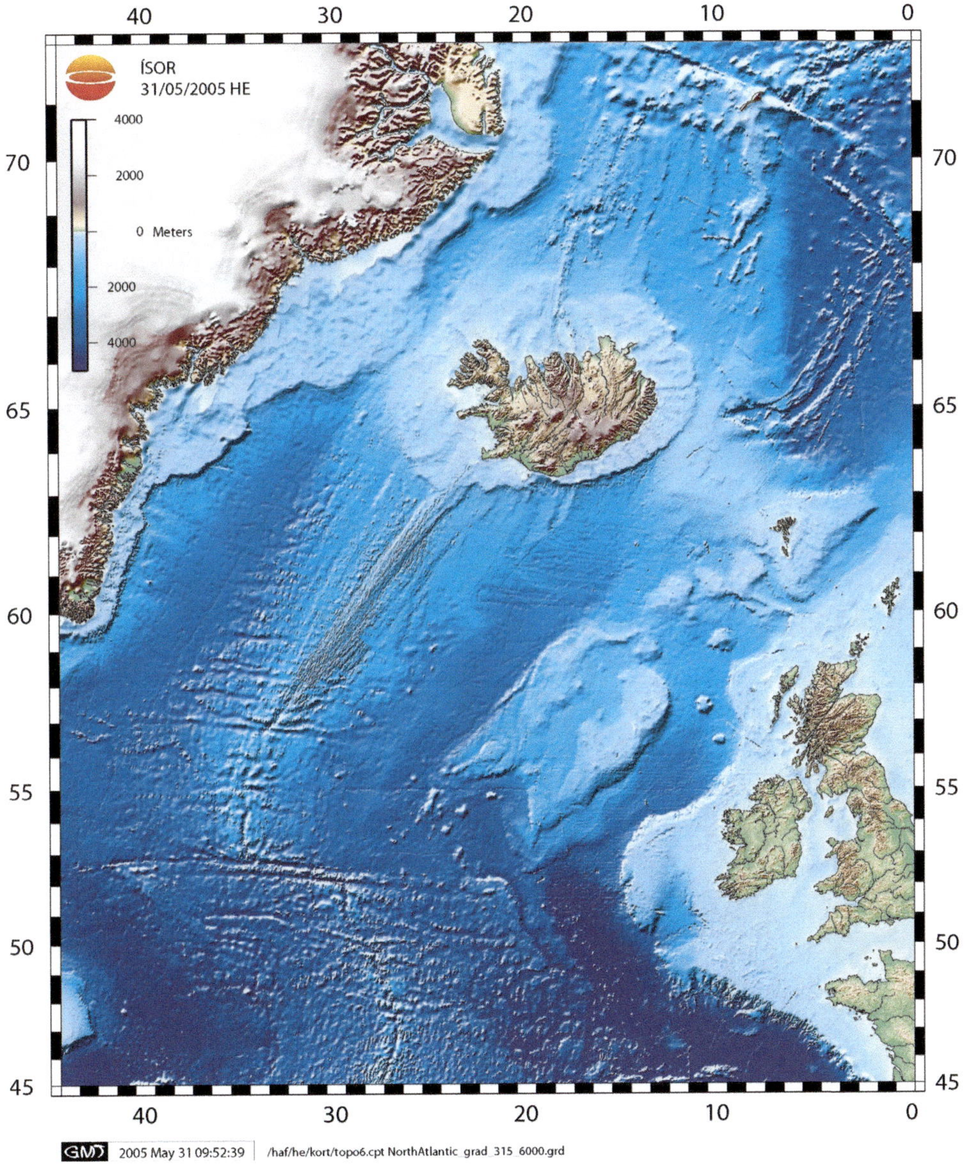

Figure 2.1. Topography of the North Atlantic showing the Iceland hotspot swell, the Mid-Atlantic Ridge (MAR), the Greenland–Scotland Ridge, and V-shaped ridges at the MAR south of Iceland.

Modified from Eysteinsson and Gunnarsson (1995). Courtesy of Hjálmar Eysteinsson, Iceland Geosurvey.

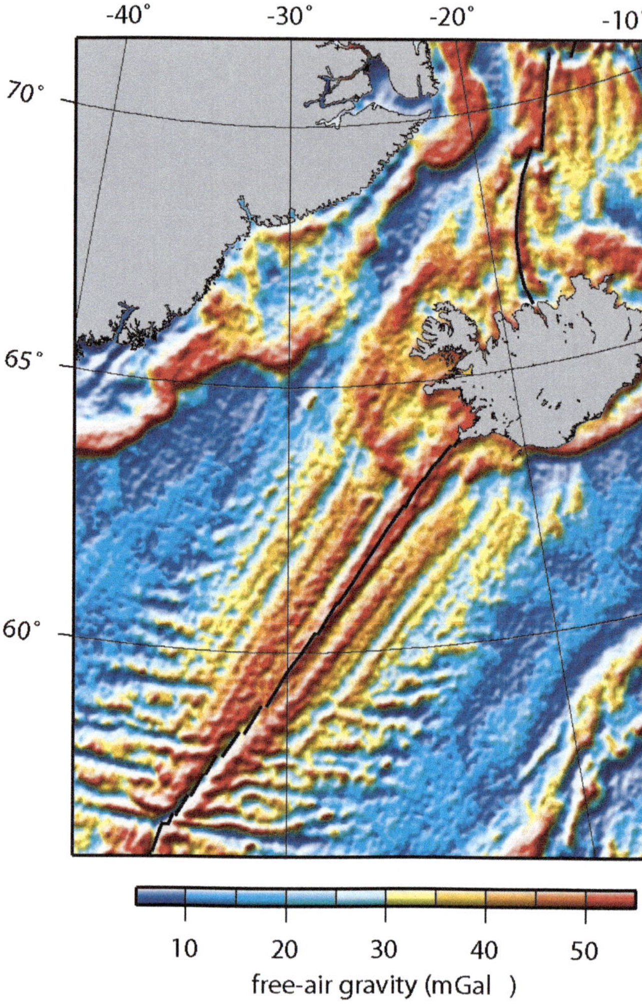

Figure 2.7. Satellite-derived free air gravity anomalies in the region surrounding Iceland based on data from Sandwell and Smith (1997). V-shaped ridges at the Reykjanes Ridge are prominent.

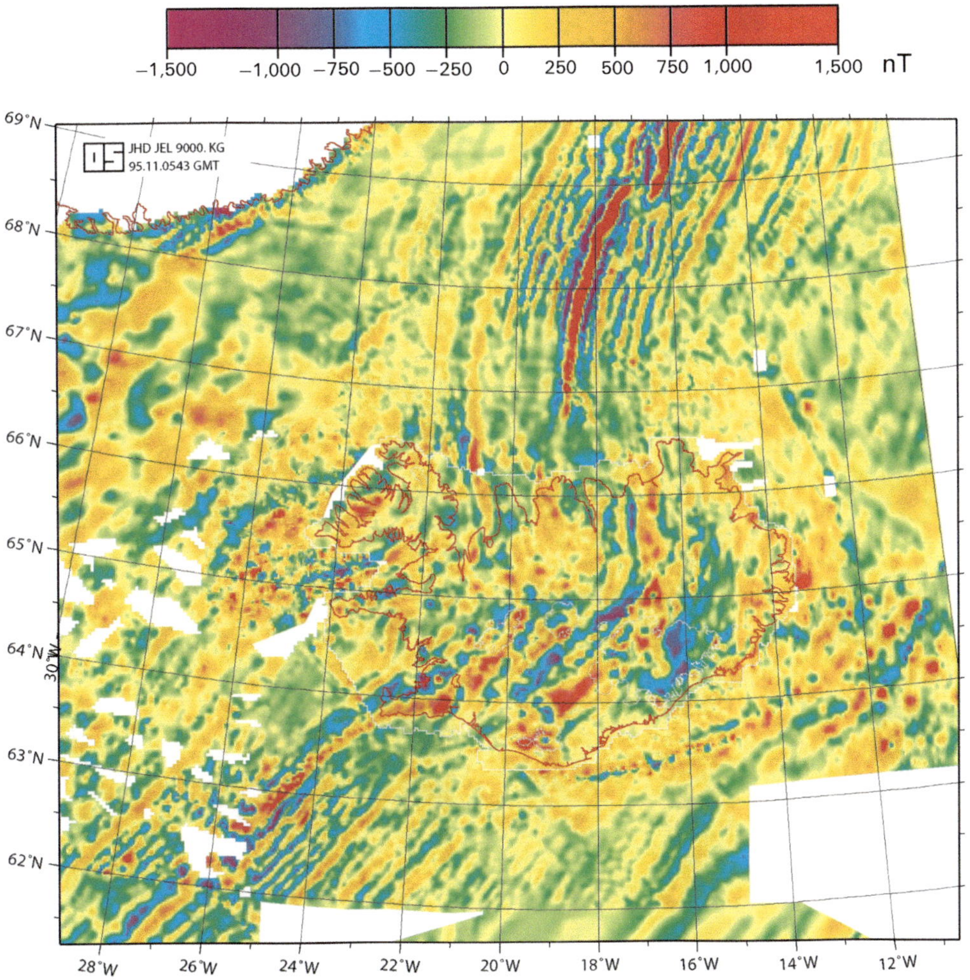

Figure 2.10. Total magnetic field anomaly map of Iceland and the North Atlantic.

Reproduced from Eysteinsson and Gunnarsson (1995) with permission of Iceland Geosurvey.

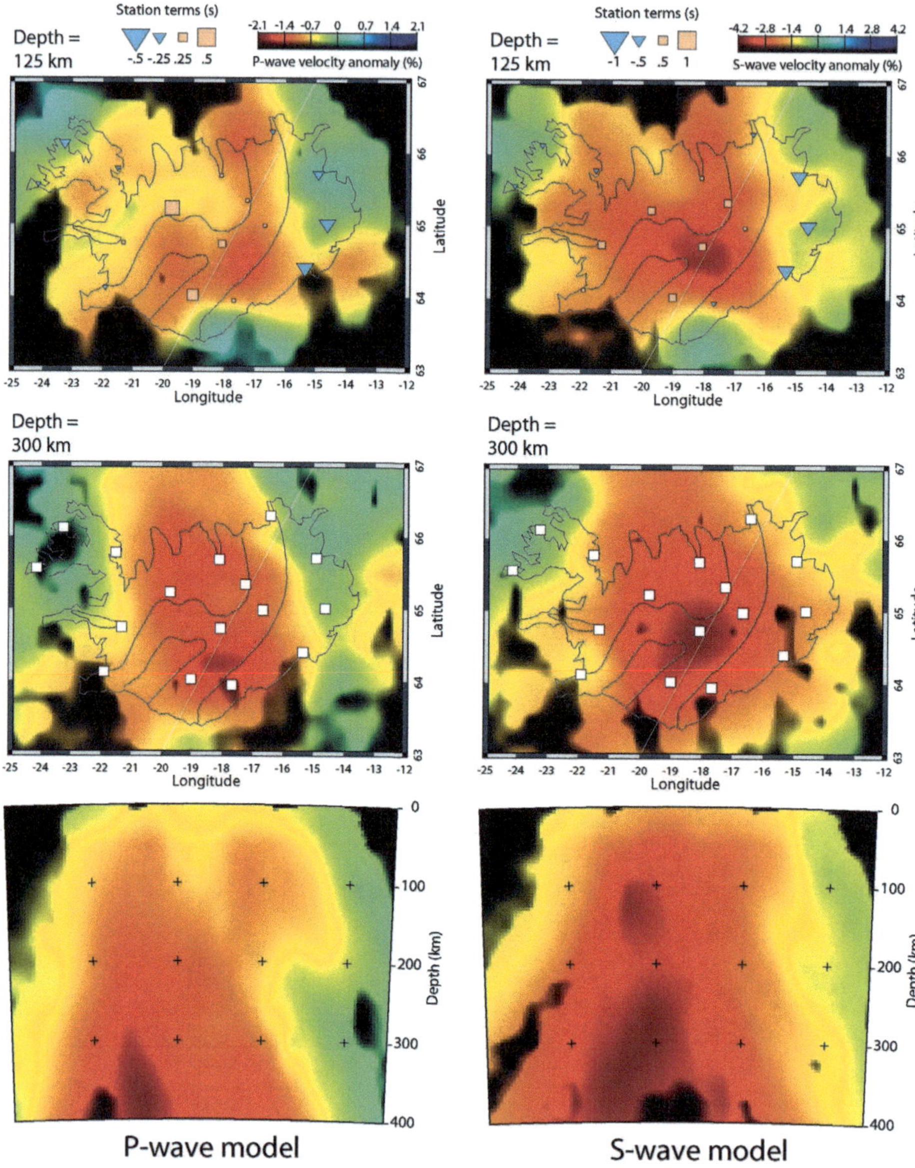

Figure 2.12. Upper mantle P-wave velocity (left) and S-wave velocity (right) anomalies under Iceland determined at depths of 125 km, 300 km, and in cross section.

Reproduced from Wolfe et al. (1997) with permission of *Nature*, London.

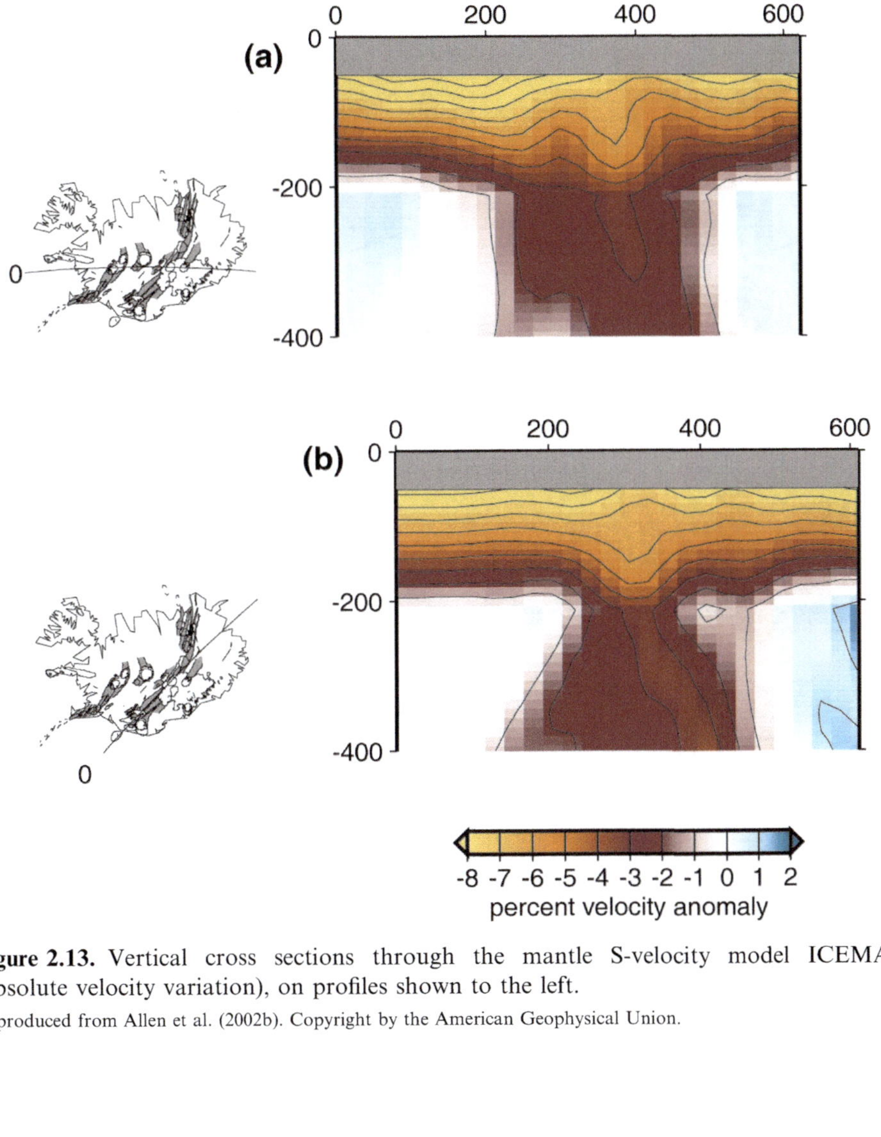

Figure 2.13. Vertical cross sections through the mantle S-velocity model ICEMAN-S (absolute velocity variation), on profiles shown to the left.

Reproduced from Allen et al. (2002b). Copyright by the American Geophysical Union.

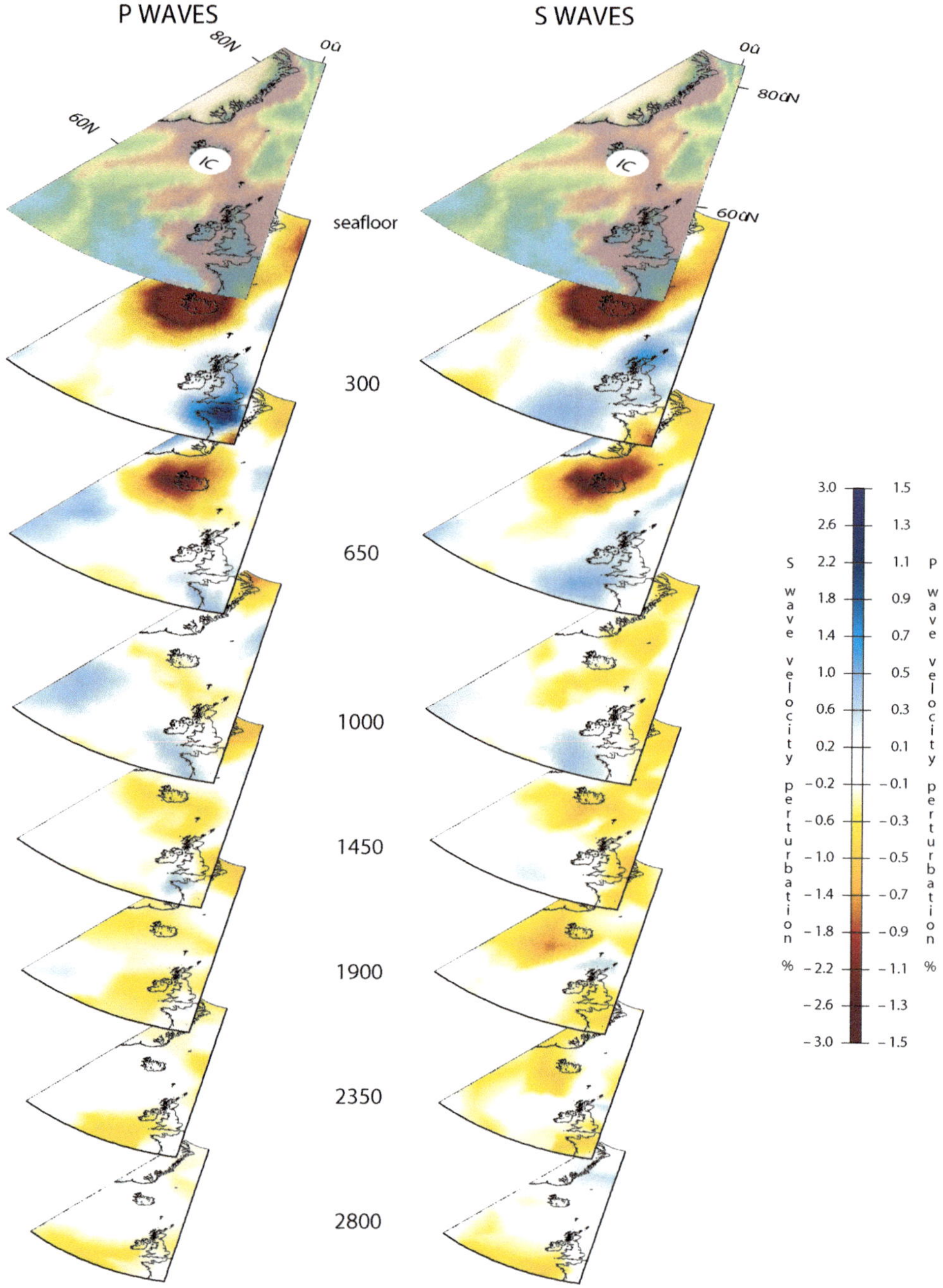

Figure 2.14. P-wave and S-wave velocity perturbations under Iceland imaged by Montelli et al. (2004c, and in preparation).

Courtesy of Rafaella Montelli, Princeton University.

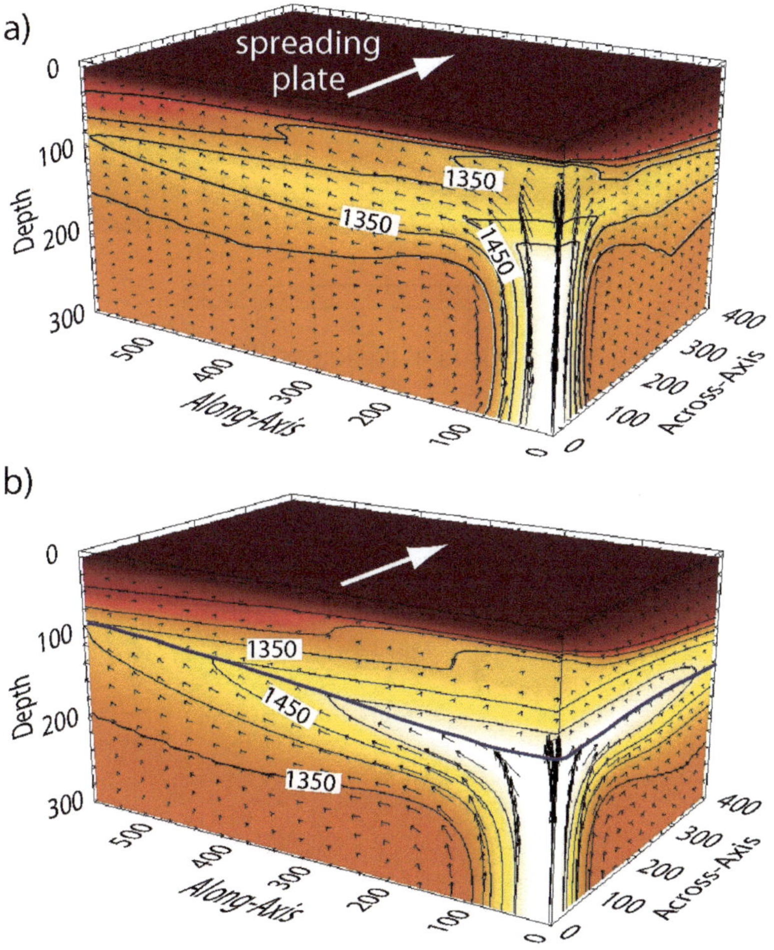

Figure 2.15. (a) Three-dimensional fluid dynamical model of a ridge-centred mantle plume. Potential temperature is coloured and contoured. Mantle flow direction and rate are shown with arrows. Maximum upwelling rate is 50 cm/yr, maximum excess temperature is 200°C. Ambient viscosity at the 250-km depth is 5×10^{19} Pa/s, minimum viscosity in the plume is 5.8×10^{18} Pa/s. (b) Same as (a), but with a viscosity increase due to extraction of water at the base of the melting zone.

Reproduced from Ito et al. (2003). Copyright by the American Geophysical Union.

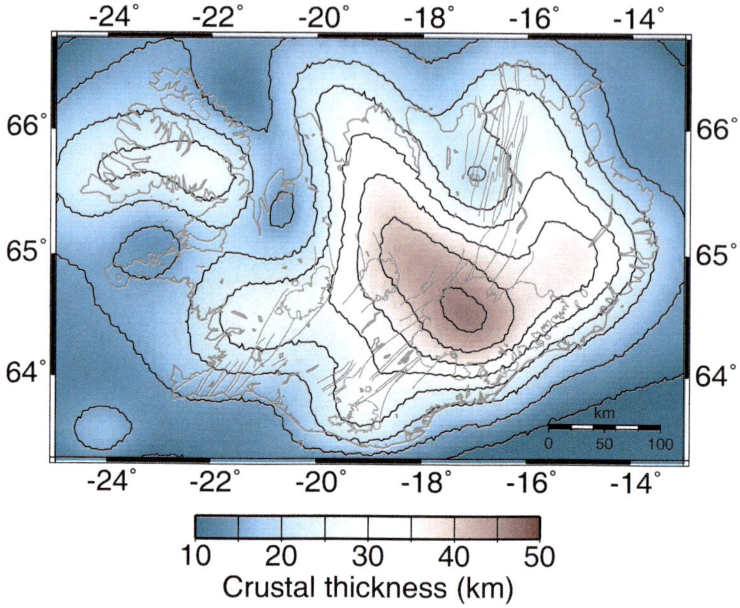

(a)

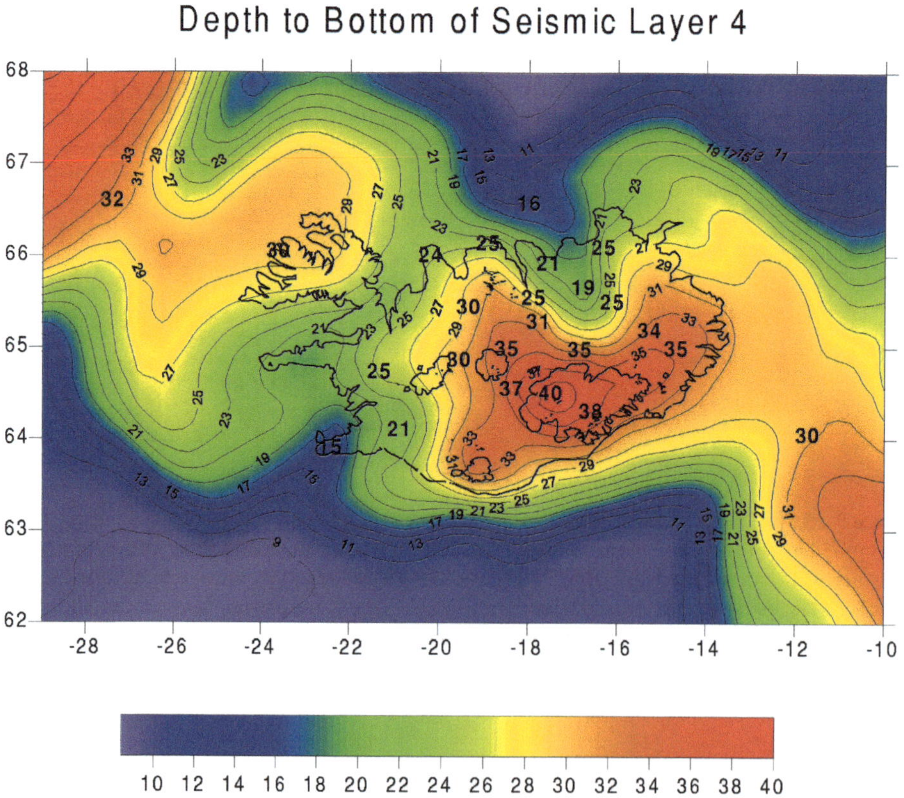

(b)

Figure 4.3. (a) Crustal thickness model ICECRTb. (b) Crustal thickness map.

(a) Reproduced from Allen et al. (2002). Copyright by the American Geophysical Union.
(b) Reproduced from Kaban et al. (2002).

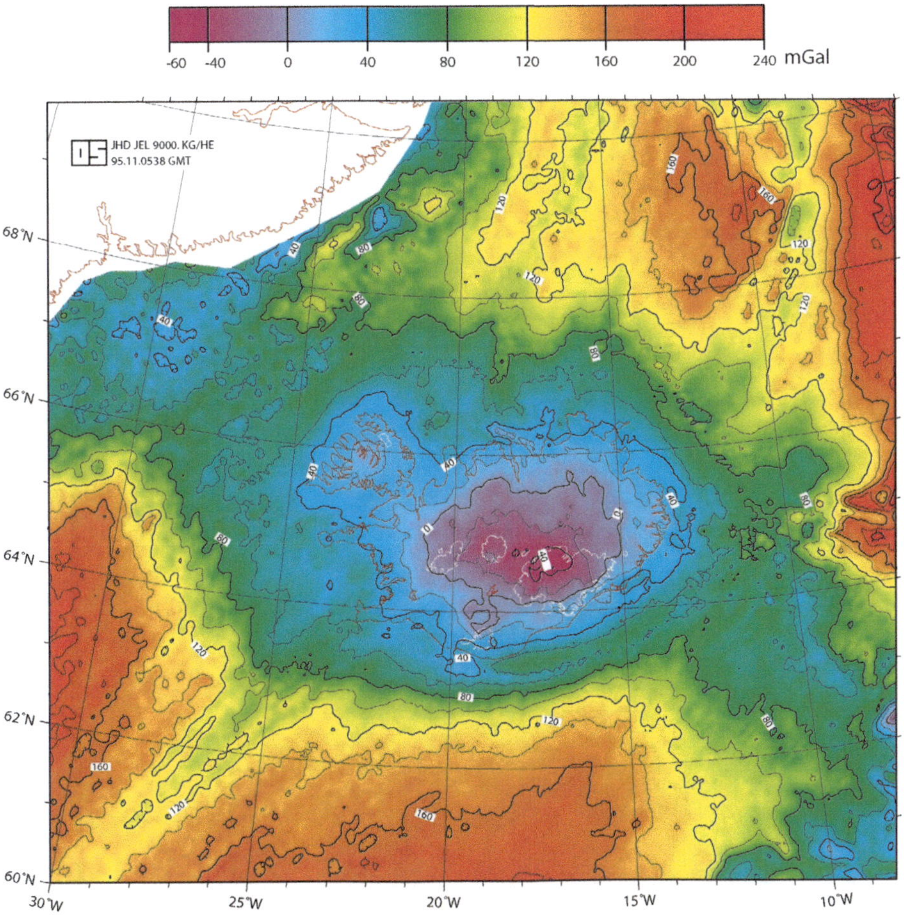

Figure 4.4. Bouguer gravity map of Iceland and surroundings.

Reproduced from Eysteinsson and Gunnarsson (1995) with permission of Iceland Geosurvey.

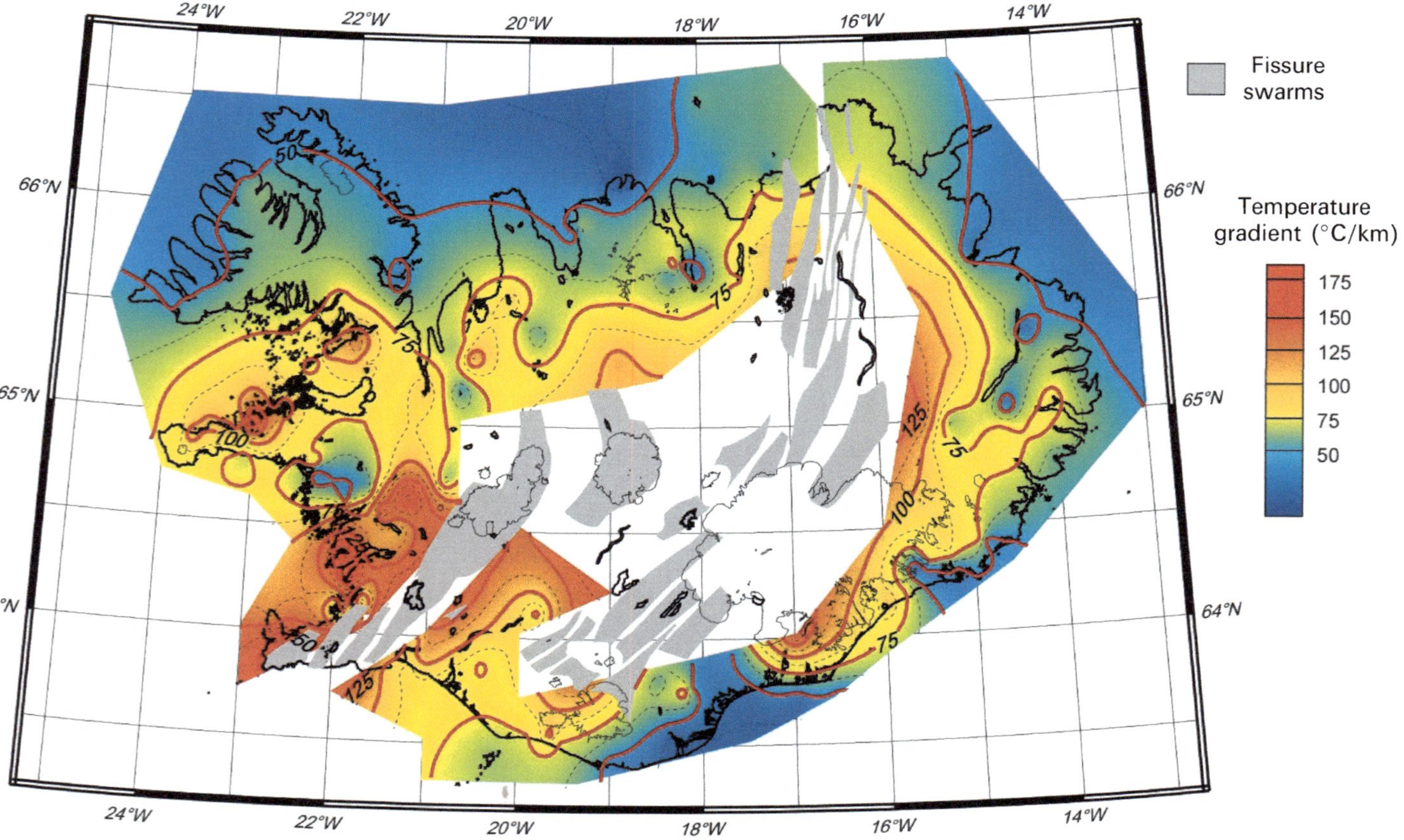

Figure 4.7. Near-surface crustal temperature gradients.
Courtesy of Ólafur Flóvenz and Kristján Sæmundsson, Iceland Geosurvey.

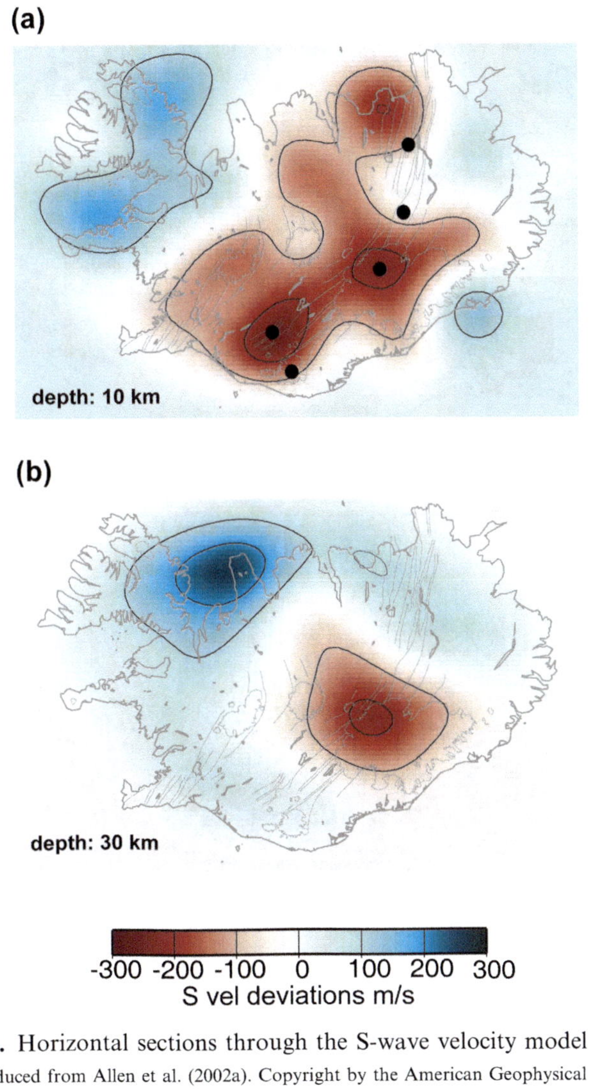

Figure 4.10. Horizontal sections through the S-wave velocity model ICECRTb.
Reproduced from Allen et al. (2002a). Copyright by the American Geophysical Union.

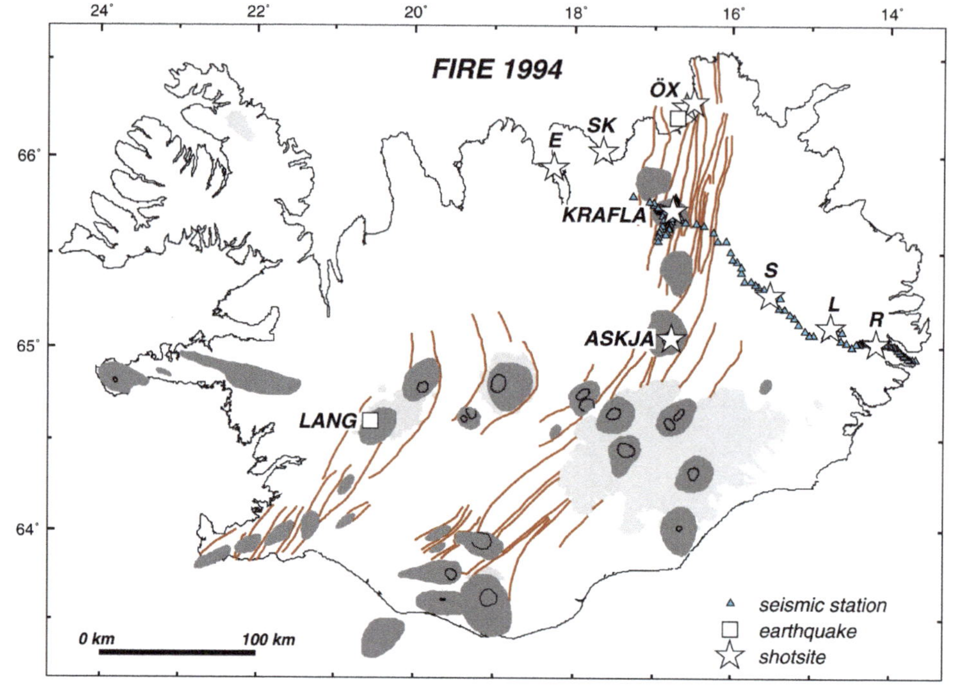

(a)

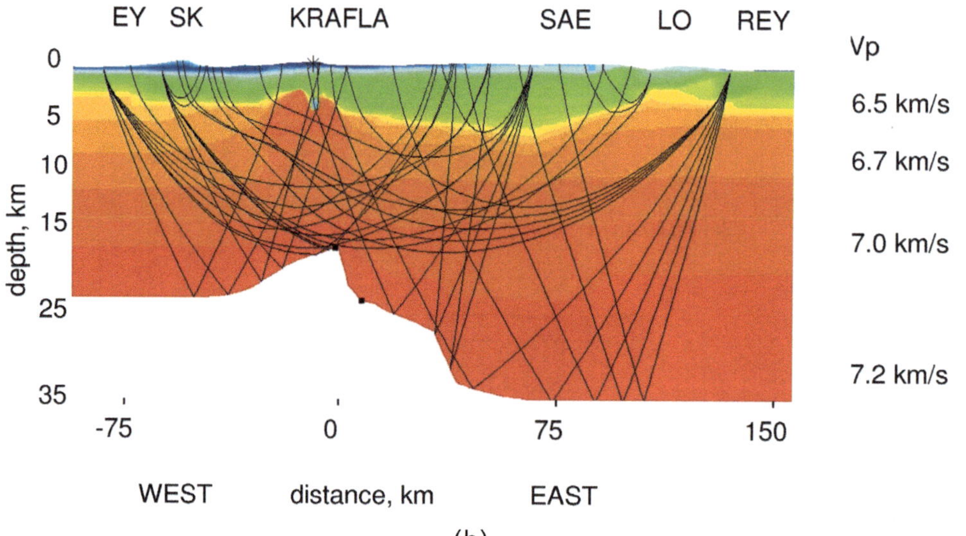

(b)

Figure 5.5. Seismic study of the Northern Volcanic Zone and the Krafla Central Volcano. (a) The seismic array. (b) P-wave velocity cross section along the profile with a low-velocity anomaly under Krafla interpreted as a magma chamber.

Modified from Brandsdottir et al. (1997). Courtesy of Bryndís Brandsdóttir.

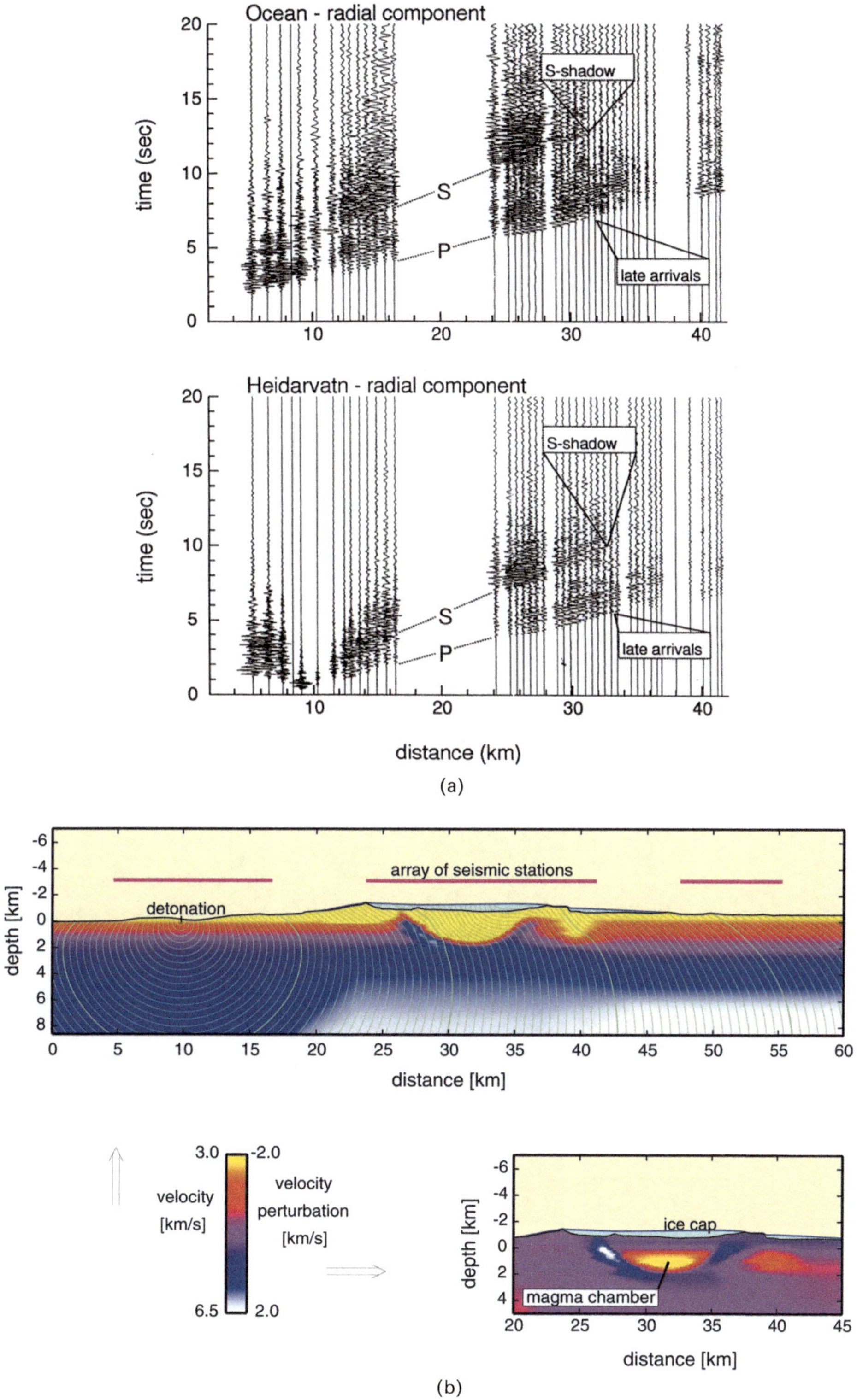

Figure 5.6. Seismic study of the Katla Volcano. (a) Seismic record section. Radial component of recorded waves for two shots from south of Katla Volcano are shown, with distance measured along a profile from the southernmost shot in the ocean. Shear-wave shadows correlate with late arrivals of P-waves in both sections. (b) Velocity model.

Figures modified from Gudmundsson et al. (1994). Courtesy of Ólafur Guðmundsson and Bryndís Brandsdóttir.

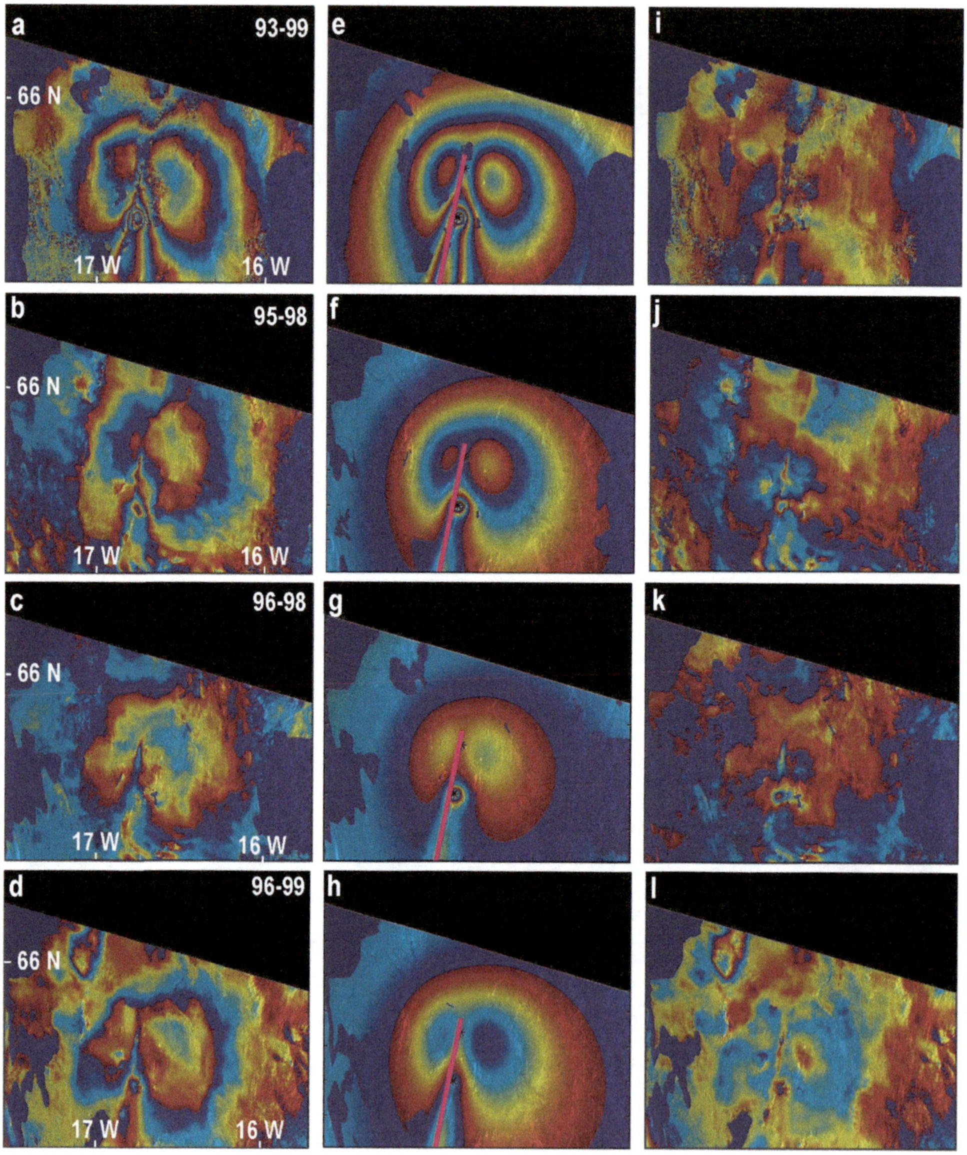

Figure 5.15. InSAR study of Krafla Volcano. Interferograms (left column), models (centre column), and residuals (right column). Each full colour cycle (fringe) corresponds to a change in range from ground to satellite of 28 mm. See text for discussion. For location, see Figure 5.14

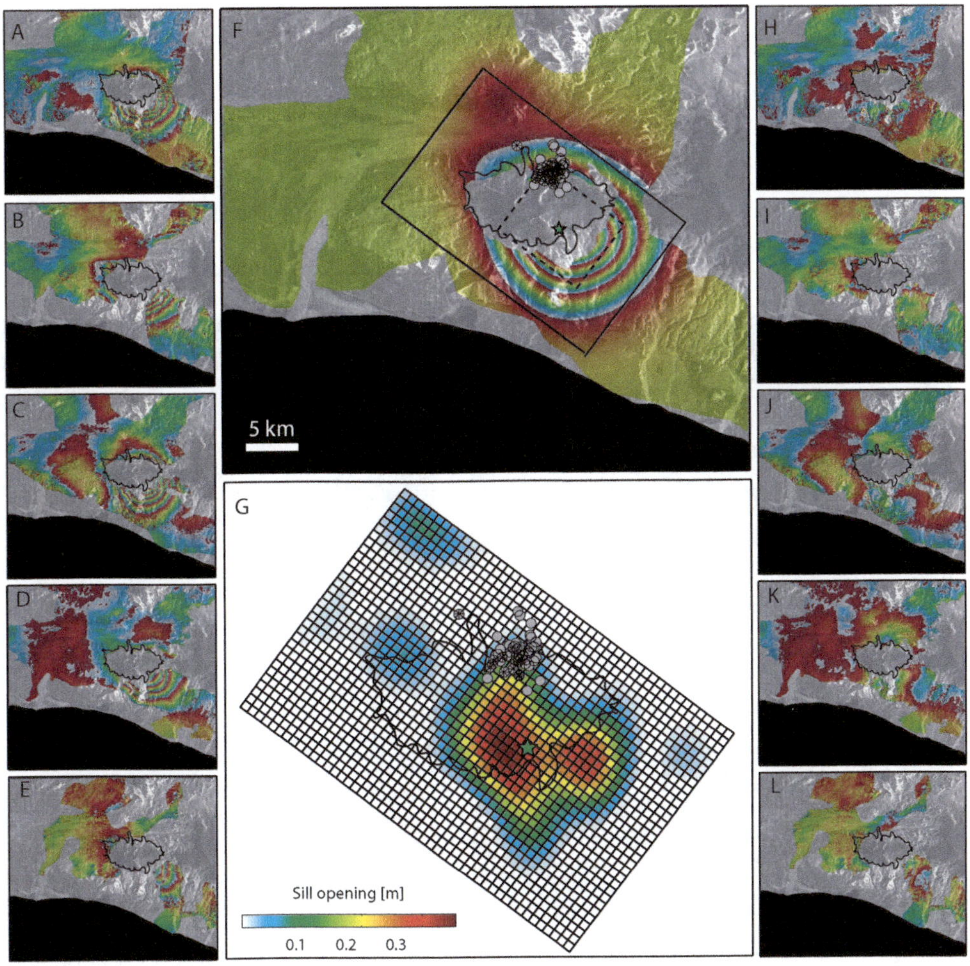

Figure 5.20. InSAR study of deformation at the Eyjafjallajökull Volcano in 1994. Interferograms (left column), model (centre), and residuals (right column). Each full colour cycle (fringe) corresponds to a change in range from ground to satellite of 28 mm. The lower centre shows the inferred variable-sill-opening model, that produces the model fringes in the upper centre panel. See text for discussion.

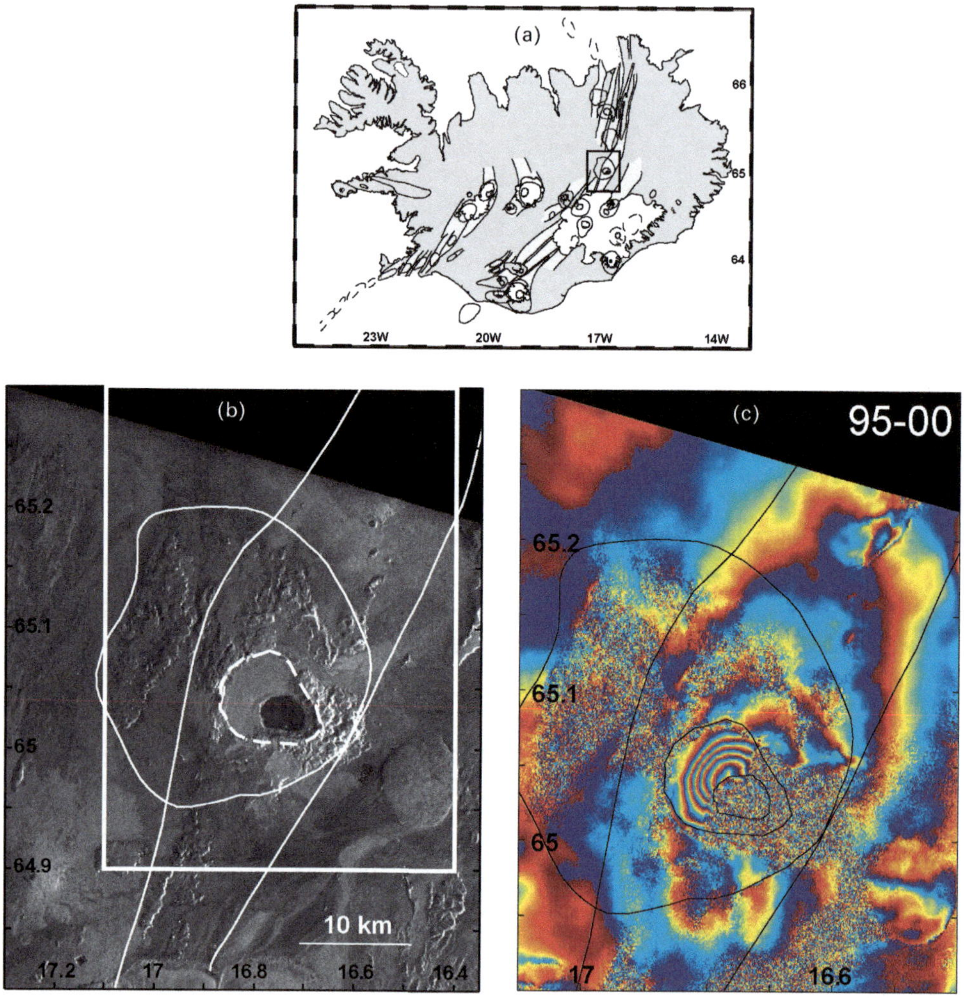

Figure 6.7. Subsidence of the Askja Volcanic System measured by InSAR. (a) Location map. The box shows the location of the InSAR amplitude image shown in (b). Overlain on it are outlines of the Askja Fissure Swarm, the Askja Central Volcano, and the Askja Caldera. (c) Interfergram spanning 1995–2000 showing fringe pattern indicative of subsidence. Each full colour cycle (fringe) corresponds to a change in range from ground to satellite of 28 mm. The partly coherent concentric fringe pattern is indicative of subsidence over a shallow magma chamber at Askja; additional elongated pattern along the fissure swarm manifests its subsidence at a rate of more than a few millimetres per year.

Courtesy of Carolina Pagli, Nordic Volcanological Centre, Institute of Earth Sciences, University of Iceland.

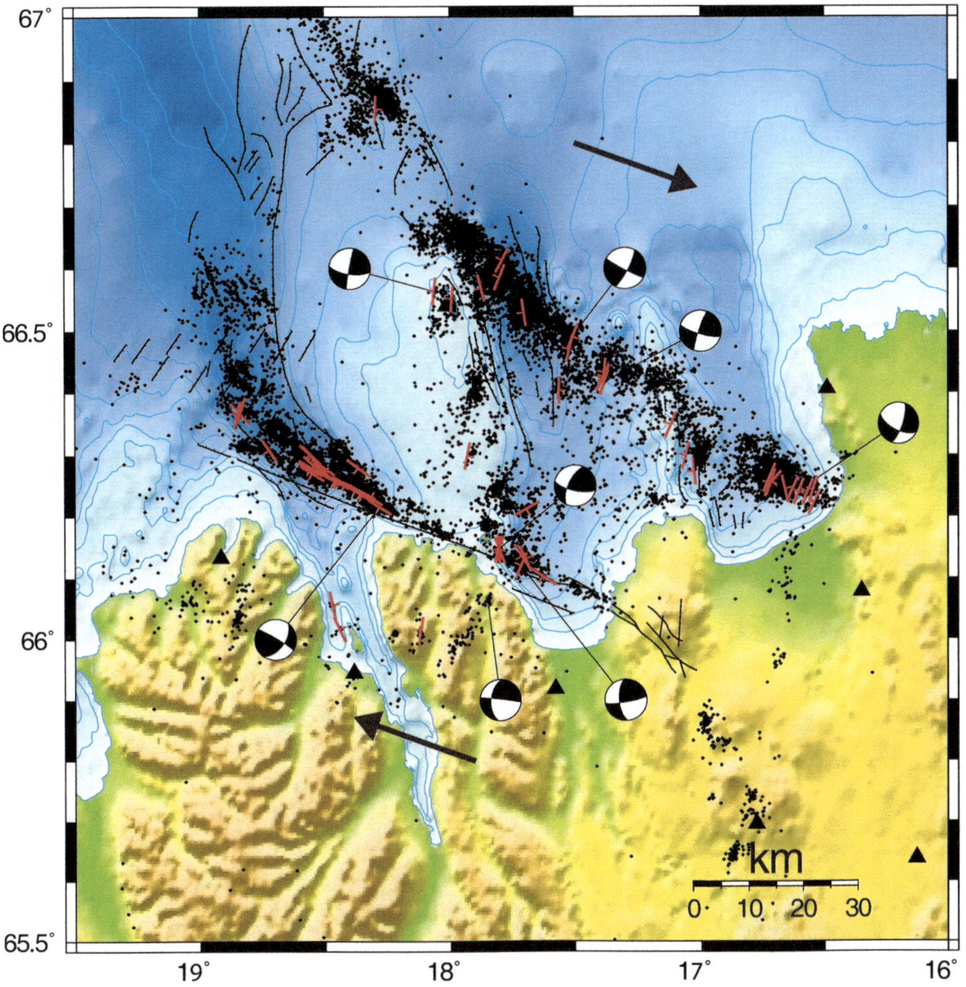

Figure 7.1. Earthquakes and faults in the Tjörnes Fracture Zone. Black dots show earthquake epicentres, 1994–2003. Red lines are active fault segments mapped using accurate relative locations of micro-earthquakes. Fault plane solutions of selected events are shown. Arrows indicate the direction of faraway plate movements. Most of the seismicity falls on two main seismic lineaments, the Grímsey Lineament (northern one) and the Húsavík–Flatey Lineament (southern one). A diffuse zone of seismicity south of the Húsavík–Flatey Lineament marks the Dalvík Lineament.

Modified from Rögnvaldsson et al. (1998) and augmented with new seismic data. Courtesy of Gunnar Guðmundsson, Icelandic Meteorological Office.

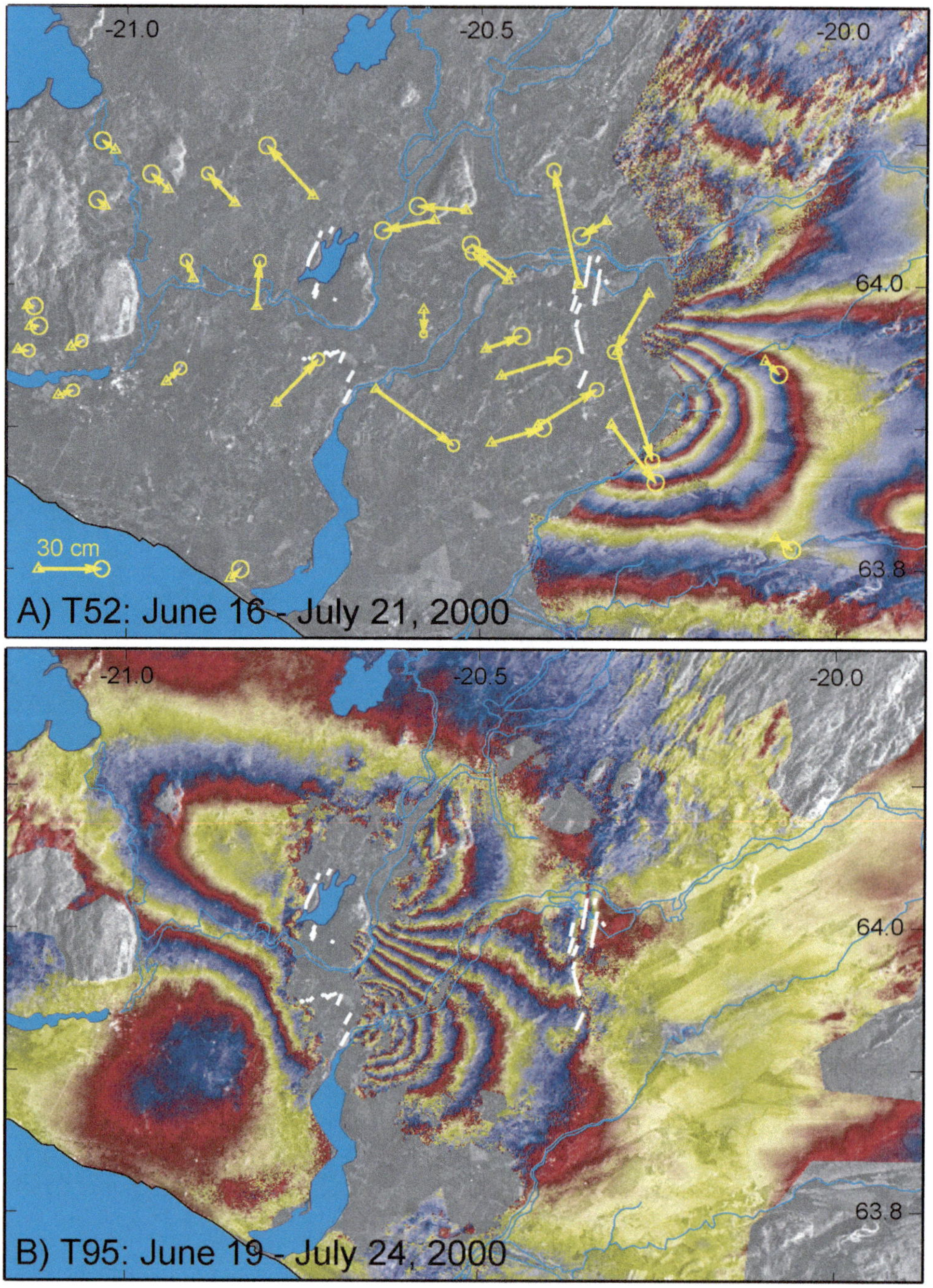

Figure 7.11. Co-seismic interferograms and horizontal GPS displacements (yellow arrows) spanning the June 17 and June 21 earthquakes in South Iceland. Surface rupture shown schematically in white. Each colour fringe in interferograms corresponds to 28.3 mm of change in range from ground to satellite. (a) Interferogram spanning June 16–July 21, 2000. It is only coherent in the area east of the June 17 earthquake. Although it includes the effects of both earthquakes, it is mostly dominated by the effects of the June 17 event. (b) Interferogram spanning June 19–July 24, 2000, capturing co-seismic deformation associated with the June 21 event, and post-seismic deformation until July 24 (local signal next to the June 17 fault trace).

Reproduced from Pedersen et al. (2003) with permission from Elsevier.

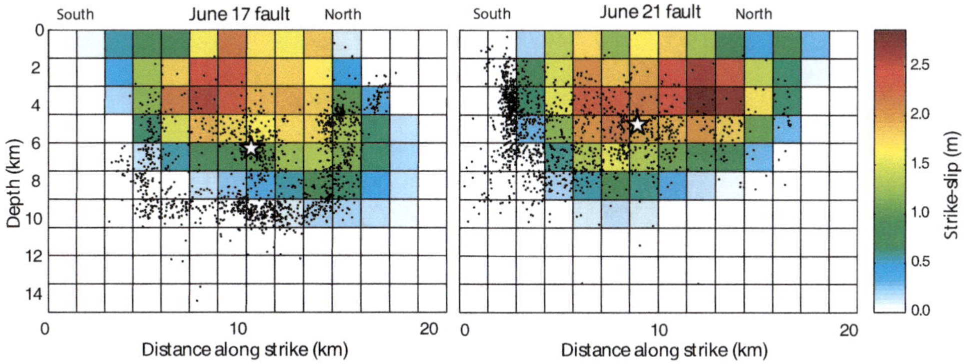

Figure 7.12. Distributed slip models for June 17 and 21, 2000, earthquakes. Colour coding shows the amount of right-lateral strike–slip movement on the fault planes. Hypocentres of main shocks are shown with white stars; black dots show aftershocks in immediate vicinity of the modelled fault planes.

Reproduced from Pedersen et al. (2003) with permission from Elsevier.

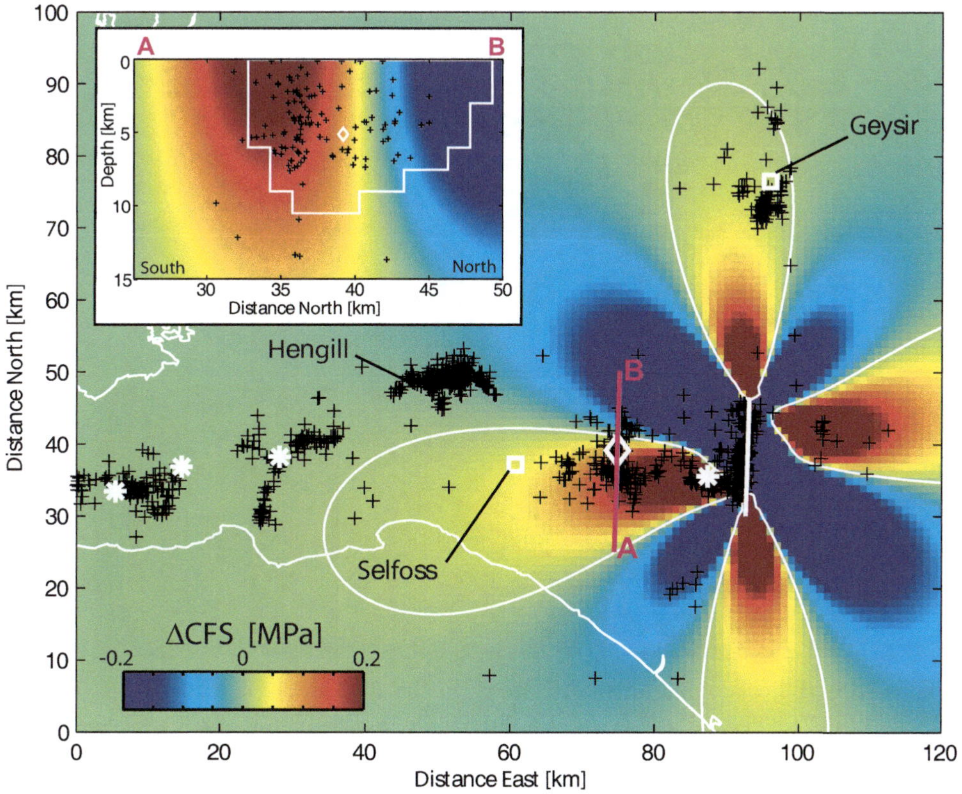

Figure 7.14. Change in static Coulomb failure stress (ΔCFS) due the June 17 main shock in South Iceland, calculated at a 5-km depth for vertical north–south faults with right-lateral strike–slip motion. Three largest triggered earthquakes on June 17 are shown with white stars, and black crosses show other well-located aftershocks June 17–21. The diamond shows the subsequent location of the June 21 fault. The red line marks the location of the cross section shown on the inset.

Reproduced from Árnadóttir et al. (2001). Copyright by the American Geophysical Union.

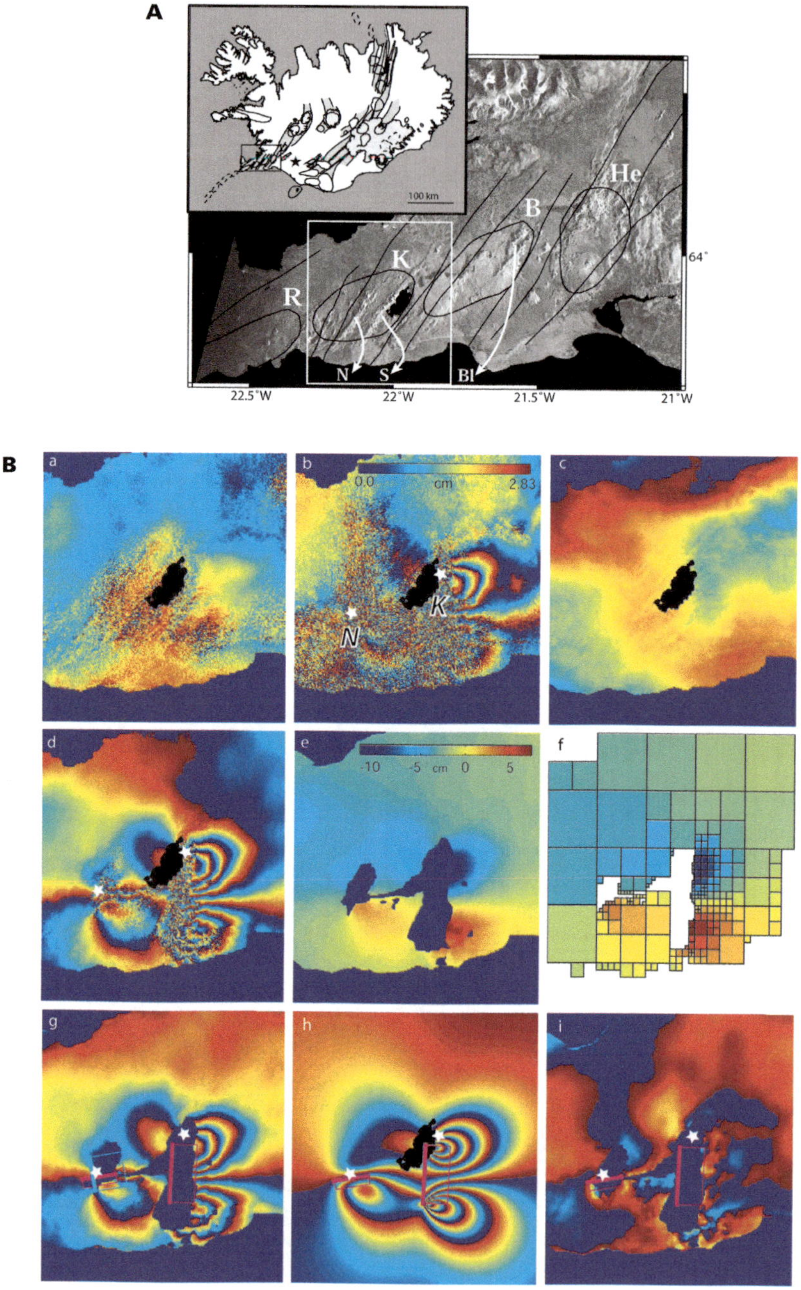

Figure 7.15. **A** Location of an InSAR study on the Reykjanes Peninsula. SAR (synthetic aperture radar) amplitude image, with a box outlining the area displayed on panels in (B). Overlaid are outlines of volcanic systems with their fissure swarms and central volcanoes: Reykjanes (R), Krísuvík (K), Brennisteinsfjöll (B), and Hengill (He). Núpshlíðarháls (N) and Sveifluháls (S) are hyaloclastite ridges. **B** Interferograms spanning (a) the pre-seismic interval, (b and d) co-seismic intervals, and (c) the post-seismic interval. (e) Unwrapped interferogram in panel (d). (f) Quad-tree division of (e). (g) Unwrapped and tilted interferogram. (h) Best fit model. (i) Residuals. White stars show epicentres of triggered earthquakes. Colour scale in panel (e) applies to the unwrapped range changes in panels (e) and (f). For other wrapped interferograms each full cycle of colour change corresponds to a change in range of 28.3 mm. Red boxes mark the outline of inferred faults that moved.

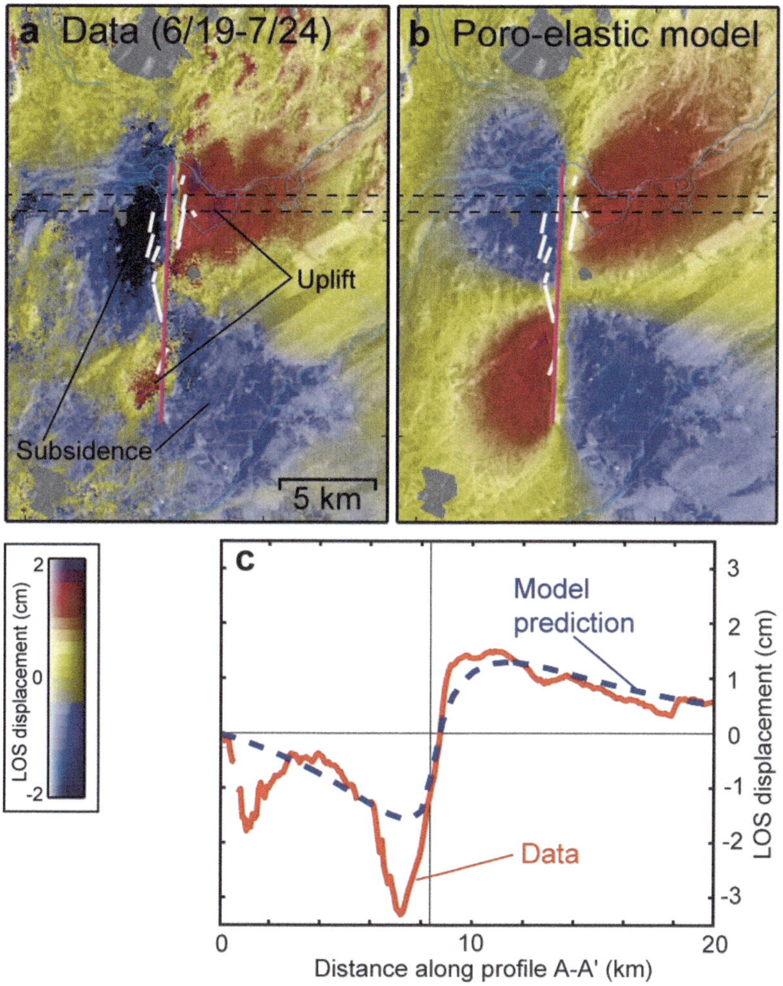

Figure 7.16. Post-seismic poro-elastic deformation at the June 17 fault trace in South Iceland. Surface rupture is shown in white, and the best fit fault plane from co-seismic geodetic data is shown as thick line. (a) Interferogram spanning June 19 to July 24, 2000. (b) Poro-elastic model prediction using an undrained Poisson's ratio of 0.31 and a drained value of 0.27. (c) Profile across the data and model.

Modified from Jónsson et al. (2003) with permission of *Nature*, London.

Printing: Mercedes-Druck, Berlin
Binding: Stein + Lehmann, Berlin